Recent Titles in This Series

159 **O. A. Ladyzhenskaya, Editor,** Proceedings of the St. Petersburg Mathematical Society, Volume II
158 **A. K. Kelmans, Editor,** Selected Topics in Discrete Mathematics: Proceedings of the Moscow Discrete Mathematics Seminar, 1972–1990
157 **M. Sh. Birman, Editor,** Wave Propagation. Scattering Theory
156 **V. N. Gerasimov, N. G. Nesterenko, and A. I. Valitskas,** Three Papers on Algebras and Their Representations
155 **O. A. Ladyzhenskaya and A. M. Vershik, Editors,** Proceedings of the St. Petersburg Mathematical Society, Volume I
154 **V. A. Artamonov et al.,** Selected Papers in K-Theory
153 **S. G. Gindikin, Editor,** Singularity Theory and Some Problems of Functional Analysis
152 **H. Draškovičová et al.,** Ordered Sets and Lattices II
151 **I. A. Aleksandrov, L. A. Bokut′, and Yu. G. Reshetnyak, Editors,** Second Siberian Winter School "Algebra and Analysis"
150 **S. G. Gindikin, Editor,** Spectral Theory of Operators
149 **V. S. Afraĭmovich et al.,** Thirteen Papers in Algebra, Functional Analysis, Topology, and Probability, Translated from the Russian
148 **A. D. Aleksandrov, O. V. Belegradek, L. A. Bokut′, and Yu. L. Ershov, Editors,** First Siberian Winter School "Algebra and Analysis"
147 **I. G. Bashmakova et al.,** Nine Papers from the International Congress of Mathematicians, 1986
146 **L. A. Aĭzenberg et al.,** Fifteen Papers in Complex Analysis
145 **S. G. Dalalyan et al.,** Eight Papers Translated from the Russian
144 **S. D. Berman et al.,** Thirteen Papers Translated from the Russian
143 **V. A. Belonogov et al.,** Eight Papers Translated from the Russian
142 **M. B. Abalovich et al.,** Ten Papers Translated from the Russian
141 **H. Draškovičová et al.,** Ordered Sets and Lattices
140 **V. I. Bernik et al.,** Eleven Papers Translated from the Russian
139 **A. Ya. Aĭzenshtat et al.,** Nineteen Papers on Algebraic Semigroups
138 **I. V. Kovalishina and V. P. Potapov,** Seven Papers Translated from the Russian
137 **V. I. Arnol′d et al.,** Fourteen Papers Translated from the Russian
136 **L. A. Aksent′ev et al.,** Fourteen Papers Translated from the Russian
135 **S. N. Artemov et al.,** Six Papers in Logic
134 **A. Ya. Aĭzenshtat et al.,** Fourteen Papers Translated from the Russian
133 **R. R. Suncheleev et al.,** Thirteen Papers in Analysis
132 **I. G. Dmitriev et al.,** Thirteen Papers in Algebra
131 **V. A. Zmorovich et al.,** Ten Papers in Analysis
130 **M. M. Lavrent′ev, K. G. Reznitskaya, and V. G. Yakhno,** One-dimensional Inverse Problems of Mathematical Physics
129 **S. Ya. Khavinson,** Two Papers on Extremal Problems in Complex Analysis
128 **I. K. Zhuk et al.,** Thirteen Papers in Algebra and Number Theory
127 **P. L. Shabalin et al.,** Eleven Papers in Analysis
126 **S. A. Akhmedov et al.,** Eleven Papers on Differential Equations
125 **D. V. Anosov et al.,** Seven Papers in Applied Mathematics
124 **B. P. Allakhverdiev et al.,** Fifteen Papers on Functional Analysis
123 **V. G. Maz′ya et al.,** Elliptic Boundary Value Problems
122 **N. U. Arakelyan et al.,** Ten Papers on Complex Analysis
121 **V. D. Mazurov, Yu. I. Merzlyakov, and V. A. Churkin, Editors,** The Kourovka Notebook: Unsolved Problems in Group Theory
120 **M. G. Kreĭn and V. A. Jakubovič,** Four Papers on Ordinary Differential Equations

(Continued in the back of this publication)

American Mathematical Society
TRANSLATIONS
Series 2 • Volume 159

Proceedings of the St. Petersburg Mathematical Society Volume II

O. A. Ladyzhenskaya
Editor

American Mathematical Society
Providence, Rhode Island

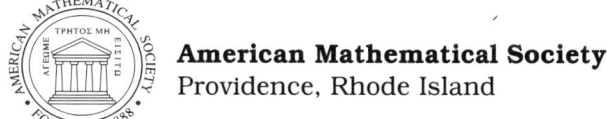

Translation edited by SIMEON IVANOV

1991 *Mathematics Subject Classification.*
Primary 01A, 11D, 11H, 11S, 16S, 16W, 26A, 35A, 35B, 35D, 35M, 35Q, 41A, 53C, 68Q, 68R, 82B; Secondary 12F, 26D, 35J.

ISBN 0-8218-7510-8
ISSN 0065-9290

⊗ The paper used in this book is acid-free and falls within the guidelines
established to ensure permanence and durability.
♻ Printed on recycled paper.
This volume was typeset using $\mathcal{A}\mathcal{M}\mathcal{S}$-TEX,
the American Mathematical Society's TEX macro system.

10 9 8 7 6 5 4 3 2 1 99 98 97 96 95 94

Contents

Amer. Math. Soc. Transl.
(2) Vol. **159**, 1994

Galois Representations in Honda's Formal Groups. Arithmetic of the Group of Points

D. G. Benois and S. V. Vostokov

In **[1]** and **[2]** an effective method was developed for studying Galois representations in the Tate module of the formal group over the ring of Witt vectors in a finite field. It is based on the explicit formulas for the norm pairing in a formal group. In the present article we generalize the results of **[1]** and **[2]** to Galois representations in formal groups over multidimensional local rings. We shall study the arithmetic of the group of points and formulate the main results on Galois representations, outlining briefly the principal ideas of the proof. Detailed proofs of the theorems formulated will be given in a forthcoming paper.

§1. Preliminaries

1. We recall that a field k is called an n-dimensional local field if there exists a chain of complete discrete valuation fields $k^{(0)} = \mathbb{F}_{q_0}, k^{(1)}, \dots, k^{(n)} = k$, in which $k^{(i)}$ is the residue class field of $k^{(i+1)}$ and $\mathbb{F}_{q_0}$ is a finite field of $q_0 = p^{f_0}$ elements (p prime).

By lifting the prime elements of the fields $k^{(i)}$ to the field k we get a system of local parameters $(t_1, t_2, \dots, t_n)$, which determines the valuation of rank n on the field k:

$$\bar{v}_k = \bar{v} = (v_1, v_2, \dots, v_n) : k^* \to \mathbb{Z}^{(n)},$$

where $v_i(a) = v_{k^{(i)}}(\overline{a/(t_n^{v_n(a)}, \dots, t_{i+1}^{v_{i+1}(a)})})$ for $1 \leqslant i < n$, $v_n(a) = v_{k^{(n)}}(a)$ (here the bar means a residue in the field $k^{(i)}$ and v_i is a valuation on this field). We assume that the group $\mathbb{Z}^{(n)}$ is lexicographically ordered (i.e., $(i_1, \dots, i_n) < (j_1, \dots, j_n)$ if for the greatest index l such that $i_l \neq j_l$ we have $i_l < j_l$).

In the case in question the valuation ring is the set of elements $a \in k$, satisfying $\bar{v}(a) \geqslant (0, \dots, 0)$. It is independent of the choice of the system of local parameters and has the only maximal ideal $\mathfrak{M} = \{a \in k : \bar{v}(a) > 0\}$. (See **[3]** for more detailed consideration of valuations on local fields.)

From Parshin's classification theorem **[4, 5]** it follows that an n-dimensional local field k of characteristic $p > 0$ is the field $\mathbb{F}_{q_0}((t_1)) \cdots ((t_n))$, and if $p = 0$ it is contained

1991 *Mathematics Subject Classification*. Primary 11S20; Secondary 12F10, 11S31.

in the field $F\{\{t_1\}\}\cdots\{\{t_{j-1}\}\}((t_j))\cdots((t_n))$, where F is a local number field (a finite extension of $\mathbb{Q}_p$). Here $F\{\{t\}\}$ denotes the set of series

$$\sum_{-\infty}^{\infty} a_i t^i, \qquad a_i \in k,$$

the coefficients a_i of which are jointly bounded with respect to the value on the field k, and $a_i \to 0$ as $i \to -\infty$. A topology corresponding to the valuation of rank n may be defined on the additive group of a multidimensional local field [**4, 6**]. Under this topology the additive group turns into a complete separable group and the multiplication is sequentially continuous: if $x_m \to x$, $y_m \to y$, then $x_m y_m \to xy$.

In this paper we consider the case of distinct characteristics of k and its residue field $k^{(n-1)}$. From the classification given above it follows that k is contained in $F\{\{t_1\}\}\cdots\{\{t_n\}\}$.

2. Series expansion of elements of multidimensional local fields. We shall call a set of multi-indices $I \subset \mathbb{Z}^{(n)}$ admissible if for any fixed set of integers $i_n, \dots, i_{l+1}$ $(1 \leqslant l \leqslant n)$ and for any set of multi-indices $\bar{r} = (r_n, r_{n-1}, \dots, r_{l+1}, r_l, \dots, r_1)$ in I such that $r_n = i_n, \dots, r_{l+1} = i_{l+1}$ the index r_l is bounded from below.

Each element $a \in k$ may be uniquely written in the form

$$a = \sum_{\bar{r} \in I} \theta_{\bar{r}} t_n^{r_n} \cdots t_1^{r_1},$$

where I is an admissible set and the coefficients $\theta_{\bar{r}}$ belong to the system of Teichmüller representatives of the field $\mathbb{F}_{q_0}$ in k (see [**7,** Lemma 2]).

The valuation of rank n on the element a coincides with the minimal multi-index $\bar{r} \in I$ for which $\theta_{\bar{r}} \neq 0$:

$$\bar{v}(a) = \min\{\bar{r} \in I : \theta_{\bar{r}} \neq 0\}.$$

We call the set

$$\bar{e}_k = \bar{v}(p) = (e_1, \dots, e_n) \tag{1.1}$$

an absolute ramification index of the field k. We denote the numbers $e_i/(p-1)$ by e_i'.

One can get a similar expansion taking the elements of the ring of Witt vectors $W = W(\mathbb{F}_{q_0})$ as coefficients:

$$a = \sum_{\bar{r} \in I} \alpha_{\bar{r}} t_n^{r_n} \cdots t_1^{r_1}, \qquad \alpha_{\bar{r}} \in W.$$

Note that in this case we lose the uniqueness of the expansion.

We also need the expansion of elements in the standard $(n+1)$-dimensional local field

$$K = T\{\{x_1\}\}\cdots\{\{x_{n-1}\}\}((x_n)),$$

where T is a quotient field of the ring of Witt vectors W and $x_1, \dots, x_n$ are independent variables. Let $\mathcal{O}_{(p)}$ be a valuation ring at prime p of K; i.e., $\mathcal{O}_{(p)}$ contains all the elements of K such that their valuation at p is nonnegative. Then each element $a(x_1, \dots, x_n)$ of $\mathscr{O}_{(p)}$ has a unique expansion

$$a(x_1, \dots, x_n) = \sum_{\bar{r} \in I} \alpha_{\bar{r}} x_n^{r_n} \cdots x_1^{r_1}, \qquad \alpha_{\bar{r}} \in W,$$

where I is an admissible set of multi-indices.

3. If a prime p is a uniformizing element with respect to a discrete valuation v_n (i.e., one can set $t_n = p$), on the field k we can introduce the Frobenius operator, raising the transcendental uniformizing elements to the pth power and acting as the ordinary Frobenius automorphism on the unramified number field. Thus, if

$$a = \sum_{\bar{r} \in I} \alpha_{\bar{r}} p^{r_n} t_{n-1}^{r_{n-1}} \cdots t_1^{r_1}$$

is the expansion of $a \in k$, then we set

$$a^{\Delta} = \sum_{\bar{r} \in I} \alpha_{\bar{r}}^{\mathrm{Fr}} p^{r_n} t_{n-1}^{pr_{n-1}} \cdots t_1^{pr_1}. \tag{1.2}$$

In the same way one can define the action of the Frobenius operator Δ on the ring $\mathscr{O}_{(p)}$.

4. The arithmetic of the multiplicative groups $\mathscr{H}_m = \mathscr{O}_{(p)}^*$, k^*-ring $\mathscr{O}_{(p)}$, and field k was studied in **[8]**. We recall the results obtained there.

A unit ε of the field K (respectively k) is called principal if its image in the last residue field $\mathbb{F}_{q_0}$ is 1. Each element $a(x_1, \dots, x_n)$ of $\mathscr{H}_m$ may be written in the form

$$a(x_1, \dots, x_n) = x_1^{r_1} \cdots x_n^{r_n} \theta \varepsilon,$$

where $(r_1, \dots, r_n) = \bar{v}(a)$, the element θ belongs to the Teichmüller system $\mathfrak{R}$, and ε is a principal unit.

On a principal ideal of the ring $\mathscr{O}_{(p)}$ one can define the function

$$E(a(x_1, \dots, x_n)) = \exp(1 + \frac{\Delta}{p} + \frac{\Delta^2}{p^2} + \cdots)(a),$$

which gives the series in $\mathscr{O}_{(p)}$ **[8,** formula (9)**]**.

The group U of principal units of the field K is generated by the elements

$$E(\theta_{\bar{r}} x_n^{r_n} \cdots x_1^{r_1}),$$

where the multi-indices $\bar{r} = (r_n, \dots, r_1)$ run over all the sets such that

$$0 < \bar{r} < \frac{p}{p-1} \bar{e}_K, \qquad p \nmid \bar{r}.$$

Let the field k contain nontrivial pth roots of 1. Then the group of principal units U_k of k together with the generators

$$\eta_{(t)}(E(\theta_{\bar{r}} x_n^{r_n} \cdots x_1^{r_1})) = E(\theta_{\bar{r}} t_n^{r_n} \cdots t_1^{r_1})$$

may contain also a generator of primary elements. To obtain this generator one should expand a primitive pth root ζ of 1 in the series $z_m(t_1, \dots, t_n)$ in the field k

and replace $t_1, \dots, t_n$ in this series by the independent variables $x_1, \dots, x_n$ (i.e., to lift to the field K). We set

$$s_m(x_1, \dots, x_n) = z_m(x_1, \dots, x_n)^p - 1.$$

Then for each $a \in W$ the element

$$\omega_m(a) = \eta_{(t)}(E(a s_m(x_1, \dots, x_n)))$$

is p-primary in k, i.e., the extension $k(\sqrt[p]{\omega_m(a)})/k$ is unramified. Thus, the group U_k of k is generated by

$$U_k = \langle E(\theta_{\bar r} t_n^{r_n} \cdots t_1^{r_1}), \omega_m(a) : 0 < \bar r < \frac{p}{p-1}\bar e_k, p \nmid \bar r \rangle.$$

5. The Cartier curves module. Let k_0 be an absolutely unramified n-dimensional local field (i.e., p is a uniformizing element with respect to the discrete valuation v_n) and let k be an n-dimensional local field containing k_0, $\mathbb{F}_{q_0}$ being the last residue field of k. As before, T is a quotient field of the ring of Witt vectors $W = W(\mathbb{F}_{q_0})$. We denote by k_1 the composite of k_0 and T.

In the fields k_0 and k we shall consider the valuation rings of rank n, denoting them by $\mathcal{O}_0$ and $\mathcal{O}$ respectively. Let $\mathfrak{M}$ be the maximal ideal of $\mathcal{O}$. Each element $\alpha \in \mathfrak{M}$ admits the following (not unique) expansion:

$$\alpha = \sum_{\substack{\bar r \in I \\ \bar r > 0}} \alpha_{\bar r} t_n^{r_n} \cdots t_1^{r_1}, \qquad \alpha_{\bar r} \in W,$$

where I is an admissible set of multi-indices (see Section 2).

Now let $F = F(x, y)$ be a one-parameter formal group over $\mathcal{O}_0$. It induces on $\mathfrak{M}$ the following structure of a formal $\mathbb{Z}_p$-module $F(\mathfrak{M})$:

$$\alpha \underset{F}{+} \beta = F(\alpha, \beta), \qquad \alpha, \beta \in \mathfrak{M},$$

$$a \underset{F}{\ } \alpha = [a](\alpha), \qquad a \in \mathbb{Z}_p, \quad \alpha \in \mathfrak{M},$$

where $[a]$ is an endomorphism of the formal group corresponding to an element a (for details see **[7**, §1**]**). One can check (see **[7**, Proposition 1**]**) that each element $\alpha \in F(\mathfrak{M})$ may be uniquely written in the form

$$\alpha = \sum_{\substack{\bar r \in I \\ \bar r > 0}}{}_{(F)} \theta_{\bar r} t_n^{r_n} \cdots t_1^{r_1}, \qquad \theta_{\bar r} \in \mathfrak{R},$$

where the series converges in $\mathfrak{M}$. Here I is some admissible set and $\sum_{(F)}$ denotes summation in the formal module $F(\mathfrak{M})$. Without requiring uniqueness, each element $\alpha \in F(\mathfrak{M})$ is representable as a series convergent in $\mathfrak{M}$:

$$\alpha = \sum_{\substack{\bar r \in I \\ \bar r > 0}} \alpha_{\bar r} t_n^{r_n} \cdots t_1^{r_1}, \qquad \alpha_{\bar r} \in W,$$

where I is an admissible set.

We consider now the standard $2n$-dimensional local field

$$L = k_1\{\{x_1\}\}\cdots\{\{x_{n-1}\}\}((x_n)),$$

where $x_1, \dots, x_n$ are independent variables and $k_1 = k_0T$. In the discrete valuation ring of the field L we consider a $\mathbb{Z}_p$-module $\mathscr{H}$ consisting of all the series of the form

$$\beta(x_1, \dots, x_n) = \sum_{\substack{\bar r \in I \\ \bar r > 0}} \beta_{\bar r} x_n^{r_n} \cdots x_1^{r_1},$$

where I is an admissible set of multi-indices and the coefficients $\beta_{\bar r}$ belong to the integral ring $\mathscr{O}_1$ of k_1. As before, we introduce the structure of a formal $\mathbb{Z}_p$-module on $\mathscr{H}$ by setting

$$\beta_1(x_1, \dots, x_n) \underset{F}{+} \beta_2(x_1, \dots, x_n) = F(\beta_1, \beta_2),$$
$$a \underset{F}{} \beta(x_1, \dots, x_n) = [a](\beta(x_1, \dots, x_n)).$$

We shall call the module thus obtained a Cartier curves module and denote it by $\mathscr{H}_F$. Each element $\beta \in \mathscr{H}_F$ can be uniquely written in the form

$$\beta = \sum_{\substack{\bar r \in I \\ \bar r > 0}}{}_{(F)} \beta_{\bar r} x_n^{r_n} \cdots x_1^{r_1}, \qquad \beta_{\bar r} \in \mathscr{O}_1.$$

§2. Arithmetic of the group of points

1. The Hilbert symbol. The fundamental result of class field theory for a multidimensional local field is that there exists a reciprocity homomorphism

$$\sigma\colon K_n(k) \to \mathrm{Gal}(k^{ab}/k)$$

from Milnor's K-group of the field k into the Galois group of its maximal abelian extension (see **[4, 6]**). Let F be a formal group of finite height h over the ring of integers $\mathscr{O}_0$ of an absolutely unramified subextension of k (see §1, paragraph 5). Assume that k contains the group μ_F of all roots of the isogeny $[p]$ of F. Then one can define the pairing

$$(\cdot,\cdot)_F\colon F(\mathfrak{M}) \times K_n(k) \to \mu_F,$$
$$(\alpha, \beta)_F = \Big(\frac{1}{[p]}\alpha\Big)^{\sigma(\beta)} \underset{F}{-} \Big(\frac{1}{[p]}\alpha\Big),$$

which we shall call the Hilbert symbol (here $\frac{1}{[p]}\alpha$ denotes γ such that $[p](\gamma) = \alpha$).

2. Using the method proposed by Honda **[9]** one can obtain a full classification of formal groups over a discrete valuation ring of the field k_0 (i.e., a valuation ring of rank 1). In particular, by this classification each class of isomorphic formal groups of finite height h is determined by the formal Artin-Hasse group F_a with logarithm

$$\lambda_a(x) = \left(1 - \frac{\mathscr{A}(\Delta)}{p}\right)^{-1}(x). \tag{2.1}$$

The coefficients $\alpha_1, \dots, \alpha_h$ of the operator $\mathscr{A}(\Delta) = \alpha_1\Delta + \cdots + \alpha_h\Delta^h$ are in the discrete valuation ring of k_0 and are such that $\alpha_1, \dots, \alpha_{h-1} \equiv 0 \bmod p$; α_h is a unit with respect to the discrete valuation. The operator Δ raises x to the pth power and acts on the coefficients according to (1.2) (see also **[1]**). In this paper we restrict ourselves to considering classes of isomorphic formal groups over $\mathcal{O}_0$ containing F_a, with the coefficients of $\mathscr{A}(\Delta)$ in $\mathcal{O}_0$, and α_h a unit of the ring $\mathcal{O}_0$ with respect to the valuation of rank n.

Thus, let F be a one-parameter formal group over $\mathcal{O}_0$ with logarithm $\lambda(x)$, and let F_a be the Artin-Hasse group with logarithm (2.1) isomorphic to it.

PROPOSITION 1. *The functions*

$$E_F(\beta) = \lambda^{-1}\left(\left(1 - \frac{\mathscr{A}(\Delta)}{p}\right)^{-1}(\beta)\right),$$

$$l_F(\beta) = \left(1 - \frac{\mathscr{A}(\Delta)}{p}\right)\lambda(\beta)$$

realize mutually inverse isomorphisms of the $\mathbb{Z}_p$-module $\mathscr{H}$ and the Cartier module.

The proof is the same as that of the corresponding proposition in **[9]**. □

3. The groups F_0 and $F_{\eta.\rho}$. In the arithmetic of the module $F(\mathfrak{M})$ the formal groups F_0 and $F_{\eta.\rho}$ (constructed for the one-dimensional case in **[1, 2]**) are of special importance. Namely, the following assertion holds:

PROPOSITION 2. *For a formal group F of height h over $\mathcal{O}_0$ there exists an isomorphic formal group F_0 with isogeny of the form*

$$[p]_0 = px + f_0(x^q),$$

where $f_0(x) = d_q x + d_{2q}x^2 + \cdots$ is a series in the ring $\mathcal{O}_0[[x]]$, invertible under superpositions (substitution of one series into another), and $q = p^h$. In addition, for any $\eta \in \mathcal{O}_0$, $\rho = 1, 2, \ldots, h-1$, there exists a formal group $F_{\eta.\rho}$ isomorphic to F with isogeny of the form

$$[p]_{\eta.\rho} = px + p\eta x^{p^\rho} + f_{\eta.\rho}(x^q)$$

for some $f_{\eta.\rho} \in \mathcal{O}_0[[x]]$. The isogeny $[p]_{\eta.\rho}$ is connected with $[p]_0$ as follows:

$$[p]_{\eta.\rho} \equiv [p]_0 + p\eta x^{p^\rho} + (d_q\eta^q - \eta d_q^{p^\rho})x^{qp^\rho} \bmod \deg(qp^\rho + 1).$$

The proof of this is much the same as that of Proposition 1.2 in **[2]**. □

Let λ_0 and $\lambda_{\eta.\rho}$ be logarithms of the formal groups F_0 and $F_{\eta.\rho}$, respectively. We set

$$\varepsilon_0(x) = \lambda^{-1} \circ \lambda_0 \circ f_0(x), \quad \varepsilon_\rho(x) = \lambda^{-1} \circ \lambda_{1.\rho} \circ f_{1.\rho}(x);$$

$\circ$ denotes superposition. The coefficients of the series $\varepsilon_\rho(x)$ are integers.

Similarly to **[2]** one can verify the congruences

$$\varepsilon_0(x) \equiv d_q x \bmod \deg 2, \tag{2.2}$$

$$\varepsilon_\rho(x) \underset{F}{-} \varepsilon_0(x) \equiv d_q x^{p^\rho} \bmod \deg(p^\rho + 1), \tag{2.3}$$

and, besides, from Proposition 2 it follows that d_q is invertible in $\mathcal{O}_0$.

4. Primary elements. We keep the notation of §1; in particular, W denotes the ring of Witt vectors with coefficients in the last residue field of $\mathbb{F}_{q_0}$. As in §2, paragraph 1, we assume that k contains the group μ_F of all roots of the isogeny $[p]$.

Let ξ be an arbitrary root of the isogeny $[p]$ of the formal group F. We expand it in a series in uniformizing elements $t_1, \dots, t_n$ with coefficients in the ring $\mathscr{O}_1$ (see §1, paragraph 5), and then replace these uniformizing elements by independent variables $x_1, \dots, x_n$. We denote the series in the $\mathbb{Z}_p$-module $\mathscr{H}$ thus obtained by $z(x_1, \dots, x_n)$; thus,

$$z(t_1, \dots, t_n) = \xi. \tag{2.4}$$

We set

$$s(x_1, \dots, x_n) = [p](z(x_1, \dots, x_n)). \tag{2.5}$$

For the sake of simplicity we denote the n-tuple of variables $(x_1, \dots, x_n)$ by X and the n-tuple of uniformizing elements by t; so

$$X = (x_1, \dots, x_n), \quad t = (t_1, \dots, t_n).$$

PROPOSITION 3. *Let $p \neq 2$, h the height of F, and a an arbitrary element in W. Let $b = a + a^{\Delta} + \dots + a^{\Delta^{h-1}}$. Then the element*

$$\omega(a) = E_F(bs(X))\big|_{X=t} \tag{2.6}$$

is p-primary in the group of points $F(\mathfrak{M})$, i.e., *the extension $k(\frac{1}{[p]}\omega(a))/k$ is unramified.*

For proof see [**2**, Proposition 2.4]. □

If $[p](x) = px + \dots + a_q x^q + \dots$ is an isogeny of a formal group F, then

$$[p] \equiv \sum_{i=1}^{\infty} a_{qi} x^{qi} \bmod p.$$

Let the valuation of rank n of p in k be $\bar{e}_p = \bar{v}_k(p)$ and $\bar{e}'_p = \bar{e}_p/(q-1)$ (see (1.1)). Then for each $\alpha \in \mathfrak{M}$ of value $\bar{r} = (r_n, \dots, r_1)$ the following congruences hold:

$$\begin{aligned} &[p](\alpha) = p\alpha + \cdots && \bar{r} > \bar{e}'_p, \\ &[p](\alpha) = a_q\alpha^q + \cdots && \bar{r} < \bar{e}'_p, \\ &[p](\alpha) = p\alpha + a_q\alpha^q + \cdots && \bar{r} = \bar{e}'_p, \end{aligned} \tag{2.7}$$

(herein dots denote the terms of higher order).

As an immediate consequence of (2.7) one has

PROPOSITION 4. *Let $\alpha \in F(\mathfrak{M})$ and $\bar{v}(\alpha) > q\bar{e}'_p$. Then $\alpha \in [p](F(\mathfrak{M}))$.*

PROPOSITION 5. *For each* $\bar{r} = (r_n, \ldots, r_1)$ *satisfying*

$$0 < \bar{r} < \bar{e}'_p, \qquad q \not| \bar{r},$$

and for each θ *in the Teichmüller system* $\mathfrak{R}$ *of the ring* $\mathcal{O}$ *let us choose an element* $\varepsilon_{\bar{r}}(\theta)$ *in* $F(\mathfrak{M})$ *such that*

$$\varepsilon_{\bar{r}}(\theta) \equiv \theta t^{\bar{r}} = \theta t_n^{r_n} \cdots t_1^{r_1} \bmod \mathfrak{M}^{\bar{r}+1}. \tag{2.8}$$

Besides, for $\bar{r} = q\bar{e}'_p$ *and* $\theta' \in \mathfrak{R}$ *we take elements*

$$\eta(\theta') \equiv \theta' t^{q\bar{e}'_p} = \theta' t_n^{qe'_n} \cdots t_1^{qe'_1} \bmod \mathfrak{M}^{q\bar{e}'_p+1}, \tag{2.9}$$

that are indivisible by the isogeny $[p]$ *in the group of points* $F(\mathfrak{M})$.

Then the elements (2.8) *and* (2.9) *generate* $F(\mathfrak{M})$, i.e., *for each* $\alpha \in F(\mathfrak{M})$ *there exists an expansion*

$$\alpha = \sum_{m.\bar{r}}{}_{(F)} [p^m](\varepsilon_{\bar{r}}(\theta_{\bar{r}.m})) +_F \sum_{m}{}_{(F)} [p^m](\eta(\theta'_m)).$$

The proof is a direct consequence of (2.7). □

By Propositions 4 and 5, taking into account the form of the isogeny $[p]$, for the primary element (2.6) we have

$$\omega(a) \equiv E_F(ba_q z^q(X))\Big|_{X=t} \bmod [p](F(\mathfrak{M})).$$

We let μ_F be the group of roots of $[p]$. This group contains h generators, which will be denoted by $\xi_1, \ldots, \xi_h$. For each ξ_j and $a \in W$ let $\omega_j(a)$ be the primary element

$$\omega_j(a) = E_F(ba_q z_j^q(X))\Big|_{X=t}. \tag{2.10}$$

5. Norm series. Let F be a formal group of finite height h over the ring of integers $\mathcal{O}_0$ of an n-dimensional local field k_0 whose characteristic may be different from that of $k_0^{(n-1)}$. Let

$$[p] = \sum_{i=1}^{\infty} a_i x^i$$

be its isogeny. We denote by $f(x)$ the series

$$f(x) = \sum_{i=1}^{\infty} a_{iq} x^i.$$

Note that $[p] \equiv f(x^q) \bmod p$ (see, for example [**10**]). By $f^{-1}(x)$ we denote the series inverse to $f(x)$ under superposition.

PROPOSITION 6. *For every* $\alpha \in F(\mathfrak{M})$ *let* k_α/k *be the extension of* k *obtained by adjoining the roots of the equation* $[p](x) = \alpha$. *Then* $f^{-1}(\alpha)$ *is a norm in this extension.*

The proof for the case of one-dimensional local fields was given in [**1**] and can be carried over to the general case without any changes. So we only outline it now.

On the set of formal power series with coefficients in $\mathcal{O}$ we introduce the operations γ and τ by

$$\gamma(\sum_{i=0}^{\infty} d_i x^i) = \sum_{i=0}^{q-1} d_i x^i,$$
$$\tau(\sum_{i=0}^{\infty} d_i x^i) = \sum_{i=q}^{\infty} d_i x^{i-q}.$$

Consider the equation $[p](x) - \alpha = 0$. From the Weierstrass preparation theorem it follows that there exists a polynomial $h(x) = x^q + \dots + u_1 x + u_0$ and an invertible (in the usual sense) series $\varepsilon(x)$ with coefficients in $\mathcal{O}$ such that $[p](x) - \alpha = h(x)\varepsilon(x)$. Let us show that u_0 is a norm in the extension k_α/k. The polynomial $h(x)$ is factorable over the field k into the product of irreducible polynomials $h_1(x), \dots, h_s(x)$, each of which generates the extension k_α/k. Hence, the constant terms of $h_1(x), \dots, h_s(x)$ are norms in k_α/k and so is their product. Thus, we have obtained that u_0 belongs to the norm subgroup of the extension k_α/k. Let us verify that $u_0 \equiv f^{-1}(\alpha) \bmod p^2$. In exactly the same way as in **[1]**, we use Manin's explicit formula **[11]**:

$$h(x) = \frac{[p] - \alpha}{\tau([p])}\left(1 + \tau \circ \frac{\gamma([p] - \alpha)}{\tau([p])}\right)^{-1}(1).$$

Computing by this formula the constant term of the series $h(x)$, we get $u_0 \equiv f^{-1}(\alpha) \bmod p^2$. Now from Proposition 4 we have that $f^{-1}(\alpha)$ belongs to the norm subgroup of k_α/k. The proposition is proved. □

It is more convenient to use this proposition in the following form:

COROLLARY 1. *An element $\alpha \in \mathfrak{M}$ is a norm in $k_{f(\alpha)}/k$.*

COROLLARY 2. *Let $\{\beta_1, \dots, \beta_n\} \in K_n(k)$, and $\beta_i \in \mathfrak{M}$ for some i. Then*

$$(f(\beta_i), \{\beta_1, \dots, \beta_n\})_F = 0.$$

PROOF. Let $\beta \in k_{f(\beta_i)}$ be such that $N_{k_{f(\beta_i)}/k}(\beta) = \beta_i$. Then the norm mapping

$$N \colon K_n(k_{f(\beta_i)}) \to K_n(k)$$

sends $\{\beta_1, \dots, \beta_{i-1}, \beta, \beta_{i+1}, \dots, \beta_n\}$ to $\{\beta_1, \dots, \beta_{i-1}, N_{k_{f(\beta_i)}/k}(\beta), \beta_{i+1}, \dots, \beta_n\} = \{\beta_1, \dots, \beta_{i-1}, \beta_i, \beta_{i+1}, \dots, \beta_n\}$ (see **[5]**). So the symbol $\{\beta_1, \dots, \beta_n\}$ belongs to the norm subgroup $N(K_n(k_{f(\beta_i)}))$. Since under the reciprocity mapping $K_n(k) \to \mathrm{Gal}(k_{f(\beta_i)}/k)$ the elements of $N(K_n(k_{f(\beta_i)}))$ transfer into the identity automorphism, $(f(\beta_i), \{\beta_1, \dots, \beta_n\})_F = 0$. □

Applying Corollary 2 to the formal groups F_0 and $F_{1.\rho}$ and taking into account that the series $\lambda^{-1} \circ \lambda_{1.\rho}$ effects the isomorphism between $F_{1.\rho}$ and F we get the following

COROLLARY 3. *Let $\{\beta_1, \dots, \beta_n\} \in K_n(k)$ be a symbol, $\beta_i \in \mathfrak{M}$. Then*

$$(\varepsilon_\rho(\beta_i), \{\beta_1, \dots, \beta_n\})_F = 0$$

for $\rho = 0, 1, \dots, h-1$.

6. A system of generators of a formal module. Herein we shall prove the main result of the section.

THEOREM 1. *Let $p \neq 2$. Denote by J the set of multi-indices $\bar{r} = (r_n, \dots, r_1)$ such that $0 < r < q\bar{e}'_p$ and $p \nmid \bar{r}$. Then the elements $\varepsilon_\rho(\theta t^{\bar{r}})$ $(\bar{r} \in J,\ 0 \leqslant \rho \leqslant h-1,\ \theta \in \mathfrak{R})$ together with the primary elements $\omega_j(a)$ $(1 \leqslant j \leqslant h,\ a \in W)$ form a system of generators of a $\mathbb{Z}_p$-module $F(\mathfrak{M})$. In addition,*

$$(\varepsilon_\rho(\theta t^{\bar{r}}), \{t_1, \dots, t_n\})_F = 0,$$
$$(\omega_j(a), \{t_1, \dots, t_n\})_F = [\operatorname{tr} a](\xi_j),$$

where $\operatorname{tr}\colon W \to \mathbb{Z}_p$ *is the trace operator.*

PROOF. The fact that the given set is a system of generators follows from Proposition 5 and identities (2.2), (2.3), and (2.10). A detailed proof for the one-dimensional case can be found in **[2]**. Now we pass to the proof of symbol properties. Let $p \nmid r_k$. Let us show that

$$(\varepsilon_\rho(\theta t^{\bar{r}}), \{t_1, \dots, t_n\})_F = 0.$$

From Corollary 3 it follows that

$$(\varepsilon_\rho(\theta t^{\bar{r}}), \{t_1, \dots, t_{k-1}, \theta t^{\bar{r}}, t_{k+1}, \dots, t_n\})_F = 0. \tag{2.11}$$

In addition,

$$\begin{aligned} \{t_1, \dots, t_{k-1}, \theta t^{\bar{r}}, t_{k+1}, \dots, t_n\} &= \{t_1, \dots, t_{k-1}, \theta, t_{k+1}, \dots, t_n\} \\ &\times \prod_{s \neq k} \{t_1, \dots, t_{k-1}, t_s, t_{k+1}, \dots, t_n\}^{r_s} \{t_1, \dots, t_{k-1}, t_k, t_{k+1}, \dots, t_n\}^{r_k}. \end{aligned} \tag{2.12}$$

It is easy to see that

$$(\varepsilon_\rho(\theta t^{\bar{r}}), \{t_1, \dots, t_{k-1}, \theta, t_{k+1}, \dots, t_n\})_F = 0$$

and if $s \neq k$, then

$$(\varepsilon_\rho(\theta t^{\bar{r}}), \{t_1, \dots, t_{k-1}, t_s, t_{k+1}, \dots, t_n\})_F = 0.$$

So, from (2.11) and (2.12) it follows that

$$(\varepsilon_\rho(\theta t^{\bar{r}}), \{t_1, \dots, t_{k-1}, t_k, t_{k+1}, \dots, t_n\}^{r_k})_F = 0.$$

Since $(r_k, p) = 1$, from the last equality we get

$$(\varepsilon_\rho(\theta t^{\bar{r}}), \{t_1, \dots, t_n\}^{r_k})_F = \left[\frac{1}{r_k}\right] \circ (\varepsilon_\rho(\theta t^{\bar{r}}), \{t_1, \dots, t_n\}^{r_k})_F = 0.$$

It was proved in **[6]** that the restriction of the automorphism $\sigma(\{t_1, \dots, t_n\})$ to an unramified extension of k coincides with the Frobenius automorphism. Then from the definition of primary elements it directly follows that

$$(\omega_j(a), \{t_1, \dots, t_n\})_F = [\operatorname{tr} a](\xi_j)$$

(see, for example, **[2, §2]**). The theorem is proved. □

§3. An explicit formula for the Hilbert symbol

In this section we define a pairing γ_V on the Cartier curves module $\mathscr{H}_F$. Its basic properties will be studied in a forthcoming paper.

1. On the elements of a multiplicative group $\mathscr{H}_m$ **[8, 3]** we define the function

$$l(\beta) = \frac{1}{p}\log \beta^p/\beta^\Delta,$$

where $\beta \in \mathscr{H}_m$, and Δ is the Frobenius operator. The series $l(\beta)$ belongs to the ring $\mathscr{O}_p$, since $\beta^p \equiv \beta^\Delta \bmod p$. Also, it is clear that

$$l(\beta_1\beta_2) = l(\beta_1) + l(\beta_2).$$

We denote the partial logarithmic derivative of the series $\beta(x_1, \dots, x_n)$ by $\delta_i(\beta) = \beta^{-1}\frac{\partial}{\partial x_i}\beta$. Let h be the height of a formal group F, $q = p^h$. On the module $\mathscr{H}_F$ we introduce the function similar to the logarithmic derivative,

$$\delta_i^F(\alpha) = \frac{\partial}{\partial x_i}\frac{\alpha_h}{q}\lambda(\alpha)^{\Delta^h},$$

where $\alpha \in \mathscr{H}_F$ and α_h is a coefficient in $\mathscr{A}(\Delta)$.

Finally, we fix a series $V(x_1, \dots, x_n) \in \mathscr{H}$ satisfying

$$\frac{\partial}{\partial x_i}V \equiv 0 \bmod q, \qquad 1 \leqslant i \leqslant n.$$

The pairing on the Cartier curves module

$$\gamma_V\colon \mathscr{H}_F \times \mathscr{H}_m^{\otimes n} \to \mathscr{O}_1 \bmod \mathfrak{M}_1$$

is defined as follows:

$$\gamma_V(\alpha, \{\beta_1, \dots, \beta_n\}) = \operatorname{res}\Phi(\alpha, \beta_1, \dots, \beta_n)/V \bmod \mathfrak{M}_1.$$

Here $\mathscr{O}_1$ is the ring of integers of the composite $k_1 = k_0T$, $\mathfrak{M}_1$ is the maximal ideal in $\mathscr{O}_1$ and Φ is of the form

$$\Phi = l_F(\alpha)D_{n+1} - l(\beta_n)^{\Delta^{h-1}}D_n + \dots + (-1)^n l(\beta_1)^{\Delta^{h-1}}D_1,$$

where $D_{n+1} = \det((\delta_i(\beta_i))_{1 \leqslant i,j \leqslant n})$,

$$D_r = \frac{1}{q^{n-r}}\begin{vmatrix} \delta_1(\beta_1) & \dots & \delta_n(\beta_1) \\ \vdots & & \vdots \\ \delta_1(\beta_{r-1}) & \dots & \delta_n(\beta_{r-1}) \\ \delta_1(\beta_{r+1}^{\Delta^h}) & \dots & \delta_n(\beta_{r+1}^{\Delta^h}) \\ \vdots & & \vdots \\ \delta_1(\beta_n^{\Delta^h}) & \dots & \delta_n(\beta_n^{\Delta^h}) \\ \delta_1^F(\alpha) & \dots & \delta_n^F(\alpha) \end{vmatrix}, \qquad 1 \leqslant r \leqslant n.$$

The division of Φ by V is performed in the field $L' = k_1\{\{x_1\}\}\cdots\{\{x_n\}\}$, and the residue is taken at x_i.

THEOREM 2. *The pairing γ_V determines the well-defined pairing (which we denote by the same letter) on the set*

$$\gamma_V : \mathscr{H}_F/[p]\mathscr{H}_F \times K_n^{\mathrm{top}}(\mathscr{H}_m)/pK_n^{\mathrm{top}}(\mathscr{H}_m) \to \mathscr{O}_1 \bmod \mathfrak{M}_1.$$

Let $u(x_1, \dots, x_n) = s(x_1, \dots, x_n)/z(x_1, \dots, x_n)$, where the series $s(x_1, \dots, x_n)$, $z(x_1, \dots, x_n)$ are those in (2.4) and (2.5).

PROPOSITION 7. *The kernel of γ_V on the left is equal to the formal $\mathbb{Z}_p$-module generated by $u(x_1, \dots, x_n)$.*

2. Pairing on the group of points $F(\mathfrak{M})$. Let J be a subgroup of $\mathscr{H}_m$ consisting of the series of the form $1 + u\psi$, where ψ is a series in $\mathscr{O}(p)$, $\psi(0) = 0$.

PROPOSITION 8. *The following isomorphism of $\mathbb{Z}_p$-modules holds:*

$$\eta_{(t)}^F : F(\mathfrak{M}) \xrightarrow{\sim} \mathscr{H}_F/I, \quad \alpha = \alpha(t_1, \dots, t_n) \mapsto \alpha(x_1, \dots, x_n) \bmod I,$$

$$\eta_{(t)} : k^* \xrightarrow{\sim} \mathscr{H}_m/J, \quad \beta = \beta(t_1, \dots, t_n) \mapsto \beta(x_1, \dots, x_n) \bmod J,$$

where $t_1, \dots, t_n$ are the uniformizing parameters of k.

The proof will be given in the next paper.

We consider a basis $\xi_1, \dots, \xi_h$ of the group μ_F of the roots of the isogeny $[p]$. For each ξ_i there are corresponding series $z_i(x_1, \dots, x_n)$ and $s_i(x_1, \dots, x_n)$ (see (2.4), (2.5)). It may be checked that there exist elements θ_{ij} in the Teichmüller system such that

$$\xi_j \equiv \theta_{ji}\xi_i \bmod \mathfrak{M}^{\frac{2q}{q-1}\bar{e}_p}.$$

In addition, in the ring of Witt vectors there are unique elements $v_1, \dots, v_n$ such that

$$\operatorname{tr}_{W/\mathbb{Z}_p}(v_i\theta_{ji}) = 0, \qquad i \neq j,$$
$$\operatorname{tr}_{W/\mathbb{Z}_p}(v_i\theta_{ii}) = 1.$$

As the series V we consecutively take the series $s_1, \dots, s_h$ and consider the pairings $\gamma_1, \dots, \gamma_h$, respectively. Thus we can define the pairing

$$\langle \cdot, \cdot \rangle_F : F(\mathfrak{M}) \times K_n(k) \to \mu_F$$

as follows:

$$\langle \alpha; \{\beta_1, \dots, \beta_n\} \rangle_F = \sum_{i=1}^{h}{}_{(F)} [\operatorname{tr}_{W/\mathbb{Z}_p}(v_i\gamma_i^F)](\xi_i),$$

where $\gamma_i^F(\alpha; \{\beta_1, \dots, \beta_n\}) = \gamma_i(\eta_{(t)}^F(\alpha), \{\eta_{(t)}(\beta_1), \dots, \eta_{(t)}(\beta_n)\})$.

THEOREM 3. *Let $p \neq 2$. The pairing $\langle \cdot, \cdot \rangle_F$ coincides with the Hilbert symbol $(\cdot, \cdot)_F$ (so, we have obtained an explicit formula for $(\cdot, \cdot)_F$).*

§4. Galois representations

The explicit formula so obtained is applicable to the computation of Galois representations on Tate modules of Honda's formal groups. Let

$$\rho : \operatorname{Gal}(\bar{k}_0/k_0) \to \operatorname{GL}(T_F) \simeq \operatorname{GL}_h(\mathbb{Z}_p)$$

be such a representation. Let k_0 contain the $(q-1)$th roots of unity. Then the extension $k_0(\mu_F)/k_0$ is a weakly ramified extension of $(q-1)$th degree, and its Galois group is isomorphic to the cyclic group $\mathbb{F}_q^*$. Let G be the image of the representation ρ. Then on G there exists a p-adic filtration

$$G_k = \{g \in G : g \equiv 1 \bmod p^k\}.$$

The Galois group $\mathrm{Gal}(k_0(\mu_F)/k_0)$ acts as conjugation on the factors $H_k = G_k/G_{k+1}$, turning them into $\mathbb{F}_q^*$-modules. We denote by M_i $(i \in \mathbb{Z}/h\mathbb{Z})$ the $\mathbb{F}_q^*$-module, isomorphic (as an abelian group) to the additive group $\mathbb{F}_q^+$, on which the group $\mathbb{F}_q^*$ operates as follows:

$$\theta * m = \theta^{1-p^i} m, \qquad \text{where } \theta \in \mathbb{F}_q^*, \quad m \in \mathbb{F}_q^+.$$

J.–M. Fontaine showed in **[12]** that for any $k \geqslant 1$ the $\mathbb{F}_q^*$-module H_k is isomorphic to the direct sum $\bigoplus_{i \in I} M_i$, where I is some subset of $\mathbb{Z}/h\mathbb{Z}$. He ascertained that $M_0 \subset H_k$ for any $k \geqslant 1$, and that if $M_i \subset H_k$, then $M_i \subset H_{k+1}$. So one can define the function

$$\nu \colon \mathbb{Z}/h\mathbb{Z} \to \mathbb{N}^+,$$
$$\nu(i) = \min\{k : M_i \subset H_k\}.$$

The problem of describing G in terms of ν has been solved so far only for some special cases (see **[12]**–**[14]**). In **[1]** an approach based on the explicit formula for the Hilbert symbol was developed for the case of one-dimensional local fields. Using Theorem 3 one can obtain similar results for multidimensional fields. Let us formulate these results more precisely. Let F be an Artin-Hasse formal group over $\mathcal{O}_0$ with the logarithm $\lambda = (1 - \frac{\mathscr{A}(\Delta)}{p})^{-1}(x)$. Denote by m the greatest integer such that all coefficients $\alpha_1, \ldots, \alpha_{h-1}$ of the operator $\mathscr{A}(\Delta) = \alpha_1\Delta + \cdots + \alpha_{h-1}\Delta^{h-1} + \alpha_h\Delta^h$ are divisible by p^m. The extension k/k_0 obtained by adjoining all the points of order p^m to k_0 is an abelian extension of k_0 of degree $(q-1)q^{m-1}$. More precisely, $H_k \simeq M_0$ for any $1 \leqslant k \leqslant m-1$. Let L be the extension of k_0 obtained by adjoining all the points of order p^{m+1} of F. The extension L/k is abelian and one has the following isomorphism

$$\varepsilon \colon K_n(k)/N_{L/k}(K_n(L)) \xrightarrow{\sim} \mathrm{Gal}(L/k).$$

Let us turn the K-group $K_n(\mathscr{H}_m)$ into an $\mathbb{F}_q^*$-module, by setting

$$\begin{aligned}\theta * \{\beta_1(x_1, \ldots, x_n), \ldots, \beta_n(x_1, \ldots, x_{n-1}, x_n)\} \\ = \{\beta_1(x_1, \ldots, x_{n-1}, \theta x_n), \ldots, \beta_n(x_1, \ldots, x_{n-1}, \theta x_n)\}\end{aligned}$$

(we denote the Teichmüller representative by the same letter θ). Further, by means of explicit formulas we construct the mapping τ from $K_n(\mathscr{H}_m)$ into the direct sum $\bigoplus_{i \in \mathbb{Z}/h\mathbb{Z}} M_i$ of Fontaine modules and prove the following result concerning τ.

THEOREM 4. *The mapping*

$$\tau \colon K_n(\mathscr{H}_m) \to \bigoplus_{i \in \mathbb{Z}/h\mathbb{Z}} M_i$$

is a homomorphism of $\mathbb{F}_q^*$*-modules. Moreover,*

$$\mathrm{Im}\,\tau = \bigoplus_{i \in I} M_i, \qquad \text{where } I = \{0\} \cup \{i : \alpha_i \not\equiv 0 \bmod p^{m+1}\}$$

(we recall that α_i is the coefficient at Δ^i in $\mathscr{A}(\Delta)$). Finally, τ induces a monomorphism

$$\varkappa\colon K_n(k)/N_{L/k}(k_n(L)) \hookrightarrow \bigoplus_{i\in\mathbb{Z}/h\mathbb{Z}} M_i.$$

The diagram

$$\begin{array}{ccc} K_n(\mathscr{H}_m) & \xrightarrow{\ \tau\ } & \bigoplus_{i\in\mathbb{Z}/h\mathbb{Z}} M_i \\ {\scriptstyle \eta_{(t)}^{-1}}\big\downarrow & & \\ K_n(k)/N_{L/k}(K_n(L)) & & \end{array}$$

proves to be commutative.

COROLLARY. *The following isomorphism takes place*:

$$H_m \simeq \bigoplus_{i\in I} M_i, \qquad \text{where } I = \{0\} \cup \{i : \alpha_i \not\equiv 0 \bmod p^{m+1}\}.$$

REMARK. For the formal groups over the ring of Witt vectors of the perfect field of characteristic p this corollary was proved by Nakamura in **[13, 14]**, but his method is quite different from ours.

References

1. D. G. Benua and S. V. Vostokov, *Norm pairing in formal groups and Galois representations*, Algebra i Analiz **2** (1990), no. 6, 69–97; English transl. in Leningrad Math. J. **2** (1991).
2. ______, *Arithmetic of the group of points of a formal group*, Zap. Nauchn. Sem. Leningrad. Otdel. Mat. Inst. Steklov. (LOMI) **191** (1991), 9–23; English transl. in J. Soviet Math. (to appear).
3. V. G. Lomadze, *On the ramification theory of two-dimensional local fields*, Mat. Sb. **109** (1979), no. 3, 378–394; English transl. in Math. USSR-Sb. **37** (1980).
4. A. N. Parshin, *Class field and algebraic K-theory*, Uspekhi Mat. Nauk **30** (1975), no. 1, 253–254. (Russian)
5. ______, *Local class field theory*, Trudy Mat. Inst. Steklov. **165** (1984), 143–170 (Russian); English transl. in Proc. Steklov. Inst. Math. **1985**, no. 3.
6. K. Kato, *A generalization of local class field theory by using K-groups*. I, J. Fac. Sci. Univ. Tokyo Sect. IA Math. **26** (1979), 303–376; II **27** (1980), 603–683.
7. S. V. Vostokov, I. B. Zhukov, and I. B. Fesenko, *Multidimensional local fields. Methods and constructions*, Algebra i Analiz **2** (1990), no. 4, 91–118; English transl. in Leningrad Math. J. **2** (1991).
8. S. V. Vostokov, *Explicit construction of the theory of class field of a multidimensional local field*, Izv. Akad. Nauk SSSR Ser. Mat. **49** (1985), no. 2, 283–308; English transl. in Math. USSR-Izv. **26** (1986).
9. T. Honda, *On the theory of commutative formal groups*, J. Math. Soc. Japan **22** (1970), 213–246.
10. A. Fröhlich, *Formal groups*, Lecture Notes in Math., vol. 74, Springer-Verlag, Berlin and New York, 1968.
11. Yu. I. Manin, *Cyclotomic fields and modular curves*, Uspekhi Mat. Nauk **26** (1971), no. 6, 7–71; English transl. in Russian Math. Surveys **26** (1971).
12. J.-M. Fontaine, *Points d'ordre fini d'un groupe formel sur une extension non ramifiée de Z_p*, Bull. Soc. Math. France Mém. **37** (1974), 75–79.
13. T. Nakamura, *Finite subgroups of formal A-modules over p-adic integer rings*, Trans. Amer. Math. Soc. **286** (1984), 765–769.
14. ______, *On torsion points of formal groups over a ring of Witt vectors*, Math. Z. **193** (1986), 397–404.

RUSSIAN STATE PEDAGOGICAL UNIVERSITY, 38, MOJKA, ST. PETERSBURG, 191065, RUSSIA

DEPARTMENT OF MATHEMATICS AND MECHANICS, ST. PETERSBURG STATE UNIVERSITY, 2 BIBLIOTECHNAYA PL., STARY PETERGOF, ST. PETERSBURG, 198904, RUSSIA

Translated by E. GOLUBEVA

Amer. Math. Soc. Transl.
(2) Vol. **159**. 1994

A Lower Bound for the Complexity of the SPLIT-FIND Problem for a Pointer Machine

D. Yu. Burago

Introduction

The task of the SPLIT-UNION-FIND algorithm (or SUF-algorithm) is to decompose a linear list $\{x_1, x_2, \dots, x_n\}$ into several intervals by means of decomposing one of the intervals already given into two (SPLIT command) or by merging two adjacent intervals (UNION command). In addition, this algorithm finds for a given element of the list the beginning of the interval containing this element (FIND command). Accordingly, the SPLIT-FIND and UNION-FIND algorithms (SF- and UF-algorithms) realize only two of the commands described here.

By complexity we mean the amortized worst-case complexity, i.e., the number of computation steps required by the algorithm (on the average) for executing one command while executing the sequence of n commands, for the "worst" sequence.

The study of complexity of SUF-algorithms is carried out for the Kolmogorov-Schönhage model [**1**], which is also called a pointer machine; but what is usually made use of is only the fact that its memory is an oriented graph with output degree of the vertices at most two.

In [**2, 3**] the notion of separating SUF-algorithms (satisfying the separation assumption) was proposed. This assumption means that x_i from different intervals lie in different components of the memory.

We give here the following table (borrowed from [**4**], where one can also get acquainted with a detailed statement of the problem), presenting the estimates of the complexity for the SUF-, SU-, and UF-algorithms.

The lower bounds obtained in this paper for the general case and for the separating algorithm differ and the question of whether this fact reflects the essence of the matter or some deficiency in the method applied for the analysis seems to be very intriguing.

Problem	General case	Separating algorithm
UF-algorithm	—	$\theta(\alpha(n))$ [2]
SF-algorithm	—	upper bound $O(\log^* n)$ [5]
SUF-algorithm	$\theta(\log\log n)$ [3]	$\theta(\log n)$ [3]

1991 *Mathematics Subject Classification*. Primary 68R10, 68Q25; Secondary 68Q30.

§1. Lower bound for the general case

1.1. We use the following formalization imposing only minimal restrictions on the computational model.

Let G be an oriented graph with n input vertices $\{x_1, x_2, \dots, x_n\}$ and n output vertices $\{y_1, y_2, \dots, y_n\}$ with some output vertices marked by the symbol $*$. Let A be an algorithm rearranging the graph G, i.e., destructing or creating some vertices or edges by the following two commands of "the experimenter":

SPLIT(y_i); in this case "the experimenter" marks by $*$ some unmarked vertex y_i and A rearranges the graph in a certain way; the cost of executing the command is the number of changes made in G;

FIND(x_i); A also rearranges G in a certain way and the cost of executing the command is the number of changes made in G plus $d(x_i, F(x_i))$, which denotes the length of the shortest path from x_i to $F(x_i)$ (where $F(x_i)$ is the output vertex having the largest number not exceeding i among all output vertices marked by $*$) in the graph G before changes are made according to this FIND command. (It is assumed that such a route from x_i to $F(x_i)$ exists, i.e., $d(x_i, F(x_i)) < \infty$.)

The pair (G, A) will be called an SF-system.

By complexity of the system we mean the smallest $Q(n)$ such that for every sequence of $m \geqslant n$ commands their total cost does not exceed $mQ(n)$ (note that here neither the complexity of the construction of the initial graph nor the cost of the work of the algorithm A computing the procedure of rearranging G are taken into account).

1.2. Here we show that one may restrict oneself to considering SF-systems having very special form.

Following **[4]** we call the graph G a k-structure if, for every i, $d(x_i, F(x_i)) < k$.

LEMMA 1. *Let a system $C1$ have the complexity $Q(n)$. Then there exists a system $C2$ with complexity no greater than $4Q(n)$ that does not rearrange the graph while executing* FIND *and in the course of its performance the graph always remains a $2Q(n)$-structure.*

PROOF. We describe the action of $C2$ when executing the SPLIT command (under our condition it does nothing after receiving a FIND command). First, $C2$ performs in the same way as $C1$ would act after receiving the SPLIT command in the same situation (it will become clear later that every graph obtained as a result of the action of $C2$ could be a result of the action of $C1$, and thus the above instruction is well-defined). Next, if the resulting graph is a $2Q(n)$-structure, the command is considered executed; otherwise, we execute FIND(x_i) (for x_i with $d(x_i, F(x_i)) > 2Q(n)$) according to the $C1$-algorithm and so on, until we obtain a $2Q(n)$-structure.

It remains to estimate the complexity of $C2$. Let $C2$ receive a sequence of commands

$$P_1 = S_1 F_1 \cdots S_{m_1} F_{m_2} \qquad (S_i - \text{SPLIT}, \quad F_i - \text{FIND}).$$

Then the graph is rearranged according to the $C2$-algorithm having received the sequence of commands

$$P_2 = S_1 F_{11} \cdots F_{1n_1} S_2 \cdots S_{m_1} F_{m_1 1} \cdots F_{m_1 n_{m_1}};$$

therefore, the cost of executing P_1 by $C2$ is not greater than the cost of executing P_2

by $C1$ plus $2Q(n)((\sum_{i-1}^{m_1} n_i) + m_2)$, since the cost of each FIND command of $C2$ is no greater than $2Q(n)$.

According to the definition of complexity, the time for executing P_2 by $C1$ does not exceed $Q(n)(m_1 + \sum_{i=1}^{m_1} n_i)$, but the cost of executing each F_{ij} in P_2 is no less than $2Q(n)$; so the number of FIND commands in P_2 is no greater than n. Therefore, the cost of executing P_2 is no greater than $Q(n)(m_1 + n)$.

Thus, the cost of executing P_1 by $C2$ does not exceed

$$Q(n)(m_1 + n) + 2Q(n)m_2 \leqslant Q(n)(2m_1 + 2m_2 + n)$$
$$\leqslant Q(n)(3(m_1 + m_2)) \leqslant 4Q(n)(m_1 + m_2). \qquad \square$$

The SF-system will be called a q-monotone system if its graph at any moment admits a strictly monotone mapping onto the segment $\{1, 2, \dots, q+1\} = N_q$ of the set of natural numbers, where x_i is mapped to 1, y_i to $q+1$, and the edge from the vertex mapped to i may go only to the vertex mapped to $i+1$. Such a mapping will be called a level.

LEMMA 2. *Let the system $C1$ have complexity $Q(n)$. Then there exists a system $C2$ with complexity not exceeding $8Q(n)^2$ that does not rearrange the graph when executing* FIND *commands; it is $2Q(n)$-monotone and its graph always remains a $2Q(n)$-structure. Moreover, the output degree of the vertices of the graph of $C2$ exceeds the output degree of $C1$ by no more than* 1.

PROOF. Consider the system S constructed as in Lemma 1 and having complexity no greater than $4Q(n)$ with the graph being a $2Q(n)$-structure. Let $V = \{v_i\}$ be the set of vertices of this graph and let $R = \{r_i\} = \{v_i^1, v_i^2\}$ be the set of its edges. Define the set of vertices of the graph $C2$ as $V \times N_{2Q(n)}$ with input vertices $(x_i, 1)$ and output vertices $(y_i, 2Q(n)+1)$ and its set of edges as

$$\{((v_1, n_1), (v_2, n_2)) \mid n_2 = n_1 + 1 \;\&\; ((v_1, v_2) \in R \vee v_1 = v_2)\}.$$

It is not hard to see that each rearrangement of the graph of the system S with cost a requires exactly $2Q(n)a$ operations for the corresponding rearrangement of the graph of $C2$; therefore the complexity of $C2$ is no greater than $8Q(n)^2$. Since the graph of S remains a $2Q(n)$-structure during the performance, the graph of $C2$ also remains a $2Q(n)$-structure. One can take the second coordinate in the product $V \times N_{2Q(n)}$ as a level. $\square$

We call an SF-system a $(d_1, d_2, \dots, d_l)$-fiber if its graph is a union of subgraphs U_i, $i = 1, 2, \dots, l$, and each U_i is an i-monotone graph (the definition of monotonicity given for the systems is carried over to graphs in an obvious way) with the output degree of x_i no greater than d_i and with the output degree of the remaining vertices no greater than three.

Obviously, a q-monotone system with output degree of the vertices not exceeding three is a $\underbrace{(0, 0, \dots, 3)}_{q}$-fiber.

For a D-fiber system, $D = (d_1, \dots, d_l)$, we define its depth as $\gamma(D) = \sum_{i=1}^{l} d_i 4^i$. For the fibration vector D we also define

$$s_1(D) = \{d_1 - 1, d_2, \dots, d_l\},$$
$$s_i(D) = \{d_1, \dots, d_{i-1} + 3, d_i - 1, \dots, d_l), \quad i > 1.$$

It is easy to check that $\gamma(s_i(D)) < \gamma(D)$, $i \in \mathbf{N}$.

Finally, we shall need the functions

$$G_0 = \exp,$$
$$G_1(n) = \exp^n(2), \quad \exp^n = \underbrace{\exp \circ \exp \circ \exp \circ \cdots \circ \exp}_{n},$$
$$\dots\dots$$
$$G_{i+1}(n) = G_i^n(2).$$

For $x = G_i(y)$ we set $l_i(x) = y$ and for all other x we define the function l_i in a piecewise-linear way. Let $\alpha(x)$ be the inverse function to $G_x(y)$, $x \in \mathbf{N}$.

1.3. THE PRINCIPAL LEMMA. *Let C be a D-fiber system. Then there exists a sequence of no more than n/p, $p = l_{\gamma(D)}(n)$,* SPLIT *commands; the cost of executing this sequence by C is no less than $n/\beta(\gamma(D))$, where $\beta(x) = 10^{10^x}$.*

The proof goes by induction on $\gamma(D)$. For convenience we extend somewhat the possibilities of the system. Namely, we allow it to destroy, while rearranging the graph, several input vertices, simultaneously adding the number of the vertices destroyed into the cost of executing this command. However, in this case we assume that the system does not rearrange the edges outgoing from the input vertices; these edges do not affect the distances from the remaining input vertices to the output ones. Thus, instead of rearranging the edge going out from the input vertex x_i the system may simply destroy this vertex.

Consider two cases.

Case 1. $d_1 \neq 0$.

In this case we have to analyze more closely the structure of the edges of U_i, i.e., of the edges directly connecting x_i with y_j. Paint red those edges (x_i, y_j) such that exactly one edge goes out from each x_i having output edges in U_i. Further we shall consider only such i that no more than one red edge leads to y_i. (No more than $n/2$ numbers remain under consideration.) Among the remaining numbers take such i that a red edge goes from x_i to y_i. If there are more than $n/4$ of such i we paint green every other corresponding x_i and for each x_i not painted with a red edge (x_i, y_i) we paint y_i green.

Otherwise, if there are less than $n/4$ red edges (x_i, y_i) we also exclude such numbers from our consideration, which leaves us with no less than $n/4$ numbers. We can choose no less than $n/12$ among these numbers in such a way that there are no red edges between x_i and y_i with these numbers. We paint x_i and y_i with the chosen numbers green.

Now we shall regard those x_i marked green as input vertices and green y_i as output vertices of a new SF-system with no less than $n/12$ input vertices (it is clear that if "the experimenter" marks by $*$ only green y_i, then the FIND operations in the old and new systems will be consistent). The graph of this system does not, however, have any red edges leading from the input vertices to the output vertices (recall that the system does not rearrange the edges going out from x_i), i.e., the new system is already $(d_1 - 1, d_2, \dots, d_l)$-fiber or, which is the same, $s_1(D)$-fiber. Thus, the inductive assumption may be applied to it. Namely, there exists a sequence of no more than

$$n/(12l_{\gamma(s_1(D))}(n/12)) < n/l_{\gamma(D)}(n)$$

SPLIT commands costing no less than

$$n/(12\beta(\gamma(s_1(D)))) \geqslant n/(12\beta(\gamma(D)-1)) \geqslant n/\beta(\gamma(D)).$$

This proves the Principal Lemma for Case 1.

Case 2. $d_1 = 0$.

Here we shall need

SUBLEMMA. *Let C be a D-fiber system, $D = (d_1, \dots, d_l)$, $\rho < n/6$. Then either there exists a sequence of $[n/\rho]$ SPLIT commands with execution cost no less than $n/3$ or, for some i, the number of vertices of U_i of level 2 is no less than $n/(3^i l\rho)$.*

PROOF. Suppose that each U_i has no more than $n/(3^i l\rho)$ level-2 vertices. Execute a sequence of $[n/\rho]$ SPLIT commands

$$P = \text{SPLIT}(y_1), \text{SPLIT}(y_{\rho+1}), \dots, \text{SPLIT}(y_{\rho[n/\rho]-\rho+1}).$$

Let m_i be the number of level-2 vertices in U_i. Then for the input vertices one can reach (travelling along the edges) no more than $\sum_{i=1}^{l} m_i 3^{i-1}$ output vertices since each vertex of U_i of level no less than 2 has no more than 3 outgoing edges. Since $m_i \leqslant n/(3^i l\rho)$, we have $\sum_{i=1}^{l} m_i 3^i \leqslant n/(3\rho)$. Therefore at least $[n/\rho] - n/(3\rho) \geqslant n/(2\rho)$ intervals among those obtained while splitting the set of the indices with asterisks put by P, may be "serviced" by the system in only one way. Namely, insofar as the edges going out from these intervals cannot be rearranged, all x_i with the numbers in these intervals must be destroyed. So the system executes no less than $n/3$ operations which proves our sublemma. □

Applying the sublemma to our system for $\rho = 2p$, we obtain some U_i having no less than $n/(2 \cdot 3^i lp)$ level-2 vertices. Take this U_i. Consider the bi-partite graph consisting of x_i and all level-2 vertices of U_i and select the largest matching.

We assume that at least one edge leads to each level-2 vertex; otherwise it is not needed by the system. Then our matching contains no less than $n/(2 \cdot 3^i lpd_i)$ pairs. In fact, since no more than d_i edges go out from x_i in U_i then, provided there were less than $n/(2 \cdot 3^i lpd_i)$ pairs in the matching, all x_i in this matching would be connected with only $n/(2 \cdot 3^i lp) - 1$ level-2 vertices altogether; this would mean that there is a level-2 vertex connected with neither x_i in this matching. However, in this case one would be able to enlarge the matching by including this vertex and an arbitrary x_i connected with it (note that at least one edge leads to this vertex).

Paint the edges of the chosen matching blue, and paint the x_j included in this matching and the y_j with the same numbers (caution: but not those that blue edges lead to) green.

Further, we shall regard, as in Case 1, the green x_i as input vertices and the green y_j as output vertices of the new system C' having no less than $n/(2 \cdot 3^i lpd_i)$ input vertices.

Consider now the system C'' defined in the following way. Take another copy of the graph U_i (denote it by U_i^*) and glue it to the graph of C' over input and output vertices. Next we erase all blue edges in U_i and destroy all the level-2 vertices in U_i^* but connect each green x_j with those level-3 vertices that were connected with the level-2 vertex where the blue edge leads from x_j. Then no more than three edges go out from each input vertex in U_i^* and the number of levels in U_i^* becomes equal to

$i-1$. Thus the system C'' is $(d_1, \dots, d_{i-1}+3, d_i-1, \dots, d_l)$-fibrated, i.e., $s_i(D)$-fibrated. The complexity of C'' has increased not more than twice as compared with the complexity of C' (now, in the case when C' would build one edge in U_i, C'' would have to build two edges, namely, in U_i and in U_i^*; note that the edges of the level $1 \to 2$ are not to be rearranged, the blue color of the edges is fixed, and the three edges that have replaced one blue edge in U_i are not to be rearranged either).

Set $q = 2d_i 3^i lp$, $r = 2 \cdot 3^i l d_i$.

Applying the inductive assumption to the system C'' and taking the double loss in the computational speed into account, we find that for the initial system C there exists a sequence of no more than

$$n/(ql_{\gamma(s_i(D))}(n/q))$$

SPLIT operations with execution cost no less than

$$n/(2q\beta(\gamma(s_i(D)))).$$

To sum up, we have that for every D-fibrated system with m inputs, $m > 6p$, we have at our disposal either (Case A) a sequence P_{A} of $m/(2p)$ commands having cost no less than $m/3$, or (Case B) a sequence P_{B} of $m/(ql_{(\gamma(D)-1)}(m/q))$ commands having cost no less than $m/(2q\beta(\gamma(D)-1))$.

In order to obtain the sequence required by the conditions of the Principal Lemma we proceed in the following way. Apply first the sequence P_{B} to the initial system C (if for C the case A takes place, the assertion of the Principal Lemma obviously holds). The asterisk set in some way decomposes the collection of the indices $1, 2, \dots, n$ into several intervals, and the x_i and y_i corresponding to each of these intervals form a new SF-system. Henceforth we do not have to study the structure of graphs; we shall use only the sequences P_{A} and P_{B} and take the possibility of destroying vertices into account. Therefore, we shall not mention the input or any other vertices but discuss only decompositions of the set of indices $1, 2, \dots, n$.

Thus, after applying the sequence P_{B}, we already have several intervals of indices; there may be not n but fewer of them, since the system may have destroyed some of the indices. We can, however, assume that we always have no less than $n/2$ indices at our disposal, since if more than $n/2$ indices had been destroyed, that would mean that the system had executed more than $n/2$ operations, which is more than the conditions of the Principal Lemma require.

At each subsequent stage we shall apply the sequence P_{B} to some of the intervals obtained after the previous stage until either $n/(100p)$ asterisks will have been set or sufficiently many intervals will have been obtained to which we can apply P_{A}. We give a more detailed description now.

Description of the stage. Before the next stage of setting asterisks we have several intervals of total length no less than $n/2$ and we have already set no more than $n/(100p)$ asterisks (and $n/(100p)+1$ intervals).

Case 1. The total length of the intervals for which the first possibility of the sublemma holds is no less than $n/4$. Then, since we have no more than $n/(100p)+1$ intervals, there exist several of them with total length no less than $n/8$, each having length no less than $12p$. The sequence provided by the first possibility of the sublemma requires the execution time indicated regardless of the order of setting asterisks; therefore, after applying the sublemma to these intervals in a certain order, we set no more than $n/(16p)$ asterisks and force the system to execute no less than $n/24$ operations. This proves the assertion of the Principal Lemma for Case 1.

Case 2. The total length of the intervals where the second possibility of the sublemma takes place (i.e., where the sequence P_{B} may be applied) is no less than $n/4$.

In this case we apply P_{B} to the longest of the intervals, then to the longest of the remaining ones and so on (we decompose no longer the intervals obtained at this stage) until the lengths of all the intervals obtained by decomposition have reached $n/64$ (or until we find ourselves in the situation of Case 1).

Thus, in Case 2 each following stage compels our system to execute no less than $n/(128q\beta(\gamma(D)-1)) \geqslant 100nr/(p\beta(\gamma(D)))$ operations (because $\beta(\gamma(D))/\beta(\gamma(D)-1) \geqslant 12800r^2$). Therefore, it remains to show that after $p/(100r)$ stages we set no more than $n/(100p)$ asterisks altogether (here we can assume that $p > 100$; otherwise the statement is obvious, since we require only one stage then).

Consider the following model subdivision process. At one stage we set equidistantly $ml_{\gamma(D)-1}(m)$ asterisks on each interval of length qm. For our model subdivision, if the length of the interval before the present stage was qm, it will become $ql_{\gamma(D)-1}(m)$ after this stage; therefore, after $p/(100r)$ stages of the model subdivision the length of the interval will become

$$q(l_{\gamma(D)-1}^{p/100}(n/pr')) > 2q,$$

since

$$l_{\gamma(D)-1}^{l_\gamma(D)^{n/pr}}(n/pr) > 2$$

and

$$l_{\gamma(D)}(n/pr) < \frac{1}{r} l_{\gamma(D)}(n/p) < 2^r l_{\gamma(D)}(n).$$

Thus it is sufficient to show that at each stage of the actual decomposition no more asterisks are set than at the corresponding stage of the model decomposition. We shall reason by induction.

Possibility 1. In Case 2 of the current stage the length of the subdivided set of indices is greater than $n/32$, i.e., only one interval of length greater than $n/32$ is subdivided. This means that we have set no more than

$$s/(ql_{\gamma(D)-1}(s/q)) \leqslant n/(ql_{\gamma(D)-1}(n/q))$$

asterisks, while there were no less than $n/(ql_{\gamma(D)-1}(n/q))$ asterisks in the model case.

Possibility 2. In Case 2 of the present stage some set of length less than $n/32$ is subdivided. Then the length of each decomposed interval is no less than $n/(8d)$ (otherwise the total length of the intervals subject to Case B would be less than $n/32 + d(n/(8d)) < n/4$), where d is the number of intervals before this stage in the model case. Therefore, in our case, we shall set no less than

$$n/(32ql_{\gamma(D)-1}(n/(8dq))) \leqslant 8n/(32ql_{\gamma(D)-1}(n/(qd)))$$

asterisks, while there will be $n/(ql_{\gamma(D)-1}(n/(qd)))$ asterisks in the model case.

Thus the Principal Lemma is proved. □

1.4. THEOREM 1. *Let C be an* SF-*system with outgoing degree of the vertices no greater than* 2. *Then its complexity is no less than $c \ln \alpha(n)$.*

PROOF. Let the complexity of C be equal to Q. Then, according to Lemmas 1 and 2, there exists a $2Q$-monotone system C' not rearranging the graph when executing FIND command and with complexity no greater than $8Q^2$.

Apply to the $\underbrace{(0,\dots,0,3)}_{2Q}$-fibrated system C', $\gamma(0,\dots,0,3)=3\cdot 4^{2Q}$, the reasoning analogous to that presented at the end of the proof of the Principal Lemma, but use here the statement of the Principal Lemma instead of the sequence P_{B} of Case B. The reasoning is simplified due to the absence of Case A and due to the possibility of destroying vertices. As a result, we obtain that the complexity of C' is no less than

$$(1/\beta(4^{Q+1}))l_{4^{2Q+1}}(n).$$

Thus $8Q^2\beta(4^{Q+1})\geqslant l_{4^{2Q+1}}(n)$. Set $q=\log_4\alpha(n)$, i.e., $l_{4^q}(n)=4^q$. If it were $2Q<q-2$ then there would also exist a $\underbrace{(0,0,\dots,0,3)}_{Q-2}$-fibrated system with complexity no greater than $8(q-2)^2$. Therefore, reasoning by contradiction, we can assume that $2Q=q-2$. Set $\delta(x)=8x^2\beta(x)$. Then $\delta((q-2)/2)\geqslant l_{4^q-1}(n)$, i.e., $G_{4^q-1}(\delta((q-2)/2))\geqslant n$ and, all the more, $G_{4^q-1}(\delta((q-2)/2))\geqslant n$. But we know that $l_{4^q}(n)=4^q$, i.e., $G_{4^q}(4^q)=n$ or $G_{4^q-1}^{4^q}(2)=n$. Then $G_{4^q-1}(\delta((q-2)/2))\geqslant G_{4^q-1}^{4^q}(2)$ or $\delta((q-2)/2)\geqslant G_{4^q-1}^{4^q-1}(2)$, which is obviously wrong for q large enough. But it follows from the definition of q that it grows infinitely with the growth of n. □

§2. Lower and upper bounds for separating systems

The notion of the separating system (satisfying the *separation* assumption) was introduced in **[2]** and was discussed later in a series of papers. An SF-system whose graph is a tree, may serve as an example of a separating system. We give here the sharp definition.

An SF-system is called *separating* if at every moment its graph consists of several connected components (regardless of orientation!) each of them containing exactly one vertex y_i marked by an asterisk and also all x_j such that $F(x_j)=y_i$ and all y_j with numbers corresponding to these x_j. Erasing of the edge is not regarded as an operation in such systems.

For separating systems the estimate of Theorem 1 can be considerably strengthened, namely, the complexity of the separating system with n inputs is no less than $c\alpha(n)$. It seems interesting, whether the estimate in Theorem 1 is exact or the presence of the logarithm in the estimate results from some deficiencies of the method applied. In any case it is clear from the sequel that the estimate in Theorem 2 is sharp.

THEOREM 2. *The complexity of the separating system with n inputs is no less than $c\alpha(n)$.*

The proof of this theorem basically coincides with that of Theorem 1 and differs only in the estimate and in the proof of the Principal Lemma. So we dwell upon this latter subject only.

THE ANALOG OF THE PRINCIPAL LEMMA FOR SEPARATING SYSTEMS. *Let C be a k-monotone separating system that is a k-structure while the outgoing degree of the input vertices of its graph is no greater than d and the outgoing degree of all remaining vertices is no greater than three. Then there exists a sequence of no more than n/p, $p=l_k^d(n)$, commands with cost of execution no less than $n/\beta(k,d)$,*

$$\beta(k,d)=10000^{k+d}d^{2k}.$$

PROOF. Similarly to the proof of the Principal Lemma we assume that the system does not rearrange the edges leading from the input vertices but it may destroy the vertices if it is necessary. We again reason by induction in k, and, for k fixed, by induction in d.

Induction step. Case 1. One can find more than n/ρ level-2 vertices where the edges from no less than $n/2$ input (level-1) vertices lead. Paint these input vertices green.

Set $*$ at the green vertex with the smallest number, skip $\rho/4$ green vertices, set $*$ again and thus decompose the green vertices into no less than $2n/\rho$ equal intervals with $\rho/4$ vertices in each. Since we have obtained twice as many intervals as level-2 vertices where the edges from green x_j come, it is clear that the outgoing degree of green x_i has decreased by one on at least half of the intervals (recall that the system destroys the edges at no cost, but instead of creating new edges it destroys input vertices). Applying the inductive assumption to these intervals we obtain a sequence of no more than

$$(n/\rho)\rho/(4l_k^{d-1}(\rho/4)) \leqslant n/(2l_k^d(n))$$

SPLIT commands with cost of execution no less than

$$n/(8\beta(k,d-1)) \geqslant 1000n/\beta(k,d).$$

Case 2. It is impossible to choose the vertices mentioned in the definition of Case 1.

Consider the largest collection of input vertices such that all level-2 vertices connected with the selected input vertices are different. This collection contains no less than $n/(d\rho)$ vertices; otherwise the set of level-2 vertices connected with the vertices in our collection and the complement of our collection in the set of all input vertices would form the set required in Case 1. Paint the selected vertices green.

Consider the system C' with graph obtained from that of the system C in the following way: green x_i are considered as input vertices, all level-2 vertices are destroyed and each green vertex is connected with those level-3 vertices that are accessible from it in the graph of C. The system C' is a $(k-1)$-monotone $(k-1)$-structure with degree of output vertices no greater than $3d$. Applying the inductive assumption to it, we obtain the sequence of no more than

$$n/(d\rho l_{k-1}^{3d}(n/(d\rho))) \leqslant n/(\rho l_{k-1}^{3d}(n/\rho))$$

SPLIT commands with execution cost no less than

$$n/(d\rho\beta(k-1,d)) > 100n/(\rho\beta(r,d)).$$

Further, the proof is concluded in just the same way as that of the Principal Lemma, the sequences of commands of Cases 1 and 2 being used instead of the sequences of commands P_A and P_B of Cases A and B, respectively.

Note that the estimate obtained in Theorem 2 is sharp: there exists a separating SF-algorithm having the complexity $c\alpha(n)$. The graph of such an algorithm is a forest with branching $\theta l_k(m)$, $1/2 \leqslant \theta \leqslant 1$, on the kth level of the tree with m leaves. □

In conclusion, the author takes the opportunity to express his gratitude to Prof. A. O. Slisenko who has acquainted him with the class of problems discussed here.

References

1. A. O. Slisenko, *Complexity problems of the theory of computations*, Uspekhi Mat. Nauk **36** (1981), no. 6, 21–103; English transl. in Russian Math. Surveys **36** (1981).
2. R. E. Tarjan, *A class of algorithms which require non linear time to maintain disjoint sets*, J. Comput. System. Sci. **18** (1979), 110–127.
3. N. Blum, *On the single-operation worst-case time complexity of the disjoint set union problem*, SIAM J. Comput. **15** (1986), 1021–1024.
4. K. Mehlhorn, S. Naher, and H. Alt, *A lower bound of the complexity of the Union-Split-Find problem*, SIAM J. Comput. **17** (1988), 1093–1102.
5. J. E. Hopcroft and J. D. Ullman, *Set merging algorithms*, SIAM J. Comput. **2** (1973), 294–303.
6. K. Mehlhorn, *Data structures and algorithms*. II. *Graph algorithms and NP-completeness*, EATCS Monogr. Theoret. Comput. Sci., Springer-Verlag, Berlin and New York, 1984.

St. Petersburg Institute for Information and Automation, Academy of Science of Russia, V.O. 39, 14 liniya, St. Petersburg, 199178, Russia

Translated by G. ROZENBLUM

Amer. Math. Soc. Transl.
(2) Vol. **159**. 1994

The Finiteness Theorem for Three-Dimensional Manifolds of Nonpositive Curvature

S. V. Buyalo

§1. Introduction

The main result of the present paper, formulated in Theorem 1.1, is related to the following theorems proven in [**1, 2**].

THEOREM A ([1]). *Let M be a complete Riemannian manifold of negative sectional curvature $-1 \leqslant K < 0$. If the volume* $\operatorname{vol}(M) < \infty$, *then M is of finite topological type, i.e., it is diffeomorphic to the interior of a compact manifold with boundary.*

The same work [**1**] contains an example of a C^∞-smooth three-dimensional manifold of finite volume and sectional curvature $-1 \leqslant K \leqslant 0$, but with infinitely generated one-dimensional homology group. There are analogous examples of arbitrary dimension.

The nonexistence of real analytic examples of this sort is shown by the following result. Let $\operatorname{InjRad} : M \to \mathbb{R}$ be the injectivity radius of the manifold M. We say that the injectivity radius tends to zero, $\operatorname{InjRad} \to 0$, if for any $\varepsilon > 0$ the set $\{p \in M : \operatorname{InjRad}(p) \geqslant \varepsilon\}$ is compact.

THEOREM B ([3]). *Let M be a real analytic complete Riemannian manifold with sectional curvatures $-1 \leqslant K \leqslant 0$ and injectivity radius* $\operatorname{InjRad} \to 0$. *Then*

(1) *M is of finite topological type.*
(2) *If the universal covering $\widetilde{M}$ does not admit an isometric decomposition of the form $\widetilde{M} = M_1 \times \mathbb{R}$, then the Betti numbers of the manifold M (with an arbitrary coefficient field) satisfy the condition $\sum_{i=0}^n b_i \leqslant C_n \operatorname{vol}(M)$ with the constant C_n depending only on the dimension n of the manifold M.*

In the following theorem the condition of real analyticity is replaced by the geometric condition of "the absence of small tori".

THEOREM C ([2], pp. 193–211). *Let M be a complete C^∞-smooth Riemannian manifold with sectional curvatures $-1 \leqslant K < 0$ and injectivity radius* $\operatorname{InjRad} \to 0$. *Let $\varepsilon > 0$ exist such that for any flat two-dimensional torus the diameter of its image via isometric totally geodesic immersion is at least ε. Then M is of finite topological type.*

The aim of this paper is to prove the following theorem generalizing and sharpening Theorem C in three dimensions.

1991 *Mathematics Subject Classification*. Primary 53C20.

1.1. THEOREM. *Let M be a complete C^∞-smooth Riemannian manifold with sectional curvatures $-1 \leqslant K \leqslant 0$, injectivity radius* InjRad $\to 0$. *Let $[F, M]_\varepsilon$, $\varepsilon > 0$, denote the set of all homotopy classes that contain an isometric totally geodesic embedding $F \to M$ of a flat torus or a flat Klein bottle of diameter less than ε ($[F, M]_\varepsilon$ may be empty). Choose one such embedding in each class of $[F, M]_\varepsilon$ and cut M along all the chosen embeddings. Then the interior of each connected component of the cut manifold M_0 is diffeomorphic to the interior of a compact manifold $W_0 \subset M_0$ with boundary. Note that any component of the boundary of W_0 is incompressible and is either a torus or a Klein bottle.*

1.2. REMARK. The manifold M cut along the embedding $i : F \to M$ should be interpreted as a metric completion of the complement $M \setminus i(F)$. The supplied limit points form a manifold F' that is either connected, if $i(F)$ is one-sided, or consists of two components, if $i(F)$ is two-sided. In the latter case F' is a two-sheeted covering of $i(F)$. If M is orientable and the conditions of the theorem hold, then F' is either a torus or a disjoint sum of two tori. The part of F' in W_0 may be considered as a component of the boundary ∂W_0. Surely, ∂W_0 may contain other components besides this kind.

1.3. REMARK. Incompressibility of the component $\partial W_0'$ of the boundary ∂W_0 means that the homomorphism of the fundamental groups $\pi_1(\partial W_0') \to \pi_1(W_0)$ induced by the embedding is monomorphic.

The proof of Theorem 1.1 goes as follows. It is shown that isometric and totally geodesic immersion of a flat torus of sufficiently small diameter is in fact a covering of some embedding, i.e., it does not have any transversal self-intersections. Hence the cut operation of the initial manifold along the sufficiently small embedded tori and Klein bottles satisfying the conditions of the theorem eliminates all the small immersed tori from all connected components of the resulting cut manifold. All the immersions turn out to be coverings of the components of the boundary ∂M_0.

The rest of the proof follows ideas of **[1, 2]**. Namely, the proper function $f : M_0 \to \mathbb{R}$ is constructed. Its critical submanifolds containing all the critical points of the function f are studied. It is proved that the number of critical submanifolds is finite, which gives the possibility of showing that all the critical points of the function f are contained in a compact set, and hence M_0 is of finite topological type.

The study of the ends of the manifold M_0 involves a detailed analysis of the set of stable points of parabolic isometry. In particular, we use the Tits metric on the sphere at infinity (cf. **[2]**).

The rest of the paper is devoted to the proof of Theorem 1.1.

§2. Immersions of small tori without transversal self-intersections

2.1. Let $\pi : \widetilde{M} \to M$ be a universal Riemannian covering of a complete Riemannian manifold M. Hereafter we consider that the sectional curvatures of the manifold M satisfy the condition $-1 \leqslant K < 0$. Recall that the group $\Gamma = \pi_1(M)$ acts on $\widetilde{M}$ discretely as a group of transformations of the covering and consists of isometries. The isometries from Γ do not have any stable points and may be of two types: those having an invariant straight line are called hyperbolic; the others are called parabolic.

Denote by $\delta_\gamma : \widetilde{M} \to \mathbb{R}$, $\delta_\gamma(x) = d(x, \gamma x)$, the displacement function of the isometry $\gamma \in \Gamma$, where d is the distance in $\widetilde{M}$. We broadly use the well-known Margulis lemma.

2.2. Lemma ([**2,** §8.3]). *There exists a constant $\mu = \mu_n$ depending only on the dimension n of the manifold M, such that for any point $x \in \widetilde{M}$ and any positive $\varepsilon \leqslant \mu$, the group $\Gamma_\varepsilon(x)$ generated by the set $\{\gamma \in \Gamma, \delta_\gamma(x) < \varepsilon\}$ is almost nilpotent, i.e., contains a nilpotent subgroup of finite index.*

2.3. A subset $U \subset \widetilde{M}$ is called precisely invariant, if for any $\gamma \in \Gamma$ either $\gamma(U) = U$ or $\gamma(U) \cap U = \varnothing$. For a precisely invariant subset U we denote by Γ_U the subgroup of Γ containing all the isometries preserving U. The natural mapping $U/\Gamma_U \to \widetilde{M}/\Gamma = M$ is an embedding in this case.

2.4. Lemma. *Let M be non-flat,* $\dim M = 3$. *Suppose that $\widetilde{M}$ contains a totally geodesically and isometrically embedded Euclidean plane E. Let E be invariant for a subgroup $H \subset \Gamma$ of the form $\mathbb{Z} \oplus \mathbb{Z}$ with generating elements $\alpha, \beta \in H$ satisfying the condition $\delta_\alpha \restriction_E, \delta_\beta \restriction_E < \mu$ with constant $\mu = \mu_3$ from the Margulis lemma. Then E is precisely invariant and $\Gamma_E \supset H$ is either of the form $\mathbb{Z} \oplus \mathbb{Z}$ or contains a subgroup of the form $\mathbb{Z} \oplus \mathbb{Z}$ of index* 2.

Proof. Since the group H equals $\mathbb{Z} \oplus \mathbb{Z}$, it acts on E as a translation group. Hence the displacement functions δ_α, δ_β are constant on E and for any $x \in E$ the group $\Gamma_\mu(x)$ contains H as a subgroup.

Suppose that an element $\gamma \in \Gamma$ exists such that $\gamma(E) \neq E$ and $\gamma(E) \cap E \neq \varnothing$. Denote $E' = \gamma(E)$, $\alpha' = \gamma\alpha\gamma^{-1}$, $\beta' = \gamma\beta\gamma^{-1}$, and let $H' = \gamma H \gamma^{-1}$ be the subgroup generated by α', β'. It is clear that $H'(E') = E'$; H, H' are of the form $\mathbb{Z} \oplus \mathbb{Z}$ and consist of hyperbolic isometries; the displacement functions $\delta_{\alpha'}$, $\delta_{\beta'}$ are constant on E', $\delta_{\alpha'} \restriction_{E'} = \delta_\alpha \restriction_E$, $\delta_{\beta'} \restriction_{E'} = \delta_\beta \restriction_E < \mu$. Take a point $x \in E' \cap E$ and consider the group $\Gamma_\mu(x)$. Then H, $H' \in \Gamma_\mu(x)$, and according to the Margulis lemma $\Gamma_\mu(x)$ contains a nilpotent subgroup N of finite index. Hence subgroups $H_1 \subset H$ and $H_1' \subset H'$ of finite index exist such that H_1, $H_1' \in N$. According to [**2,** §7.4(3)], H_1, H_1' lie in the center of N, and hence generate a free abelian subgroup H_0 of rank k. Since $\widetilde{M}$ is non-flat and three-dimensional, $k < 3$. Since the ranks of H_1, H_1' equal 2, the rank $k = 2$. Hence H_1, H_1', and also $H_1 \cap H_1'$, are of finite index in H_0; in particular, the rank of $H_1 \cap H_1'$ is 2. But any element $h \in H_1 \cap H_1'$ preserves the straight line $E \cap E'$. Hence $\operatorname{rank}(H_1 \cap H_1') \leqslant 1$. This contradiction implies that the plane $E \subset \widetilde{M}$ is precisely invariant. It remains to show that Γ_E has the stated form.

Since the action of the group Γ in $\widetilde{M}$ is free, the action of the subgroup $\Gamma_E \subset \Gamma$ in E is effective, i.e., any isometry from Γ_E different from the identity is nontrivial in E. Hence Γ_E may be considered as a group of isometries of the Euclidean plane. By the conditions $\Gamma_E \supset H = \mathbb{Z} \oplus \mathbb{Z}$, it acts freely in E, and hence has the stated form. □

2.5. Corollary. *If a three-dimensional non-flat Riemannian manifold M with sectional curvatures $-1 \leqslant K < 0$ admits an isometric totally geodesic immersion $i : T^2 \to M$ of a flat torus of diameter* $\operatorname{diam} T^2 < \mu/2$, *then i is a finite-sheeted covering of either an embedded torus or of an embedded Klein bottle in M.*

PROOF. By the condition there is a commutative diagram

$$\begin{array}{ccc} E^2 & \xrightarrow{\ \tau\ } & \widetilde{M} \\ \downarrow & & \downarrow \pi \\ T^2 & \xrightarrow{\ i\ } & M \end{array}$$

where the vertical arrows are universal Riemannian covering mappings and the horizontal arrows are isometric totally geodesic mappings. Hence the group Γ contains a subgroup H of the form $\mathbb{Z}\oplus\mathbb{Z}$ that preserves the plane $E=\tau(E^2)$ and $E/H=T^2$. Since the longer diagonal of the parallelogram exceeds its sides, the displacements in the plane E for the suitable generators $\alpha, \beta \in H$ are less then $2\operatorname{diam} T^2 < \mu$. Now the assertion follows from Lemma 2.4. □

§3. The cut operation along embedded small tori and Klein bottles

Fix a positive number $\nu \leqslant \mu/8$, where $\mu=\mu_3$ is the constant from the Margulis lemma. According to the conditions of Theorem 1.1 we choose one isometric and totally geodesic embedding $F\to M$ of a flat torus or a flat Klein bottle F for every class in $[F, M]_{2\nu}$. Arguments similar to those in the proof of Lemma 2.4 show that the images of all the chosen embeddings are disjoint. Hence M can be cut along all the chosen embeddings. We can consider M to be orientable, because as we pass to the two-sheeted oriented covering the finiteness of the topological type is preserved.

Let M_0 be a connected component of the resulting cut manifold. As has already been mentioned in Remark 1.2, all the components of the boundary ∂M_0 are flat totally geodesic tori. Denote by $X\subset\widetilde{M}$ one of the components of the preimage of M_0 in the universal covering $\widetilde{M}$ of the manifold M. By construction and Corollary 2.5, X possesses the following properties.

3.1. THEOREM. *X is a closed convex subset of $\widetilde{M}$ and every component of its boundary is a plane, i.e., it is an isometric totally geodesic embedding of a Euclidean plane E^2.*

3.2. *The action of the group Γ on X is discrete, it consists of orientation-preserving isometries, and $X/\Gamma = M_0$.*

3.3. *If a subgroup $G\subset\Gamma$ preserves a plane $E\subset X$, the action of G on E is uniform, and G is generated by isometries whose displacements at some point of E are less than 2ν, then E is parallel to some component of the boundary ∂X. In particular, G preserves the orientation of E and is of the form $\mathbb{Z}\oplus\mathbb{Z}$.*

The function $\delta_\Gamma : X\to\mathbb{R}$, $\delta_\Gamma(x)=\inf\{\delta_\gamma(x):\gamma\in\Gamma-\mathrm{id}\}$, is Γ-invariant and determines a function $M_0=X/\Gamma\to\mathbb{R}$, twice the injectivity radius $\mathrm{InjRad}: M_0\to\mathbb{R}$ of the manifold M_0. Let $\pi: X\to M_0$ be the projection. Then $\mathrm{InjRad}(p)=\frac{1}{2}\delta_\Gamma(x)$, where $x\in\pi^{-1}(p)$, and the value $\delta_\Gamma(x)$ does not depend on the choice of the point $x\in\pi^{-1}(p)$. The last property of X and Γ to be stated here is the following.

3.4. THEOREM. $\mathrm{InjRad}\to 0$, *i.e., for any $\varepsilon>0$ the set $\{p\in M_0 : \mathrm{InjRad}(p)\geqslant\varepsilon\}$ is compact.*

This property is the consequence of $\mathrm{InjRad}\to 0$ on the initial manifold M.

3.5. REMARK. For any component E of the boundary ∂X consider all the planes in X parallel to E. The union of these planes is isometric to either $[0, a] \times \mathbb{R}^2$, $0 \leqslant a < \infty$, or $[0, \infty) \times \mathbb{R}^2$. The latter case is excluded. Suppose that $[0, \infty) \times \mathbb{R}^2$ is flat. The group Γ preserves the boundary $\partial X = \{0\} \times \mathbb{R}^2$, and hence $\Gamma = \mathbb{Z} \oplus \mathbb{Z}$, and M_0 is isometric to $[0, \infty) \times T^2$, which contradicts the condition InjRad $\to 0$. If for different components of the boundary ∂X the described sets $[0, a] \times \mathbb{R}^2$ intersect, then M_0 is obviously isometric to $[0, a] \times T^2$. This contradicts the fact that the tori chosen for the cut operation are in different homotopy classes. Hence in the sequel we can assume that $a = 0$, i.e., that there are no planes in X parallel to any component of the boundary.

§4. The subgroups of Γ generated by small isometries

4.1. REMARK. If the fundamental group of an orientable surface H contains a subgroup isomorphic to $\mathbb{Z} \oplus \mathbb{Z}$, then H is homeomorphic to a torus.

Indeed, passing to the covering (and using the orientability), we can assume that $\pi_1(H) = \mathbb{Z} \oplus \mathbb{Z}$. Suppose that H is not compact. Then, since its fundamental group is finitely generated, it is homeomorphic to the interior of a compact surface with boundary. Hence $\pi_1(H)$ is free, which contradicts the conditions. Hence H is compact and has no boundary, and then the assertion is obvious.

Similar arguments show that if $\pi_1(H)$ is nilpotent, then it is a free abelian group of rank at most 2.

We call the set $\mathrm{MIN}(\gamma) = \{x \in X : \delta_\gamma(x) = \inf \delta_\gamma\}$ the minimal set of the isometry γ.

Since the function δ_γ is convex (cf. **[2]**), the minimal set is convex. It is known (**[2]**) that the minimal set of a parabolic isometry is empty, and for a hyperbolic isometry it consists of invariant straight lines and is isometric to the product $D \times \mathbb{R}$, and the isometry acts on it as (id, translation).

4.2. LEMMA. *Let $\varepsilon \leqslant 2\nu$. Let the group $\Gamma_\varepsilon(x) \subset \Gamma$ be nontrivial and consist of hyperbolic isometries. Then it is one of the following*:

(1) $\Gamma_\varepsilon(x) = \mathbb{Z}$, *the set* $\mathrm{MIN}(\Gamma_\varepsilon(x)) = \cap\{\mathrm{MIN}(\gamma) : \gamma \in \Gamma_\varepsilon(x)\}$ *is nonempty and admits an isometric decomposition* $D \times \mathbb{R}$. *The isometries from* $\Gamma_\varepsilon(x)$ *act on* $\mathrm{MIN}(\Gamma_\varepsilon(x))$ *as* (id, translation).

(2) $\Gamma_\varepsilon(x) = \mathbb{Z} \oplus \mathbb{Z}$, *the set* $\mathrm{MIN}(\Gamma_\varepsilon(x))$ *is nonempty, is isometric to* E^2, *and is a component of the boundary* ∂X. *The isometries from* $\Gamma_\varepsilon(x)$ *act on* $\mathrm{MIN}(\Gamma_\varepsilon(x))$ *as translations.*

PROOF. Since $\varepsilon \leqslant \mu/4$, according to Lemma 2.2 the group $\Gamma_\varepsilon(x)$ contains a nilpotent subgroup N of finite index. Since $\Gamma_\varepsilon(x)$ consists of hyperbolic isometries, $N = \mathbb{Z}^s$, $s = 0, 1, 2, 3$ (cf. **[2,** §7.4.(3)**]**). The group $\Gamma_\varepsilon(x)$ is not trivial, so that $s > 0$, and since $\widetilde{M}$ is not flat, we have $s < 3$. The set X contains an isometrically and totally geodesically embedded Euclidean space E^s invariant for $\Gamma_\varepsilon(x)$ (cf. **[2,** §7.4.(2)**]**). If $s = 1$, then $\Gamma_\varepsilon(x) = \mathbb{Z}$, and the set $\mathrm{MIN}(\Gamma_\varepsilon(x))$ possesses the stated properties. Let $s = 2$. Then $\Gamma_\varepsilon(x)$ may be regarded as a group of isometries of the Euclidean plane $E^2 \subset X$. The displacements for the elements of $\Gamma_\varepsilon(x)$ at the point $x' \in E^2$ closest to x are not greater than at x. Hence the group $\Gamma_\varepsilon(x)$ is generated by the isometries with displacements less than $\varepsilon \leqslant 2\nu$ at some point in E^2. By Property 3.3, $\Gamma_\varepsilon(x) = \mathbb{Z} \oplus \mathbb{Z}$, and by Remark 3.5, the plane E^2 is a component of the boundary ∂X. The elements of $\Gamma_\varepsilon(x)$ act on E^2 as translations; so $\mathrm{MIN}(\Gamma_\varepsilon(x)) = E^2$. □

Recall that points of the absolute $X(\infty)$ are defined for a convex set $X \subset \widetilde{M}$ as classes of asymptotically equivalent rays in X. For any point $z \in X(\infty)$ the corresponding Busemann function $b_h : X \to \mathbb{R}$ is $b_h(x) = \lim_{t\to\infty}(d(x,h(t)) - t)$, where $h : [0,\infty) \to X$ is a ray, $h \in z$. Different Busemann functions for the same $z \in X(\infty)$ differ by a constant. The level sets of the Busemann function b_h are called horospheres with center $z \in X(\infty)$. The Busemann function is convex; so the horoballs (i.e., its sublevels) are convex. For more details see **[2]**.

4.3. LEMMA. *For any $\varepsilon \leqslant 2v$ and any $x \in X$ the group $\Gamma_\varepsilon(x)$ is a free abelian group of rank not greater than two. If $\Gamma_\varepsilon(x)$ contains a parabolic isometry γ then there exists a point $z \in X(\infty)$ such that $\Gamma_\varepsilon(x)$ preserves z and all its horospheres, and the displacement function δ_γ is strictly decreasing along any ray $h \in z$.*

PROOF. If $\Gamma_\varepsilon(x)$ consists of hyperbolic isometries, then the assertion is a corollary of Lemma 4.2. If the group $\Gamma_\varepsilon(x)$ contains a parabolic element, then according to **[2,** §7.9.(3)**]** there exists a point $z' \in X(\infty)$ such that $\Gamma_\varepsilon(x)$ preserves z' and all its horospheres. Since the group Γ preserves orientation of X and $\Gamma_\varepsilon(x)$ preserves horoballs of the point z', $\Gamma_\varepsilon(x)$ also preserves orientation of the horospheres. Since the horospheres are simply connected, the group $\Gamma_\varepsilon(x)$ is the fundamental group of an orientable surface $S/\Gamma_\varepsilon(x)$, where S is a horosphere of the point z'. Let $N \subset \Gamma_\varepsilon(x)$ be a nilpotent subgroup of finite index. Then S/N is a finite-sheeted covering surface for $S/\Gamma_\varepsilon(x)$ and by Remark 4.1 it is a torus, or N is isomorphic to $\mathbb{Z}$. In any case $\Gamma_\varepsilon(x)$ is a free abelian group of rank at most 2.

Let $\gamma \in \Gamma_\varepsilon(x)$ be a parabolic isometry. There exists $z \in X(\infty)$ such that any ray $h \in z$ minimizes the displacement function δ_γ. Since $\Gamma_\varepsilon(x)$ is abelian, the function δ_γ is $\Gamma_\varepsilon(x)$-invariant. According to **[2,** §3.9**]**, the group $\gamma_\varepsilon(x)$ preserves z and all its horospheres. □

§5. The proper function on M_0 and its critical points

5.1. Let the function $g : (0,\infty) \to [0,\infty)$ be of class C^∞ and possess the following properties: $g(t) = 0$ for $t \geqslant v$; $g(t) \to \infty$ for $t \to 0+$; $g'(t) < 0$ for $0 < t < v$. Define the function $F : X \to \mathbb{R}$, $F(x) = \sum_{\gamma\in\Gamma-\mathrm{id}} g \circ \delta_\gamma$. It is clear that F is smooth and Γ-invariant. Hence its projection is a smooth function $f : M_0 \to \mathbb{R}$.

Note that the sum in the definition of F contains only the addends with $\delta_\gamma < v$ and inside M_0 there are no immersed tori with generatrix of length less than $2v$; this will be used below.

5.2. LEMMA. *The function f is proper, i.e., the preimage of any compact set in $\mathbb{R}$ is compact in M_0.*

PROOF. Let $t > 0$. Choose $\varepsilon > 0$ to satisfy $g(2\varepsilon) = t$. Clearly, $0 < 2\varepsilon < v$. If $p \in f^{-1}([0,t])$ and $x \in \pi^{-1}(p)$ then $F(x) \leqslant t$. Hence $\delta_\gamma(x) \geqslant 2\varepsilon$ for all $\gamma \in \Gamma - \mathrm{id}$. Then $\mathrm{InjRad}(p) = \frac{1}{2}\delta_\Gamma(x) \geqslant \varepsilon$. Thus the set $f^{-1}([0,t])$ is contained in the compact set $\{p \in M_0 : \mathrm{InjRad}(p) \leqslant \varepsilon\}$, and the assertion becomes obvious. □

Denote $\widetilde{\Gamma}_t(x) = \{\gamma \in \Gamma : \delta_\gamma(x) < t\}$. Due to the properties of g we have

$$F(x) = \sum_{\gamma\in\widetilde{\Gamma}_v(x)-\mathrm{id}} g \circ \delta_\gamma(x)$$

and

$$\operatorname{grad} F = \sum_{\gamma \in \widetilde{\Gamma}_\nu(x) - \mathrm{id}} g'(\delta_\gamma(x)) \cdot \operatorname{grad} \delta_\gamma(x).$$

5.3. LEMMA. *Let x be a critical point of the function F. Then the group $\Gamma_\nu(x)$ does not contain any parabolic isometries and $x \in \mathrm{MIN}(\Gamma_\nu(x))$.*

PROOF. By Lemma 4.3, the group $\Gamma_\nu(x)$ is free abelian. Hence if the generating set $\widetilde{\Gamma}_\nu(x)$ contains no parabolic isometries, then $\Gamma_\nu(x)$ contains no parabolic isometries. Suppose that for the critical point x the group $\Gamma_\nu(x)$ contains a parabolic isometry γ_0. We can assume that $\gamma_0 \in \widetilde{\Gamma}_\nu(x)$. Choose the point $z \in X(\infty)$ as was done in Lemma 4.3. Let $S = b_z^{-1}(b_z(x))$ be the horosphere passing through the point z, where b_z is the Busemann function of the point z. Since S is a boundary of a convex horoball invariant for $\Gamma_\nu(x)$, $\langle \operatorname{grad} \delta_\gamma(x), \operatorname{grad} b_z(x)\rangle \geqslant 0$ for all $\gamma \in \Gamma_\nu(x)$. Hence

$$0 = \langle \operatorname{grad} F(x), \operatorname{grad} b_z(x)\rangle = \sum_{\gamma \in \widetilde{\Gamma}_\nu(x) - \mathrm{id}} g'(\delta_\gamma(x))\langle \operatorname{grad} \delta_\gamma(x), \operatorname{grad} b_z(x)\rangle \leqslant 0,$$

and hence $\langle \operatorname{grad} \delta_\gamma(x), \operatorname{grad} b_z(x)\rangle = 0$ for all $\gamma \in \widetilde{\Gamma}_\nu(x)$, which contradicts the fact that δ_{γ_0} is strictly decreasing along the ray $h : [0, \infty) \to X$, $h(0) = x$, $h(\infty) = z$.

So, the group $\Gamma_\nu(x)$ consists of hyperbolic isometries. If it is trivial, then the assertion of Lemma 5.3 is obvious. Suppose that $\Gamma_\nu(x)$ is not trivial. According to Lemma 4.2, $\Gamma_\nu(x) = \mathbb{Z}^s$, $s = 1$ or 2. Let the function $b(y) = d(y, \mathrm{MIN}(\Gamma_\nu(x)))$ be smooth outside of the compact set $\mathrm{MIN}(\Gamma_\nu(x))$. If $x \notin \mathrm{MIN}(\Gamma_\nu(x))$ then by the same argument we get $\langle \operatorname{grad} \delta_\gamma(x), \operatorname{grad} b(x)\rangle = 0$ for all $\gamma \in \widetilde{\Gamma}_\nu(x)$. Hence δ_γ is constant over the interval $[x, x']$, where $x' \in \mathrm{MIN}(\Gamma_\nu(x))$ is the closest point to x, and the quadrangle $xx'\gamma(x')\gamma(x)$ is a flat totally geodesic rectangle in X.

If $s = 1$, then there exists a straight line passing through the point x and invariant for some generator $\gamma_0 \in \widetilde{\Gamma}_\nu(x)$ of the group $\Gamma_\nu(x) = \mathbb{Z}$, and hence invariant for the whole group $\Gamma_\nu(x)$. Hence $x \in \mathrm{MIN}(\Gamma_\nu(x))$.

Let $s = 2$. Then by Lemma 4.2 the set $E = \mathrm{MIN}(\Gamma_\nu(x))$ is a component of the boundary ∂X. Any point $\gamma(x)$, $\gamma \in \Gamma$, will be critical for F together with x. Let $\alpha, \beta \in \widetilde{\Gamma}_\nu(x)$ be the generators of the group $\Gamma_\nu(x) = \mathbb{Z} \oplus \mathbb{Z}$. Then the projections of the quadrangle $x\alpha(x)(\beta\alpha(x) = \alpha\beta(x))\beta(x)$ in E are totally geodesic rectangles. Hence the sum of its angles is 2π and it bounds a flat totally geodesic parallelogram. Hence x belongs to a plane in X parallel to E, and hence $x \in E$. □

5.4. COROLLARY. *A point $x \in X$ is critical for the function F if and only if the group $\Gamma_\nu(x)$ contains no parabolic isometries and $x \in \mathrm{MIN}(\Gamma_\nu(x))$.*

PROOF. The necessity is proven in Lemma 5.3. Let us prove the sufficiency. By the condition and Lemma 4.2 the group $\Gamma_\nu(x) = \mathbb{Z}^s$, $s = 0, 1, 2$, consists of hyperbolic isometries and the set $\mathrm{MIN}(\Gamma_\nu(x))$ is nonempty. Since $x \in \mathrm{MIN}(\Gamma_\nu(x))$, it follows that $\operatorname{grad} \delta_\gamma(x) = 0$ for all $\gamma \in \Gamma_\nu(x)$. Hence $\operatorname{grad} F(x) = 0$. □

5.5. COROLLARY. *Every point $x \in \partial X$ is critical for the function F.*

PROOF. Let E be a component of the boundary ∂X, $x \in E$, and Γ_E a subgroup of Γ preserving E. The diameter of any torus or a Klein bottle $T \in [F, M]_{2\nu}$ chosen for the cut operation is less than $2\nu \leqslant \mu/4$. The corresponding component of the boundary ∂M_0 is a torus of diameter not greater than $2 \operatorname{diam} T < \mu/2$. Hence the group Γ_E, since it is isomorphic to $\mathbb{Z} \oplus \mathbb{Z}$ and consists of hyperbolic isometries, has generators with displacements in the plane E not greater than $4 \operatorname{diam} T < \mu$. Hence the group Γ_E is a finite-index subgroup of $\Gamma_\mu(x)$ and there exists a plane $E' \subset X$ invariant for both of them. Hence $E' \subset \mathrm{MIN}(\Gamma_E)$, is parallel to E, and by Remark 3.5 coincides with E. So $\Gamma_\mu(x) = \Gamma_E$. Since $\Gamma_\nu(x) \subset \Gamma_\mu(x)$, the group $\Gamma_\nu(x)$ contains no parabolic isometries and $E \subset \mathrm{MIN}(\Gamma_\nu(x))$. The rest is derived from Corollary 5.4. □

§6. Critical manifolds

6.1. For any point $x \in X$ critical for the function F the set $U_x = \mathrm{MIN}(\Gamma_\nu(x))$ is called the critical manifold of the point x. By Lemma 5.3, $x \in U_x$. This set may not consist of points critical for F, but sometimes it does. For example, if $\dim U_x = 1$ or $x \in \partial X$, and $\Gamma_\nu(x)$ is isomorphic to $\mathbb{Z} \oplus \mathbb{Z}$, then U_x consists of critical points.

Since for the critical point x the group $\Gamma_\nu(x) = \mathbb{Z}^s$, $s = 0, 1, 2$, we distinguish three possibilities:

(1) $s = 0$, $\Gamma_\gamma(x) = \mathrm{id}$. So $U_x = X$.

(2) $s = 1$, $\Gamma_\gamma(x) = \mathbb{Z}$. In this case $U_x = D \times \mathbb{R}$.

(3) $s = 2$, $\Gamma_\gamma(x) = \mathbb{Z} \oplus \mathbb{Z}$. In this case $U_x \subset \partial X$ is a component of the boundary.

The action of the isometries from $\Gamma_\nu(x)$ onto $U_x = D \times \mathbb{Z}^s$ is of the form (id, translation) and hence for $y \in U_x$ the inclusion $\Gamma_\nu(x) \subset \Gamma_\nu(y)$ is valid. If y is critical for F, then $U_y \subset U_x$. These inclusions can be strict. But if $U_y = U_x$, then $x \in U_y$ and hence $\Gamma_\nu(x) = \Gamma_\nu(y)$. So the critical submanifold as a subset of X unambiguously determines its decomposition and the group $\Gamma_\nu(x)$, where x is a critical point, and hence the following definition is meaningful.

Critical submanifolds U_x, U_y are said to be conjugate if $U_y = \gamma(U_x)$ for some $\gamma \in \Gamma$.

The following proposition is the principal one.

6.2. Proposition. *Up to a conjugation there exists only a finite number of critical submanifolds.*

6.3. Corollary. *There exists $\varepsilon > 0$ such that all the points critical for f are contained in the set $\{p \in M_0 : \mathrm{InjRad}(p) \geqslant \varepsilon\}$.*

Proof. Suppose the contrary; then there exists a sequence p_i of points critical for f, $\mathrm{InjRad}(p_i) \to 0$. Let $x_i \in X$, $\pi(x_i) = p_i$, and let $U_i = U_{x_i}$ be the corresponding critical submanifolds. Passing to a subsequence and using Proposition 6.2 we can assume that $x_i \in U_i = U_0$. Since x_i is critical, $\delta_\gamma(x_i) \geqslant \nu$ for all parabolic $\gamma \in \Gamma$. If $\delta_\gamma(x_i) < \nu$, then $\gamma \in \Gamma_\nu(x_i) = \Gamma_\nu(x_0)$ acts on $U_0 = D \times \mathbb{R}^s$ as (id, translation). Hence $\delta_\gamma(x_0) = \delta_\gamma(x_i)$, which contradicts the fact that $\frac{1}{2}\delta_\Gamma(x_i) = \mathrm{InjRad}(p_i) \to 0$. □

6.4. Corollary. *The interior of the manifold M_0 is diffeomorphic to the interior of a compact manifold $W_0 \subset M_0$ with boundary.*

Proof. By Corollary 6.3, all the points critical for $f : M_0 \to \mathbb{R}$ lie in the compact set $\{\mathrm{InjRad} \geqslant \varepsilon\}$. So the set $f^{-1}([0, \eta])$ with $\eta > \max\{f(p) : \mathrm{InjRad}(p) \geqslant \varepsilon\}$ may be taken as W_0. □

§7. The proof of Proposition 6.2

7.1. LEMMA. *If the dimension of the minimal set* $\mathrm{MIN}(\gamma) = D \times \mathbb{R}$ *of a hyperbolic isometry* γ *is at least two, then this set is maximal in the sense that any straight line in* X *parallel to some (and hence to any) line of the form* $d \times \mathbb{R} \subset \mathrm{MIN}(\gamma)$ *lies in* $\mathrm{MIN}(\gamma)$.

PROOF. The isometry γ acts on $D \times \mathbb{R}$ as (id, translation). Since $\dim D \geqslant 1$, $\dim X = 3$, and γ preserves the orientation, any straight line parallel to a line of the form $d \times \mathbb{R}$ is invariant for γ. So it lies in $\mathrm{MIN}(\gamma)$. □

7.2. LEMMA. *Let* γ_1, γ_2 *be hyperbolic isometries such that they do not belong to the same cyclic subgroup of* Γ, *their minimal sets do not lie in* ∂X, *and* $\inf \delta_{\gamma_i} < 2v$, $i = 1, 2$. *Then their minimal sets are disjoint.*

PROOF. Suppose the contrary and let $z \in \mathrm{MIN}(\gamma_1) \cap \mathrm{MIN}(\gamma_2)$. Since the isometries γ_1, γ_2 do not belong to the same cyclic subgroup, their invariant straight lines passing through the point z are different. The group $\Gamma_{2v}(z)$ is abelian due to Lemma 4.3; so its subgroup generated by the isometries γ_1, γ_2 is isomorphic to $\mathbb{Z} \oplus \mathbb{Z}$ and consists of hyperbolic isometries. So there is a plane E passing through z invariant for this group such that $E \subset \mathrm{MIN}(\gamma_1) \cap \mathrm{MIN}(\gamma_2)$. According to §3.3 and Remark 3.5 the plane E should be a component of the boundary ∂X, and there are no other planes in X parallel to E. Since there exists a plane parallel to E passing through any point of $\mathrm{MIN}(\gamma_1) \cap \mathrm{MIN}(\gamma_2)$, one of the minimal sets for γ_1, γ_2 should coincide with $E \subset \partial X$, which contradicts the conditions. □

Let U_x be a critical manifold with group $\Gamma_v(x) \neq \mathrm{id}$. The set $V_x = \bigcup\{\mathrm{MIN}(\gamma) : \gamma \in \widetilde{\Gamma}_v(x) - \mathrm{id}\}$ will be called the extension of U_x.

7.3. LEMMA. *For any critical manifold* U_x *with nontrivial group* $\Gamma_v(x)$ *its extension is convex and coincides with one of the minimal sets for the isometries from* $\widetilde{\Gamma}_v(x) - \mathrm{id}$ *which has the maximal dimension.*

PROOF. If $\Gamma_v(x) = \mathbb{Z} \oplus \mathbb{Z}$ then U_x is a component of the boundary ∂X and by Remark 3.5 the minimal sets for the isometries from $\Gamma_v(x)$ may differ from U_x only within the same cyclic subgroup. If $\Gamma_v(x) = \mathbb{Z}$, then all the invariant straight lines for the isometries of the group are parallel. The rest is derived from Lemma 7.1. □

7.4. LEMMA. *Let the extension* V_x *of the critical manifold* U_x *be three-dimensional, and let* A *be the union of the components of the boundary* ∂V_x *different from the components of* ∂X. *There exists a point* $y \in A$ *such that for any isometry* $\sigma \in \Gamma \setminus \Gamma_v(x)$ *the displacement* $\delta_\sigma(y) \geqslant v$.

PROOF. Let $V_x \neq X$ and, hence, A be nonempty. Suppose that the statement is wrong and that for any point $y \in A$ there exists an isometry $\sigma_y \in \Gamma \setminus \Gamma_v(x)$ with $\delta_{\sigma_y}(y) < v$. Let $y \in A$ be the nearest to x. Consider the geodesic ray $h : [0, \infty) \to \widetilde{M}$, $h(0) = x$, passing through the point y, $y = h(t_0)$. Since $U_x \subset V_x$, $x \in V_x$. By the supposition the critical point $x \notin A$ and hence $t_0 > 0$. It is clear that $h([0, t_0]) \subset V_x$ and the ray h is orthogonal at the point y to the corresponding component of the boundary ∂V_x. Let $\gamma \in \widetilde{\Gamma}_v(x)$ be the isometry for which $V_x = \mathrm{MIN}(\gamma)$. Since $\inf \delta_\gamma < v$, by Lemma 4.3 the isometry σ_y commutes with γ. Hence σ_y preserves the convex set V_x and its decomposition $V_x = D \times \mathbb{R}$. Hence the displacement function δ_{σ_y} does not decrease along $h([t_0, \infty))$. Let t_1 be the point of minimum of

$\delta_{\sigma_y}(t) = \delta_{\sigma_y} \circ h(t)$, $0 \leqslant t < \infty$. Then $t_1 \in [0, t_0]$. Suppose that $t_1 > 0$. Then the interval $[h(t), \sigma_y \circ h(t_1)]$ is orthogonal to the rays h and $\sigma_y \circ h$ at its ends. Since the isometry respects the decomposition $V_x = D \times \mathbb{R}$, the straight line l in $\widetilde{M}$ passing through the points $h(t_1)$ and $\sigma_y \circ h(t_1)$ is invariant for σ_y. The displacement for the isometry σ_y along l is not greater than $\delta_{\sigma_y}(t_0) < v$. The straight line l cannot be parallel to the component $\mathbb{R}$ of the decomposition $V_x = D \times \mathbb{R}$, because then either σ_y in V_x is a translation along $\mathbb{R}$, or the distance $d(\sigma_y(x), A)$ is less than $d(x, y)$. The former case contradicts $\sigma_y \notin \Gamma_v(x)$, and the latter contradicts the choice of y as the point in A nearest to x. So the plane E passing through l parallel to the factor $\mathbb{R}$ is invariant for the group generated by the isometries σ_y and γ. According to §3.3 and Remark 3.5 the plane E is a component of the boundary ∂X. This contradicts the fact that $h(t_1) \in E$ lies inside of V_x if $t_1 < t_0$, or $h(t_1) \in A$ if $t_1 = t_0$. So $t_1 = 0$ and hence $\delta_{\sigma_y}(x) \leqslant \delta_{\sigma_y}(y) < v$, which contradicts the condition $\sigma_y \notin \Gamma_v(x)$. □

7.5. Lemma. *If the extension V_x of the critical manifold U_x does not coincide with X, then there exist a point $x_1 \in X$ and an isometry $\gamma_x \in \widetilde{\Gamma}_v(x)$ such that $\delta_\Gamma(x_1) \geqslant v$ and $\delta_{\gamma_x}(x_1) = v$.*

Proof. If $\dim V_x = 3$ then take A and $y \in A$ as in Lemma 7.4 and consider a geodesic ray $h : [0, \infty) \to \widetilde{M}$, $h(0) = x$, $h(t) \in \widetilde{M} \setminus V_x$ for $t > 0$, orthogonal to the corresponding component of the boundary ∂V_x. If $\dim V_x \leqslant 2$, then consider a ray $h : [0, \infty) \to \widetilde{M}$, $h(0) = x$, $h(t) \in \widetilde{M} \setminus V_x$ for $t > 0$, intersecting X and orthogonal to V_x. We consider these cases simultaneously.

Since the ray h lies outside of V_x, the function $\delta_\gamma(t) = \delta_\gamma \circ h(t)$ is strictly increasing for all $\gamma \in \widetilde{\Gamma}_v(x) - \mathrm{id}$. Hence there exists $t_0 > 0$ such that $\delta_\gamma(t_0) \geqslant v$ for all $\gamma \in \Gamma_v(x) - \mathrm{id}$ and $\delta_\gamma(t_0) = v$ for some isometry $\gamma_x \in \Gamma_v(x)$. We can assume that $\gamma_x \in \widetilde{\Gamma}_v(x)$. Let us prove that $x_1 = h(t_0) \in X$. Suppose not. Then there exists a positive $t_1 < t_0$ such that $h(t_1) \in \partial X$. Note that $\delta_{\gamma_x}(t_1) < v$. Hence $h(t_1)$ is a critical point and the isometry $\gamma_x \in \Gamma_v(h(t_1))$ and hence it acts as translation on the corresponding component of the boundary ∂X. This contradicts the fact that $h(t_1) \notin V_x \supset \mathrm{MIN}(\gamma_x)$. So $x_1 \in X$. Suppose that $\delta_\gamma(x_1) < v$ for some nontrivial isometry $\gamma \in \Gamma$. By the choice of x_1 the isometry $\gamma \notin \Gamma_v(x)$. On the other hand, $\gamma, \gamma_x \in \Gamma_{2v}(x)$ commute according to Lemma 4.3. Hence γ preserves the set $\mathrm{MIN}(\gamma_x)$ and its decomposition. If $\mathrm{MIN}(\gamma_x) \neq V_x$ then by Lemma 7.1 $\dim V_x \geqslant 2$ and the set V_x is maximal. Since the isometric image of a pair of parallel straight lines is a pair of parallel straight lines, and γ preserves $\mathrm{MIN}(\gamma_x) \subset V_x$, γ also preserves the convex set V_x. Hence δ_γ does not decrease along the ray $h : \delta_\gamma(h(0)) \leqslant \delta_\gamma(x_1) < v$. If $\dim V_x \leqslant 2$ then $h(0) = x$, which contradicts $\gamma \notin \Gamma_v(x)$. If $\dim V_x = 3$ then $h(0) = y \in A$, which contradicts Lemma 7.4. Hence $\delta_\Gamma(x_1) \geqslant v$. □

Different critical manifolds may have the same extension. Let γ be a hyperbolic isometry with $\dim \mathrm{MIN}(\gamma) \geqslant 2$ and $\inf \delta_\gamma < v$. If the critical manifold U_x is a proper subset of $\mathrm{MIN}(\gamma)$ then either $\dim U_x = 1$ and U_x is a straight line invariant for γ, or U_x is a component of the boundary ∂X. In the latter case $\dim \mathrm{MIN}(\gamma) = 3$, because $\gamma \in \Gamma_v(x)$ and by Lemma 7.1.

7.6. Lemma. *Let the set V be an extension of some critical manifold and $\bar{\gamma}$ a hyperbolic isometry such that $V = \mathrm{MIN}(\bar{\gamma})$, with $\rho = \inf \delta_{\bar{\gamma}} < v$ being minimal among all the hyperbolic isometries with the same minimal set V. Then for any critical manifold $U_x \subset V$ there exist $x_1 \in X$ and an isometry $\gamma_x \in \widetilde{\Gamma}_v(x)$ with $\mathrm{MIN}(\gamma_x) = U_x$ such that $\delta_\Gamma(x_1) \geqslant \rho$ and $\delta_{\gamma_x}(x_1) < v$.*

PROOF. The argument is analogous to the proof of Lemma 7.5. If $U_x = V$ then we take the point x as x_1 and the isometry $\bar{\gamma}$ as γ_x. So we consider that U_x is a proper subset of V. Just before the formulation of Lemma 7.6 it was explained that U_x is an invariant for $\bar{\gamma}$ a straight line or a component of the boundary ∂X. In any case there exists a ray $h : [0, \infty) \to \widetilde{M}$, $h(0) = x$, orthogonal to U_x and intersecting X. The group $\Gamma_v(x)$ preserves U_x; so the displacement functions for all isometries from $\gamma_v(x)$ do not decrease along h and strictly increase for the isometries $\gamma \in \Gamma_v(x) - \mathrm{id}$ with $\delta_\gamma(x) < \rho$ as a consequence of $U_x \subset V$ and Lemma 7.1. Hence there exists a point $x_1 = h(t_0)$, $t_0 \geqslant 0$ such that $\delta_\gamma(x_1) \geqslant \rho$ for all $\gamma \in \Gamma_v(x) - \mathrm{id}$ and $\rho \leqslant \delta_{\gamma_x}(x_1) < v$ for some isometry $\gamma_x \in \widetilde{\Gamma}_v(x)$ with $\mathrm{MIN}(\gamma_x) = U_x$.

The inclusion $x_1 \in X$ can be proven as in Lemma 7.5. If the displacement for a nontrivial isometry $\sigma \in \Gamma$ at the point x_1 is less than ρ, then $\sigma \notin \Gamma_v(x)$. On the other hand, by Lemma 4.3, σ commutes with γ_x and hence preserves its minimal set $\mathrm{MIN}(\gamma_x) = U_x$. So the displacement function δ_σ does not decrease along h and hence $\delta_\sigma(x) \leqslant \delta_\sigma(x_1) < \rho < v$, which contradicts the condition $\sigma \notin \Gamma_v(x)$. So $\delta_\Gamma(x_1) \geqslant \rho$. □

Two extensions V_x, V_y of critical manifolds are called conjugate if $\gamma(V_x) = V_y$ for some isometry $\gamma \in \Gamma$.

7.7. LEMMA. *Let U_x, U_y be critical manifolds. Let their extensions V_x, V_y be nonconjugate. Choose points x_1, y_1 and isometries γ_x, γ_y as was done in Lemma 7.5. Then the distance between the orbits of the points x_1 and y_1 via the action of the group Γ is not less than $v/2$.*

PROOF. Suppose that the distance between the orbits of the points x_1 and y_1 is less than $v/2$. We can assume that $d(x_1, y_1) < v/2$ and hence

$$\begin{aligned}\delta_{\gamma_x}(y_1) &\leqslant d(y_1, x_1) + \delta_{\gamma_x}(x_1) + d(\gamma_x(x_1), \gamma_x(y_1)) \\ &= 2d(y_1, x_1) + \delta_{\gamma_x}(x_1) < v + v = 2v.\end{aligned}$$

Since $\delta_{\gamma_x}(y_1) < v$, the isometries γ_x, γ_y lie in the group $\Gamma_{2v}(y_1)$. By Lemma 4.3, the isometries commute and hence their minimal sets intersect. If both sets lie in ∂X, then they both coincide with the corresponding component of the boundary. Since the minimal sets of the isometries γ_x, γ_y lie correspondingly in V_x, V_y, we have $V_x = V_y$ which contradicts the conditions. If one of them, say $\mathrm{MIN}(\gamma_x)$, lies in ∂X, then it coincides with U_x and is a component of the boundary ∂X. Then $\mathrm{MIN}(\gamma_y)$ is three dimensional and is an extension of both U_x and U_y. This contradicts the fact that V_x and V_y are nonconjugate. But if neither of $\mathrm{MIN}(\gamma_x)$ and $\mathrm{MIN}(\gamma_y)$ lies in ∂X, by Lemma 7.2 the isometries γ_x, γ_y belong to the same cyclic subgroup, and hence the sets V_x, V_y coincide, which contradicts the conditions. □

7.8. LEMMA. *Let U_x, U_y be nonconjugate critical manifolds that are proper subsets of their common extension V. Choose a positive $\rho < v$, points x_1, y_1 and isometries γ_x, γ_y as was done in Lemma 7.6. Then the distance between the orbits of the points x_1, y_1 via the action of the group Γ is not less than $\rho/2$.*

PROOF. If we assume the contrary, then, as in the proof of Lemma 7.7, we find that the minimal sets of the isometries γ_x, γ_y intersect. On the other hand, by Lemma 7.6 these minimal sets coincide with the corresponding critical manifolds U_x, U_y. Since U_x, U_y are proper subsets of V, according to the remark before Lemma 7.6 they coincide, which contradicts the condition that V_x and V_y are nonconjugate. □

7.9. Corollary. *The number of critical manifolds up to a conjugation is finite.*

Proof. First we prove that there exist only a finite number of mutually nonconjugate extensions of critical manifolds. Suppose the contrary; then by Lemmas 7.5 and 7.7 there exists an infinite sequence $\{p_i\}$ of points in $M_0 = X/\Gamma$, any two of them at a distance not less than $\nu/2$, and the injectivity radius $\mathrm{InjRad}(p_i) \geqslant \nu/2$, which contradicts $\mathrm{InjRad} \to 0$. Similarly by Lemmas 7.6 and 7.8, every extension can contain only a finite number of mutually nonconjugate critical manifolds. So the number of critical manifolds up to a conjugation is finite. □

Now Proposition 6.2 is proven. It remains to prove that all the components of the boundary ∂W_0 (see §6.4) are incompressible tori or Klein bottles.

§8. The components of the boundary ∂W_0

After the passage to a two-sheeted oriented covering, any component of the boundary ∂W_0 is orientable. The components of ∂W_0 contained in ∂M_0, i.e., created in M due to the cut operation, are totally geodesic tori, and for them the assertion of Theorem 1.1 is obvious. If M_0 is compact then $M_0 = W_0$ and ∂W_0 does not have any other components. So we assume that M_0 is not compact, and a component H of the boundary ∂W_0 is not a component of ∂M_0. According to Corollary 6.4 we can assume that $H = H_\eta$ is a connected component of the level set $f^{-1}(\eta)$. Denote by N_η the component of the set $f^{-1}([\eta, \infty))$ containing H_η. The function f has no critical points in N_η; so all of its level sets $H_t \subset f^{-1}(t)$, $t \geqslant \eta$, contained in N_η are homeomorphic to H_η and N_η is homeomorphic to $H \times [0, 1)$. Denote by N_t, $t \geqslant \eta$, the part of the set $f^{-1}([t, \infty))$ contained in N_η. Fix one of the components of $\pi^{-1}(N_\eta)$ in X, where $\pi : X \to M_0$ is the projection, and denote it by $\widetilde{N}_\eta$. Denote by $\widetilde{N}_t$, $t \geqslant \eta$, the component of the preimage $\pi^{-1}(N_t)$ contained in $\widetilde{N}_\eta$, $\widetilde{H}_t = \partial \widetilde{N}_t$. It is clear that $\widetilde{N}_t \subset F^{-1}([t, \infty))$, $\widetilde{H}_t \subset F^{-1}(t)$. Denote by Γ_H the subgroup of Γ preserving $\widetilde{N}_\eta$. Then $\widetilde{H}_\Gamma/\Gamma_H = H$. It is clear that the sequence

$$1 \to \pi_1(\widetilde{H}_\eta) \to \pi_1(H) \to \Gamma_H \to 1$$

is exact and $\Gamma_H = \pi_1(H)$ if and only if H is incompressible.

8.1. Lemma. *For any point $x \in \widetilde{N}_\eta$ the group $\Gamma_\nu(x)$ preserves $\widetilde{N}_\eta$.*

Proof. Let $x \in \widetilde{N}_\eta$. For $\gamma \in \widetilde{\Gamma}_\nu(x)$ consider the interval $[x, \gamma(x)]$. Since the group $\Gamma_\nu(x)$ is abelian, we have $\delta_{\gamma'}(\gamma(x)) = \delta_{\gamma'}(x)$ for $\gamma' \in \Gamma_\nu(x)$. So for any point $x' \in [x, \gamma(x)]$ we have $\delta_{\gamma'}(x') \leqslant \delta_{\gamma'}(x)$ and hence $F(x') \geqslant F(x)$. So the whole interval $[x, \gamma(x)] \subset \widetilde{N}_\eta$ and the group $\Gamma_\nu(x)$ preserves $\widetilde{N}_\eta$. □

8.2. Lemma. *For any point $x \in \widetilde{N}_\eta$ the group $\Gamma_\nu(x)$ is isomorphic to $\mathbb{Z}$ or the surface H is homeomorphic to a torus.*

Proof. For an isometry $\gamma \in \Gamma$ the set $C_\gamma^a = \{y \in X : \delta_\gamma(y)) \leqslant a\}$ is convex. Since the group $\Gamma_\nu(x)$ is abelian, the set $A_x = \cap\{C_\gamma^{\delta_\gamma(x)} : \gamma \in \Gamma_\nu(x)\}$ is convex, nonempty ($x \in A_x$), and $\Gamma_\nu(x)$-invariant, and for $x \in \widetilde{N}_\eta$ it is contained in $\widetilde{N}_\eta$. Suppose that for the point $x \in \widetilde{N}_\eta$ the group $\Gamma_\nu(x)$ is isomorphic to $\mathbb{Z} \oplus \mathbb{Z}$. By Lemma 8.1,

$\Gamma_\nu(x) \subset \Gamma_H$; so we have a continuous mapping $A_x/\Gamma_\nu(x) \to \widetilde{N}_\eta/\Gamma_H = N_\eta$ induced by the inclusion $A_x \subset \widetilde{N}_\eta$. The corresponding homomorphism

$$\mathbb{Z} \oplus \mathbb{Z} = \pi_1(A_x/\Gamma_\nu(x)) \to \pi_1(N_\eta) = \pi_1(H)$$

is an injection by construction. So by Remark 4.1, H is homeomorphic to a torus. □

8.3. LEMMA. *Let $G \subset \Gamma$ be a cyclic subgroup consisting of parabolic isometries. There exists a point $z \in X(\infty)$ such that any isometry $\gamma \in G$ preserves z and all of its horospheres, and the displacement function $\delta_\gamma \circ h(t) \to \inf \delta_\gamma$ as $t \to \infty$ along any ray $h \in z$. In particular, $\operatorname{grad} \delta_\gamma \circ h(t) \to 0$ as $t \to \infty$. In addition, if $\alpha \in \Gamma$ commutes with all $\gamma \in G$, then $\alpha(z) = z$.*

PROOF. Let γ_0 be a generator of group G. There exists a point $z \in X(\infty)$ such that any ray $h \in z$ minimizes the displacement function δ_{γ_0}. Similarly to Lemma 4.3 we find that the group G preserves z and all of its horospheres and $\alpha(z) = z$. Let $\gamma \in G$. Then $\gamma = \gamma_0^n$, n an integer. For $t \to \infty$ we have

$$\delta_\gamma \circ h(t) \leqslant |n|\delta_{\gamma_0} \circ h(t) \to |n| \inf \delta_{\gamma_0} = \inf \delta_{\gamma_0^n} = \inf \delta_\gamma,$$

and hence $\lim_{t\to\infty} \delta_\gamma \circ h(t) = \inf \delta_\gamma$ for any ray $h \in z$. □

8.4. LEMMA. *Suppose that for some subset $A \subset \widetilde{N}_\eta$, intersecting $\widetilde{N}_t$ for all $t \geqslant \eta$, all the groups $\Gamma_\nu(x)$, $x \in A$, lie in the same cyclic subgroup $G \subset \Gamma$. Then any isometry $\gamma \in G - \operatorname{id}$ is parabolic and $\inf \delta_\gamma = 0$.*

PROOF. Since A intersects all the $\widetilde{N}_t$, $t \geqslant \eta$, the group G contains arbitrarily short isometries: for any $\varepsilon > 0$ there exists $\gamma_\varepsilon \in G - \operatorname{id}$ with $\inf \delta_{\gamma_\varepsilon} < \varepsilon$. Let γ_0 be a generator of the group G. Let us prove that $\inf \delta_{\gamma_0} = 0$. According to [2, Lemma 6.6(2)] we have $\inf \delta_{\gamma_0} = \lim_{k \to \infty} \frac{1}{k} \delta_{\gamma_0^k}(x_0)$ for arbitrary $x_0 \in X$. Let $\gamma_\varepsilon = \gamma_0^n$, $n \neq 0$. Then

$$\inf \delta_{\gamma_0} = \lim_{k\to\infty} \frac{1}{k} \delta_{\gamma_0^k}(x_0) = \lim_{k\to\infty} \frac{1}{k|n|} \delta_{\gamma_0^{kn}}(x_0)$$

$$= \frac{1}{|n|} \lim_{k\to\infty} \delta_{\gamma_\varepsilon^k}(x_0) = \frac{1}{|n|} \inf \delta_{\gamma_\varepsilon} < \frac{\varepsilon}{|n|}.$$

Hence $\inf \delta_{\gamma_0} = 0$. But then $\inf \delta_\gamma = 0$ for any isometry $\gamma \in G$. In particular, G consists of parabolic isometries. □

8.5. LEMMA. *Suppose that all groups of the form $\Gamma_\nu(x)$, $x \in \widetilde{N}_t$, for some $t \geqslant \eta$ lie in the same cyclic subgroup $G \subset \Gamma$. Then the surface $\widetilde{H}_\eta \subset X$ is simply connected and, in particular, the component H of the boundary ∂W_0 is incompressible and $\pi_1(H) = \gamma_H$.*

PROOF. We can assume that the subgroup $G \subset \Gamma$ is the smallest among those satisfying the conditions. Since for $\gamma \in \Gamma_H$ the relation $\Gamma_\nu(\gamma x) = \gamma \Gamma_\nu(x) \gamma^{-1}$ is valid, we have $\gamma G \gamma^{-1} = G$, i.e., G is a normal subgroup in γ_H. Moreover, since G is cyclic, for the generator γ_0 we have $\gamma \gamma_0 \gamma^{-1} = \gamma_0^{\pm 1}$, $\gamma \in \Gamma_H$. So the centralizer $\mathbf{Z}_G$ is a subgroup of index $\leqslant 2$ in γ_H.

By Lemmas 8.3 and 8.4, there exists a point $z \in X(\infty)$ such that the group $\mathbf{Z}_G$ preserves z and the function δ_γ is strictly decreasing along any ray $h \in z$ for any isometry $\gamma \in G - \operatorname{id}$. So the angle $\angle(\operatorname{grad} b_z, \operatorname{grad} \delta_\gamma) < \frac{\pi}{2}$, where b_z is the Busemann function of the point z.

Hence for any point $x \in \widetilde{H}_t$ we have $\alpha(x) = \angle(-\operatorname{grad} F(x), \operatorname{grad} b_z(x)) < \frac{\pi}{2}$, because $\operatorname{grad} F(x) = \sum_{\gamma \in \Gamma_\nu(x) - \operatorname{id}} g'(\delta_\gamma(x)) \operatorname{grad} \delta_\gamma(x)$, and $\Gamma_\nu(x) \subset G$. Since $\mathbf{Z}_G$

preserves the point z, the field grad b_z is $\mathbf{Z}_G$-invariant. Hence the function $\alpha = \alpha(x)$ is also $\mathbf{Z}_G$-invariant. The group $\mathbf{Z}_G$ has a finite index in Γ_H and hence $\widetilde{H}_t/\mathbf{Z}_G$ is compact and $\alpha(x) \leqslant \alpha_0 < \frac{\pi}{2}$ for all $x \in \widetilde{H}_t$. So the projection of the surface $\widetilde{H}_t$ onto any horosphere of the point z along the gradient lines grad b_z is a homeomorphism. Since the horospheres are simply connected, the surface $\widetilde{H}_t$ is simply connected as well, and hence $\widetilde{H}_\eta$ is simply connected. □

8.6. COROLLARY. *If a component H of the boundary ∂W_0 is homeomorphic to a torus, then it is incompressible.*

PROOF. Let K be the kernel of the inclusion homomorphism $\pi_1(H) \to \pi_1(W_0)$. Then $K = \pi_1(\widetilde{H}_\eta)$ and $\Gamma_H = \pi_1(H)/K$. Since the group Γ_H is abelian and does not contain any finite-order elements, $\pi_1(H) = K \oplus \Gamma_H$. So, if K is nontrivial, then Γ_H is a cyclic group. Then by Lemmas 8.1 and 8.5 the component H is incompressible, which contradicts the condition $K \neq 1$. □

Hereafter in accordance with Lemma 8.2 and Corollary 8.6 we assume that for any point $x \in \widetilde{N}_\eta$ the group $\Gamma_v(x)$ is isomorphic to $\mathbb{Z}$.

8.7. LEMMA. *For any point $x \in \widetilde{N}_\eta$ the group $\Gamma_v(x)$ consists of parabolic isometries.*

PROOF. Assume it is not so. Since $\Gamma_v(x)$ is a cyclic group, it consists of hyperbolic isometries. Let $y \in \mathrm{MIN}(\Gamma_v(x))$ be the nearest to x. Then $\delta_\gamma(y) \leqslant \delta_\gamma(x)$ for $\gamma \in \Gamma_v(x)$. Hence $F(y) \geqslant F(x) \geqslant \eta$ and $y \in \widetilde{N}_\eta$. By the assumption, $\Gamma_v(x)$ is a cyclic group and $\Gamma_v(x) \subset \Gamma_v(y)$. Hence $\Gamma_v(y)$ contains no parabolic isometries. If $y \in \mathrm{MIN}(\Gamma_v(y))$ then by Corollary 5.4 the point y is critical for F. This contradicts the definition of η. Hence $y \notin \mathrm{MIN}(\Gamma_v(y))$ and hence by Lemma 7.1, $\dim \mathrm{MIN}(\Gamma_v(y)) = 1$. Analogous arguments show that the point $y' \in \mathrm{MIN}(\Gamma_v(y))$ nearest to y lies in $\widetilde{N}_\eta$ and is critical for the function F. The contradiction proves the lemma. □

8.8. LEMMA. *Any increasing sequence $\{G_i\}$ of cyclic subgroups in Γ, $G_i \subset G_{i+1}$ for all $i \geqslant 1$, terminates, i.e., $G_i = G_{i_0}$ when $i \geqslant i_0$ for some i_0.*

PROOF. Let $G = \varinjlim G_i$ be the direct limit of the sequence $\{G_i\}$. Then G is a free abelian subgroup in Γ, and since Γ is discrete, the group G is cyclic. Let α, α_i be the generators of G, G_i. If we assume that the statement is wrong, then $\alpha_1 = \alpha_i^{k_i}$, $\alpha_i = \alpha^{l_i}$, where $|k_i|, |l_i| \to \infty$. Obviously, this is impossible. □

A parabolic isometry $\gamma \in \Gamma$ is said to be strictly parabolic if $\inf \delta_\gamma = 0$. Our first goal is to show that for any point $x \in \widetilde{N}_\eta$ the group $\Gamma_v(x)$ consists of strictly parabolic isometries.

8.9. LEMMA. *Let $x \in \widetilde{N}_\eta$, $z \in X(\infty)$ as was in* Lemma 8.3 *for the group $\Gamma_v(x)$, the ray $h \in z$, $h(0) = x$. Then $\Gamma_v(x) \subset \Gamma_v(h(s))$ for $s \geqslant 0$. If $\Gamma_v(x) = \Gamma_v(h(s))$ for all $s \geqslant 0$, then the ray h intersects all the sets $\widetilde{N}_t$, $t \geqslant \eta$.*

PROOF. According to Lemma 8.3 the displacement functions δ_γ do not increase along h for any $\gamma \in \Gamma_v(x)$. Hence $h([0,\infty)) \subset \widetilde{N}_\eta$ and $\Gamma_v(x) \subset \Gamma_v(h(s))$ for $s \geqslant 0$. Suppose that $\Gamma_v(x) = \Gamma_v(h(s))$ for all $s \geqslant 0$ and $h([0,\infty)) \cap \widetilde{N}_t = \varnothing$ for some $t > \eta$. Then there exists a constant $c > 0$ such that $\delta_\gamma \circ h(s) \geqslant c$ for all $\gamma \in \Gamma_v(x) - \mathrm{id}$ and $s \geqslant 0$

and hence $\operatorname{card}(\widetilde{\Gamma}_v(h(s))) \leqslant C$ for some $C > 0$. By Lemma 8.3, $\operatorname{grad}\delta_\gamma(h(s)) \to 0$ as $s \to \infty$ for all $\gamma \in \Gamma_v(x) - \mathrm{id}$; so for $s \to \infty$

$$|\operatorname{grad} F(h(s))| \leqslant |g'(c)| \sum_{\gamma \in \widetilde{\Gamma}_v(h(x)) - \mathrm{id}} |\operatorname{grad}\delta_\gamma(h(s))| \to 0.$$

Hence $\operatorname{grad} f(\pi \circ h(s)) \to 0$ as $s \to \infty$. By the supposition the geodesic $\pi \circ h$ is contained in the compact set $\overline{N_\eta \setminus N_t} = (\overline{\widetilde{N}_\eta \setminus \widetilde{N}_t})/\Gamma_H$. So the function f has critical points in this compact set, which contradicts the definition of η. □

8.10. LEMMA. *For any point $x_0 \in \widetilde{N}_\eta$ the group $\Gamma_v(x_0)$ consists of strictly parabolic isometries.*

PROOF. Let the point $z \in X(\infty)$ and let the ray $h \in z$, $h(0) = x_0$ be as in Lemma 8.9. If $\Gamma_v(x_0) = \Gamma_v(h(s))$ for all $s \geqslant 0$, then by Lemmas 8.9 and 8.4, the group $\Gamma_v(x_0)$ consists of strictly parabolic isometries. So we assume that $\Gamma_v(x) \subsetneqq \Gamma_v(x_1)$, $x_1 = h(s_1)$ for some $s_1 > 0$. If the statement is wrong, then, by the same arguments we find a sequence $\{x_i\} \subset \widetilde{N}_\eta$ such that for the sequence of cyclic groups $\{\Gamma_v(x_i)\}$ the inclusion $\Gamma_v(x_i) \subsetneqq \Gamma_v(x_{i+1})$ is valid, which contradicts Lemma 8.8. □

8.11. LEMMA. *A strictly parabolic isometry $\gamma \in \Gamma$ preserves all the horospheres of any of its stable points $z \in X(\infty)$.*

PROOF. If γ does not preserve the horospheres of the stable point z, then it translates them all by the same distance $T_z(\gamma) > 0$ ([**2**, §3.7])[1]. Let $0 < \varepsilon < T_z(\gamma)$. Along the ray $h : [0, \infty) \to X$, $h(\infty) = z$, $h(0) \in C_\gamma^\varepsilon$, the displacement function δ_γ does not increase. Hence the isometry γ translates any point of the ray h by a distance not greater than $\varepsilon < T_z(\gamma)$, which contradicts the definition of $T_z(\gamma)$. □

Consider the Tits metric on $X(\infty)$ and the corresponding metric topology (cf. [**2**]). Then the group Γ acts on $X(\infty)$ isometrically.

8.12. PROPOSITION. *Let X be a three-dimensional Hadamard manifold with sectional curvatures $-1 \leqslant K \leqslant 0$, and let $\gamma : X \to X$ be a strictly parabolic isometry. Then the set of its stable points $\operatorname{fix}(\gamma) \subset X(\infty)$ is homeomorphic to an interval, possibly degenerated (in the metric topology on $X(\infty)$). The interval is the unique shortest line in $X(\infty)$ (in the Tits metric) between any two of its interior points.*

The proposition will be proven in §9.

8.13. LEMMA. *For any group $\Gamma_v(x)$, $x \in \widetilde{N}_\eta$, there exists a unique maximal cyclic group $\Gamma(x) \subset \Gamma_H$ that contains it.*

PROOF. The group Γ acts on $X(\infty)$ isometrically. Hence if the set $\operatorname{fix}(\gamma)$ of strictly parabolic isometry $\gamma \in \Gamma$ is not a single point, then $\operatorname{fix}(\gamma) = \operatorname{fix}(\gamma^n)$ for all $n \neq 0$. This follows from Lemma 8.12. Suppose that for the generator $\gamma \in \Gamma_v(x) - \mathrm{id}$, $x \in \widetilde{N}_\eta$, $\gamma_1^n = \gamma = \gamma_2^k$, $n, k \neq 0$, $\gamma_1, \gamma_2 \in \Gamma$. Then γ_i is a strictly parabolic isometry and $z \in \operatorname{fix}(\gamma_i) \subset \operatorname{fix}(\gamma)$, $i = 1, 2$, where z is the midpoint of the interval $\operatorname{fix}(\gamma)$. If neither of the sets $\operatorname{fix}(\gamma_i)$, $i = 1, 2$ is a singleton, then by the previous considerations they coincide with $\operatorname{fix}(\gamma)$. In any case, by Lemma 8.3, a point $z' \in \operatorname{fix}(\gamma_1) \cap \operatorname{fix}(\gamma_2)$ may be found such that at least for one of the isometries, say γ_1, the displacement function $\delta_{\gamma_1} \to 0$ along any ray $h \in z'$. Moreover, since $z' \in \operatorname{fix}(\gamma_2)$, there exists a

[1] *Editor's note.* There the term *invariant* is used for *stable*.

ray $h' : [0,\infty) \to X$, $h'(\infty) = z'$ such that $\delta_{\gamma_2} \circ h'(t) \leqslant \nu$ for all $t \geqslant 0$. Choose $t_0 \geqslant 0$ so that $\delta_{\gamma_1} \circ h'(t_0) \leqslant \varepsilon$, where $g(\varepsilon) \geqslant \eta$. Then $h'(t_0) \in \widetilde{N}_\eta$ and the isometries $\gamma_1, \gamma_2 \in \Gamma_\nu(h'(t_0)) \subset \Gamma_H$ lie in the same cyclic subgroup. Since an increasing sequence of cyclic subgroups in Γ_H terminates, any group $\Gamma_\nu(x)$, $x \in \widetilde{N}_\eta$, is contained in a unique maximal cyclic subgroup in Γ_H. □

8.14. LEMMA. *All the groups $\Gamma_\nu(x)$, $x \in \widetilde{N}_\eta$, are contained in the same maximal cyclic subgroup $G \subset \Gamma_H$.*

PROOF. Let $x \in \widetilde{N}_\eta$, and let $\Gamma(x)$ be the unique cyclic subgroup in Γ_H containing the group $\Gamma_\nu(x)$. Denote by A the set of points $x' \in \widetilde{N}_\eta$ for which $\Gamma_\nu(x') \subset \Gamma(x)$. Since $\widetilde{\Gamma}_\nu(x) \subset \widetilde{\Gamma}_\nu(x')$ for all x' sufficiently close to x, the set A is open. Let $x_0 \in \widetilde{N}_\eta$ be a limit point of A. Then $\widetilde{\Gamma}_\nu(x_0) \subset \widetilde{\Gamma}_\nu(x')$ for all x' sufficiently close to x_0, in particular, for $x' \in A$. Hence $\Gamma_\nu(x_0) \subset \Gamma_\nu(x') \subset \Gamma(x)$. So the set A is closed and nonempty. Hence $A = \widetilde{N}_\eta$. □

8.15. COROLLARY. *The surface $\widetilde{H}_\eta$ is connected, the component H of the boundary ∂W_0 is incompressible, and $\Gamma_H = \pi_1(H)$.*

This follows from Lemmas 8.14 and 8.5.

8.16. COROLLARY. *Let the subgroup $G \subset \Gamma_H$ be as in Lemma 8.14. Then its centralizer $\mathbf{Z}_G$ is a subgroup of Γ_H of index no greater than two.*

The proof is analogous to that of Lemma 8.5.

8.17. COROLLARY. *The component H of the boundary ∂W_0 is homeomorphic to a torus.*

PROOF. If the group $\pi_1(H) = \Gamma_H$ contains no subgroups isomorphic to $\mathbb{Z} \oplus \mathbb{Z}$, then according to Corollary 8.16 the centralizer $\mathbf{Z}_G = G$ is a cyclic subgroup in Γ_H of index not greater than two, which contradicts the compactness of H. Hence $\mathbb{Z} \oplus \mathbb{Z} \subset \pi_1(H)$ and by Remark 4.1, H is homeomorphic to a torus. □

Incompressibility of H is proven in Corollary 8.6. So the proof of Theorem 1.1 is complete.

8.18. REMARK. The group $\Gamma_H = \pi_1(H)$ contains parabolic isometries. Indeed, by Lemma 8.1, $\Gamma_\nu(x) \subset \Gamma_H$ for all $x \in \widetilde{N}_\eta$, and hence $\inf\{\inf \delta_\gamma : \gamma \in \Gamma_H - \mathrm{id}\} = 0$. Since $\Gamma_H = \mathbb{Z} \oplus \mathbb{Z}$, this means that it cannot consist of hyperbolic isometries only.

8.19. EXAMPLE. On the product $Y = [0,\infty) \times T^2$ there exists a metric of negative sectional curvature $K \geqslant -1$ and injectivity radius InjRad $\to 0$ (and even of finite volume) such that all the isometries from the group $\pi_1(Y) = \mathbb{Z} \oplus \mathbb{Z}$ of the universal covering $\widetilde{Y} = [0,\infty) \times \mathbb{R}^2$ (except the identity) are parabolic, but neither of them is strictly parabolic.

Let (r, u, v) be the coordinates in $\widetilde{Y}$, where $r \geqslant 0$ and u, v are the standard coordinates in $\mathbb{R}^2$. Take $ds^2 = dr^2 + e^{-2r}du^2 + (e^{-r} + \varepsilon)^2 dv^2$, where $\varepsilon > 0$. The group Γ acts on $\widetilde{Y} = [0,\infty) \times \mathbb{R}^2$ as (id, Γ'), where Γ' is a lattice of translations in $\mathbb{R}^2$ obtained from the lattice of the integers by a rotation about the origin through an angle with irrational tangents. It is clear that Γ acts on $(\widetilde{Y}, ds^2)$ discretely and isometrically. It is easily verified that $Y = \widetilde{Y}/\Gamma$ possesses the necessary properties.

According to 8.18, all the hyperbolic isometries from the group Γ_H lie in the same cyclic subgroup G. Suppose that G is nontrivial. Then the group Γ_H preserves the set $A = \mathrm{MIN}(G)$ and its decomposition $A = D \times \mathbb{R}$. Hence $\dim A = 3$, the group Γ_H acts on ∂A uniformly, and hence $\widetilde{N}_t \subset A$ for some $t \geqslant \eta$. So we can assume that the universal covering of the end N_η of the manifold M is isometric to the product of the surface by a straight line. Nevertheless even in this case we cannot state that the group Γ_H contains strictly parabolic isometries.

8.20. Example. On the product $Y = [0, \infty) \times T^2$ there exists a metric with sectional curvatures $-1 \leqslant K \leqslant 0$ and of finite volume such that in the group $\pi_1(Y) = \mathbb{Z} \oplus \mathbb{Z}$ of isometries of the universal covering $\widetilde{Y}$ there are hyperbolic isometries and neither of the parabolic isometries is strictly parabolic.

Let $\gamma' : H^2 \to H^2$ be a parabolic isometry of a hyperbolic plane H^2, and let $B \subset H^2$ be a horoball invariant for γ'. Let the group Γ of isometries of the product $\widetilde{Y} = \mathbb{R} \times B$ be generated by the isometries $\gamma_1 = (t \mapsto t + p, \mathrm{id})$, $\gamma_2 = (t \mapsto t + q, \gamma')$, where p/q is irrational. It is clear that Γ acts on $\widetilde{Y}$ discretely and that $Y = \widetilde{Y}/\Gamma$ possesses the necessary properties.

§9. Stable points of a strictly parabolic isometry

This section is devoted to the proof of Proposition 8.12. The proof is essentially based on the following geometrical fact.

9.1. Lemma. *There exists a constant $c_0 > 0$, such that for any point $z \in X(\infty)$ and any horosphere $S_z(t) = b_z^{-1}(t)$ the normal curvatures of the horosphere (with respect to the "outer" normal ∇b_z) are nonnegative and do not exceed c_0.*

Proof. It is derived from the uniform boundedness of the sectional curvatures of X and the fact that for any point of the horosphere $S_z(t)$ one can find a tangent sphere contained in X with arbitrary given radius, for instance, a unit sphere, contained in the horoball $b_z^{-1}((-\infty, t])$, i.e., to one side from $S_z(t)$. □

Let the angles formed by any two of the vectors $v_0, v_1, v_2 \in T_xX$ not exceed $\pi/2$. Denote by $|v_0v_1v_2|$ the area of the spherical triangle on the unit sphere in T_xX with vertices v_0, v_1, v_2. Since $\dim X = 3$, from Lemma 9.1 we immediately derive

9.2. Corollary. *There exists a function $\varepsilon = \varepsilon(\omega)$, monotone in $[0, \pi/2]$, $\varepsilon(0) = 0$, positive in $(0, \pi/2)$, with the following property.*

Let $x \in X$, the horospheres S_0, S_1, S_2 pass through the point x, and the angles formed by their unit normals v_0, v_1, v_2 at x not exceed $\pi/2$. Then inside a ball of radius $\varepsilon(|v_0v_1v_2|)$ with center x the horospheres have the only common point x.

Let $z \in X(\infty)$, $x \in X$. Denote by $v_z(x) \in T_xX$ the unit outer normal to the horosphere with center z, passing through x. Hence for the ray $h(t) = \exp_x(-tv_z(x))$, $t \geqslant 0$, we have $h(\infty) = z$.

9.3. Corollary. *Let $\gamma : X \to X$ be a strictly parabolic isometry, let the points $z_0, z_1, z_2 \in \mathrm{fix}(\gamma)$ be separated by Tits distances not exceeding $\pi/2$, and let $\omega \in (0, \pi/2]$.*

If the isometry γ shifts a point $x \in X$ by a distance less than $\varepsilon(\omega)$, then the area of the spherical triangle $|v_0v_1v_2| < \omega$, where $v_i = v_{z_i}(x)$, $i = 0, 1, 2$.

Proof. According to **[2, §4]** and the condition of the Corollary, the angles between v_0, v_1, v_2 are at most $\pi/2$. Suppose that $|v_0v_1v_2| \geqslant \omega$. Then, as $\varepsilon(\omega) \leqslant$

$\varepsilon(|v_0v_1v_2|)$, by Corollary 9.2 the horospheres S_0, S_1, S_2 have a unique common point x in the ball $B_{\varepsilon(\omega)}(x)$ of radius $\varepsilon(\omega)$ with center x. By Lemma 8.11 the isometry γ preserves any horosphere of any of its stable points. So $x, \gamma(x) \in B_{\varepsilon(\omega)}(x)$ are at least two different common points of the horospheres S_0, S_1, S_2. Hence $|v_0v_1v_2| < \omega$. □

According to **[2,** Appendix 3] any two stable points $z, w \in X(\infty)$ of the parabolic isometry γ are separated by Tits distances at most π. So they can be connected by a shortest line lying in $X(\infty)$. If the distance $Td(z,w) < \pi$, the shortest is unique. In this case it is entirely contained in $\operatorname{fix}(\gamma)$.

9.4. LEMMA. *Let $\gamma : X \to X$ be a strictly parabolic isometry. For any triple of points $z_0, z_1, z_2 \in \operatorname{fix}(\gamma)$, pairwise at Tits distances not exceeding $\pi/2$, in case of proper enumeration*

$$Td(z_0, z_2) = Td(z_0, z_1) + Td(z_1, z_2).$$

PROOF. We can consider that all the distances $d_i = Td(z_{i+1}, z_{i+2})$ are positive. Let $v_i = v_{z_i}$, $\varphi_i(x) = \measuredangle(v_{i+1}(x), v_{i+2}(x))$ (enumerated mod 3). Let $\omega(x) = |v_0(x)v_1(x)v_2(x)|$ be the area of the spherical triangle $v_0(x)v_1(x)v_2(x)$.

Suppose first that not all of the distances d_0, d_1, d_2 are equal. For definiteness let $\sigma = d_1 - d_0 > 0$.

Since $z_i \in \operatorname{fix}(\gamma)$ for $i = 0, 1, 2$, the function δ_γ does not decrease along the ray $h_i \in z_i$ with arbitrary starting point $x \in X$. Since the isometry γ is strictly parabolic, for any $\varepsilon > 0$ there exists a point $x \in X$ with $\delta_\gamma(x) < \varepsilon$. Hence by Corollary 9.3 and Lemmas 4.2 and 4.7 from **[2]**, we find a sequence of points $x_k \in X$ with the following properties (for $k \to \infty$):

(1) $\omega(x_k) \to 0$;
(2) $0 \leqslant d_0 - \varphi_0(x_k) \to 0$;
(3) $0 \leqslant d_1 - \varphi_1(x_k) \to 0$.

Passing, if neccessary, to a subsequence, from property 1) we derive that the angles of the spherical triangles $v_0(x_k)v_1(x_k)v_2(x_k)$ tend, as $k \to \infty$, to either 0 or π. If the value of the angle $\alpha_2(x_k)$ corresponding to the vertex $v_2(x_k)$ tends to π, then from the positivity of the distances d_0, d_1 and from the properties 2) and 3) we derive that the difference between $\varphi_2(x_k)$ and $\varphi_0(x_k) + \varphi_1(x_k)$ is arbitrarily small. The sum on its turn is arbitrarily close to $d_0 + d_1$. Since $d_2 \geqslant \varphi_2(x_k)$, this implies $d_2 \geqslant d_0 + d_1$, and hence $d_2 = d_0 + d_1$.

Let us consider henceforth that $\alpha_2(x_k) \to 0$ as $k \to \infty$. Fix a positive number $\omega \ll \sigma$. Choose k large enough to ensure that the area of the spherical triangle $\omega(x_k) < \omega$, and that the angles $\varphi_0(x_k)$, $\varphi_1(x_k)$ are as close as desired to d_0, d_1 correspondingly. We can assume that $\delta_\gamma(x_k) < \varepsilon(\omega)$.

The function $\delta_\gamma \circ h(t)$ does not decrease along the ray $h \in z_1$, $h(0) = x_k$; so the area $\omega \circ h(t) < \omega$. According to Lemma 4.2 from **[2]** the angles $\varphi_0 \circ h(t)$, $\varphi_2 \circ h(t)$ do not decrease and tend to d_0, d_2 as $t \to \infty$, correspondingly. Since $\omega \ll \sigma = d_1 - d_0$, the angle $\varphi_1 \circ h(t)$ in comparison with $\varphi_1(x_k)$ may diminish by a value of order $o(\omega)$. So the distance d_i may be approximated by φ_i, $i = 0, 1, 2$, and φ_1 is arbitrarily close to $\varphi_0 + \varphi_2$. Hence $d_1 = d_0 + d_2$.

To complete the proof of the lemma we have to consider the case $d_0 = d_1 = d_2 = d$. Since $d \leqslant \pi/2$, the shortest $[z_0z_1] \subset \operatorname{fix}(\gamma)$. Choose a point z_0' on the shortest line close enough to z_0. Applying the above considerations to z_0', z_1, z_2 we arrive at a contradiction. □

9.5. Proof of Proposition 8.12. Let the points $z_0, z_1, z_2 \in \operatorname{fix}(\gamma)$ be chosen so that $\max_{0 \leqslant i,j \leqslant 2} Td(z_i, z_j) \leqslant \pi/2$. By Lemma 9.4 the points z_0, z_1, z_2 belong to the same shortest line in $\operatorname{fix}(\gamma)$. The lemma also implies that the shortest lines do not branch in $\operatorname{fix}(\gamma)$. Thence the assertion of Proposition 8.12 is obvious.

References

1. M. Gromov, *Manifolds of negative curvature*, J. Differential Geom. **13** (1978), 223–230.
2. W. Ballmann, M. Gromov, and V. Schroeder, *Manifolds of nonpositive curvature*, Progr. Math., vol. 61, Birkhäuser, Boston, 1985.

St. Petersburg Branch of Steklov Mathematical Institute, 27, Fontanke, St. Petersburg, 191011, Russia

Translated by M. PANKRATOV

Amer. Math. Soc. Transl.
(2) Vol. **159**, 1994

On Two Conjectures of Linnik

E. P. Golubeva and O. M. Fomenko

§1. Introduction. Formulation of the main results

The analysis of the lattice point equidistribution on a large three-dimensional sphere has led Yu. V. Linnik to the following conjecture [**1**]:

CONJECTURE 1. *Let* $n \equiv 1, 2, 3, 5, 6 \pmod 8$; *then the equation*

$$x_1^2 + x_2^2 + x^2 = n$$

is solvable for some $x = O(n^\varepsilon)$.

Linnik stated also a more general conjecture on the existence of lattice points in convex domains on a large three-dimensional sphere with small area, but here we consider only the aforementioned particular case. The following weaker conjecture was also given in [**1**].

CONJECTURE 2. *The Diophantine inequality*

$$|n - x_1^2 - x_2^2| < n^\varepsilon$$

is solvable for any large n.

In both conjectures ε is an arbitrary positive fixed number. Verifying the conjectures seems to be very difficult, especially the first one.

We begin by discussing Conjecture 2. It is easy to show that the Diophantine inequality

$$|n - x_1^2 - x_2^2| < n^{1/4} \tag{1}$$

is solvable. Note the related problem on the number of lattice points in a planar annulus. A recent result on the circle problem [**2**] implies the following fact: for $h < y$

$$\sum_{y<n\leqslant y+h} r_2(n) = \pi h + O(y^{(7/22)+\varepsilon}), \tag{2}$$

where $r_2(n)$ is the number of representations of n by sums of squares of two integers, ε being positive and arbitrarily small.

Even if as a remainder in (2) we take $O_\varepsilon(y^{1/4+\varepsilon})$, which corresponds to the unsolved circle problem, we have a weaker result for Conjecture 2 than the trivially

1991 *Mathematics Subject Classification*. Primary 11H06, 11D75.

obtained (1). However, Conjecture 2 is a consequence of the following assumption on the number of integers in a short segment representable by a sum of squares of two integers (see [3]). Let

$$M(x,h) = \sum_{x<s<x+h} 1,$$

s being a number representable by a sum of squares of two integers; then one assumes that the asymptotics

$$M(x,h) \asymp \frac{h}{\sqrt{\log x}}. \tag{3}$$

holds in the domain $x^{\varepsilon} < h < x$. Note that combining the reasons of Hooley [3] with (2) one can establish (3) for $x^{(7/22)+\varepsilon} < h < x$.

Now we pass to Conjecture 1. Linnik [1] developed an original method for investigating the lattice point equidistribution on second-order surfaces. He studied in detail the case of the three-dimensional sphere and the simplest two-sheeted hyperboloid

$$H: \quad ac - b^2 = D; \qquad D > 0.$$

This method of proof required some additional assumptions. In particular, for Conjecture 1 Linnik obtained the following result: for all sufficiently large $n \equiv 1, 2, 3, 5, 6 \pmod 8$ the equation $x_1^2 + x_2^2 + x^2 = n$ is solvable with some $x \ll n^{1/2}/(\log n)^{\beta}$ and constant $\beta > 0$ if one supposes the extended Riemann hypothesis or its weakened version (see [1]) to be valid.

For a long time Linnik's results related to the lattice point equidistribution on a large three-dimensional sphere and, particularly, the aforementioned result related to Conjecture 1 have been the only contribution to the problem.

Thanks to the paper [4] by Iwaniec on estimates of Fourier coefficients of half-integral weight parabolic forms, it became possible to establish the lattice point equidistribution on a large sphere without any additional assumptions and in more precise form.

In the present paper we consider only Conjecture 1. In [5] to prove the lattice point equidistribution on a large three-dimensional sphere the authors used an analogue of the Iwaniec estimate (uniform with respect to weight) and estimates of coefficients in the expansion by spherical harmonics of the smoothed characteristic function of a domain on the sphere. We act in a similar fashion. However, we use a more precise uniform estimate for the Fourier coefficients of the cusp form of half-integral weight obtained in [6] (for another variant of the estimate see [7]). Besides, we take into account the specific character of the domain on the sphere considered, that is, a spherical belt.

Let us state the main result of this paper.

THEOREM 1. *Let $r_3(n)$ be the number of representations of a positive integer n by a sum of squares of three integers; $n \equiv 1, 2, 3, 5, 6 \pmod 8$,*

$$r(n,B) = \sum_{-B \leqslant x/\sqrt{n} \leqslant B} r_2(n - x^2) = \sum_{\substack{x_1^2+x_2^2+x^2=n \\ -B \leqslant x/\sqrt{n} \leqslant B}} 1.$$

Then the asymptotic formula

$$r(n,B) = Br_3(n) + O_{\varepsilon}(n^{(1/2)-(7/705)+\varepsilon}) \tag{4}$$

holds, where $n \to \infty$, $1 > B > 0$, the positive ε being arbitrarily small.

COROLLARY. *The equation*

$$x_1^2 + x_2^2 + x^2 = n$$

is solvable for some $x = O_\varepsilon(n^{(1/2)-(7/705)+\varepsilon})$.

Indeed, the last assertion immediately follows from (4) and the estimate $r_3(n) \underset{\varepsilon}{\gg} n^{(1/2)-\varepsilon}$, which is a consequence of the known Siegel theorem.

Note that in the case of a spherical belt Theorem 1 refines the result of our paper **[6]**.

For $n = dl^2$, with a constant integer d, we can significantly refine the results of Theorem 1. To this end we apply the Shimura lifting and reduce the problem to estimating the Fourier coefficients of a cusp form of integral weight, and these estimates (in contrast to the case of half-integer weight) are very precise. The Shimura lifting in implicit form appeared in the paper **[8]** by Linnik as a lifting of ternary theta series to quaternary ones.

THEOREM 2. *Let* $n = dl^2$ *be a positive integer,* $n \equiv 1, 3, 5 \pmod 8$, d *a positive integer constant. Then*

$$r(dl^2, B) = BC(d)l + O_\varepsilon(l^{(3/4)+\varepsilon}) \tag{5}$$

as $l \to \infty$. *Here* $1 > B > 0$; $C(D) > 0$ *is a constant dependent on* d; $\varepsilon > 0$ *arbitrarily small. From now on we do not indicate the dependence of constants on* d.

COROLLARY. *The equation* $x_1^2 + x_2^2 + x^2 = dl^2$ *is solvable for some* $x = O_\varepsilon(l^{(3/4)+\varepsilon})$.

The remainder term $O_\varepsilon(n^{(1/4)+\varepsilon})$ in (4) is assumed to be unimprovable, as is $O_\varepsilon(l^{(1/2)+\varepsilon})$ in (5).

§2. Proof of Theorem 1

First we state two lemmas. The first one deals with estimates of Fourier coefficients of cusp forms of half-integer weight uniform with respect to the weight.

Let $k = \frac{3}{2} + m$ with some integer $m \geqslant 1$. Denote by $S_k(\Gamma)$ for $\Gamma = \Gamma_0(N)$ the linear space of cusp forms $f(z)$ of weight k (and the identity character) for Γ. $S_k(\Gamma)$ is a finite-dimensional Hilbert space with inner product

$$(f, g) = \int_{\Gamma\backslash H} f(z)\overline{g(z)}y^{k-2}\,dx\,dy, \qquad H = \{z = x + iy \mid y > 0\}.$$

We write the Fourier expansion for f

$$f(z) = \sum_{n=1}^{\infty} \widehat{f}(n)\exp(2\pi i n z).$$

LEMMA 1. *Represent* n *as* tn_0^2, *where* t *is squarefree. Let* $k = \frac{3}{2} + m$, $m \geqslant 1$, *and* $f(z) \in S_k(\Gamma)$. *Then for any* n *subject to* $(n_0, N) = 1$ *one has*

$$\widehat{f}(n) \underset{\varepsilon.N}{\ll} \frac{(4\pi)^{m/2}n^{m/2}}{(\Gamma(m+\frac{1}{2}))^{1/2}}\{m^{\frac{3}{16}}n^{\frac{1}{2}-\frac{1}{40}+\varepsilon} + m^{\frac{85}{32}}n^{\frac{1}{2}-\frac{7}{160}+\varepsilon}\}(f, f)^{1/2},$$

with an arbitrarily small $\varepsilon > 0$.

The proof is based on Iwaniec results **[4]** and can be found in the paper **[6]**.

LEMMA 2. *Let δ be a small positive number, $\chi(t)$ the characteristic function of $[-B, B]$, $0 < B < 1$. Then there exists a "smoothed characteristic function" $\chi_\delta(t)$ having the properties:*

$$\begin{aligned} \chi_\delta(t)\ (|t| < \infty) &\ \text{is an even function;} \\ \chi_\delta(t) = 1, &\quad \text{for } |t| \leqslant B; \\ \chi_\delta(t) = 0, &\quad \text{for } |t| \geqslant B + \delta; \\ 0 \leqslant \chi_\delta(t) \leqslant 1, &\quad \text{if } B < |t| < B + \delta; \end{aligned}$$

$\chi_\delta(t)$ can be expanded in a series of Legendre polynomials:

$$\chi_\delta(t) = \sum_{m=0}^{\infty} \alpha_m P_m(t),$$

where $\alpha_0 = B + O(\delta)$, $\alpha_m \underset{r_0}{\ll} 1/[(\delta_m)^{2r_0-1} m^{1/2}]$, $r_0 = 1, 2, \dots$.

The proof of a more general result can be found in **[5]**.

Now we prove Theorem 1. We shall use the method of **[5]**. By Lemma 2

$$\begin{aligned} r(n, B) &= (B + O(\delta)) r_3(n) + \sum_{m=1}^{\infty} \alpha_m \sum_{x_1^2+x_2^2+x^2=n} P_m(x/\sqrt{n}) \\ &= (B + O(\delta)) r_3(n) + \sum_{m=1}^{\infty} \alpha_m R_m(n). \end{aligned}$$

It is known that

$$c_m(n) = n^{m/2} R_m(n) = n^{m/2} \sum_{x_1^2+x_2^2+x^2=n} P_m(x/\sqrt{n})$$

is a Fourier coefficient of the theta series with spherical function:

$$\vartheta(z) = \sum_{n=1}^{\infty} c_m(n) \exp(2\pi i n z),$$

where $\vartheta(z) \in S_{3/2+m}(\Gamma)$, with $\Gamma = \Gamma_0(4)$. By Lemma 1

$$c_m(n) \underset{\varepsilon}{\ll} \frac{(4\pi)^{m/2} n^{m/2}}{(\Gamma(m+\frac{1}{2}))^{1/2}} \{m^{\frac{3}{16}} n^{\frac{1}{2}-\frac{1}{40}+\varepsilon} + m^{\frac{85}{32}} n^{\frac{1}{2}-\frac{7}{160}+\varepsilon}\} (\vartheta, \vartheta)^{1/2}. \tag{6}$$

Let us now estimate (ϑ, ϑ). First we estimate $R_m(n)$, and, simultaneously, $c_m(n)$. Recall the following facts (see **[9]**):

$$|P_m(\cos\theta)| (\sin\theta)^{1/2} < (2/\pi)^{1/2} m^{-1/2} \qquad (0 \leqslant \theta \leqslant \pi), \tag{7}$$

$$P_m(1) = 1; \tag{8}$$

besides, $r_2(k) \ll k^\varepsilon$. Summing $\sum_{x_1^2+x_2^2+x^2=n} P_m(x/\sqrt{n})$ over x and taking into account (7) and (8) we obtain

$$R_m(n) \ll \sum_{0 \leqslant x < \sqrt{n}} \frac{1}{m^{1/2}} \frac{1}{(1 - x^2/n)^{1/2}} r_2(n - x^2) + 1.$$

Hence

$$R_m(n) \underset{\varepsilon}{\ll} n^{(1/2)+\varepsilon} m^{-1/2} + 1,$$

whence

$$c_m(n) \underset{\varepsilon}{\ll} n^{m/2}(n^{(1/2)+\varepsilon} m^{-1/2} + 1). \tag{9}$$

Let us establish the inequality

$$(\vartheta, \vartheta) \underset{\varepsilon}{\ll} \frac{1}{(4\pi)^m} \Gamma(m + 1 + \varepsilon). \tag{10}$$

Using (9) one has

$$\begin{aligned}
(\vartheta, \vartheta) &\underset{\varepsilon}{\ll} \int_{\sqrt{3}/2}^{\infty} y^{m-\frac{1}{2}} \left(\sum_{n=1}^{\infty} e^{-4\pi n y} (n^{m+1+\varepsilon}/m + n^m) \right) dy \\
&\underset{\varepsilon}{\ll} \frac{1}{(4\pi)^m} \left[\int_{\sqrt{3}/2}^{\infty} \frac{y^{m-\frac{1}{2}} y^{-m-2}}{m} \, dy \int_{4\pi y}^{\infty} e^{-t} t^{m+1+\varepsilon} \, dt \right. \\
&\qquad \left. + \int_{\sqrt{3}/2}^{\infty} y^{m-\frac{1}{2}} y^{-m-2} \, dy \int_{4\pi y}^{\infty} e^{-t} t^{m+\varepsilon} \, dt \right] \\
&\underset{\varepsilon}{\ll} \frac{1}{(4\pi)^m} \left[\frac{1}{m} \int_0^{\infty} e^{-t} t^{m+1+\varepsilon} \, dt + \int_0^{\infty} e^{-t} t^{m+\varepsilon} \, dt \right] \\
&\underset{\varepsilon}{\ll} \frac{\Gamma(m+1+\varepsilon)}{(4\pi)^m}.
\end{aligned}$$

If we substitute (10) in (6) and apply Lemma 2, we obtain that

$$\begin{aligned}
\alpha_m R_m(n) &\underset{\varepsilon . r_0}{\ll} \frac{m^{(1/4)+\varepsilon}}{(m\delta)^{2r_0-1} m^{1/2}} \{ m^{\frac{3}{16}} n^{\frac{1}{2}-\frac{1}{40}+\varepsilon} + m^{\frac{85}{32}} n^{\frac{1}{2}-\frac{7}{160}+\varepsilon} \} \\
&\underset{\varepsilon . r_0}{\ll} \frac{1}{(m\delta)^{2r_0-1}} \{ m^{-\frac{1}{16}+\varepsilon} n^{\frac{1}{2}-\frac{1}{40}+\varepsilon} + m^{\frac{77}{32}+\varepsilon} n^{\frac{1}{2}-\frac{7}{160}+\varepsilon} \}.
\end{aligned} \tag{11}$$

Decompose $\sum_{m=1}^{\infty} \alpha_m R_m(n)$ into two sums taking in the first one $m \leqslant 1/\delta^{1+\varepsilon} = U$. It follows from Lemma 2 that by the appropriate choice of r_0 the second sum

$$\Sigma_2 = \sum_{m>U} \alpha_m R_m(n)$$

can be made small in comparison with $\delta r_3(n)$ (below, $\delta = n^{-c}$ where c is a positive constant).

To estimate the first sum

$$\Sigma_1 = \sum_{m \leqslant U} \alpha_m R_m(n)$$

it suffices to use (11) and set $\delta = n^{-7/705}$. Theorem 1 is proved.

§3. Proof of Theorem 2

We need the following

LEMMA 3. *Let $k \geqslant 2$ be even, $f(z) \in S_k(\Gamma_0(N))$,* i.e., *$f(z)$ is a $\Gamma_0(N)$-cusp form of weight k (and of identity character), $f(z) = \sum_{n=1}^{\infty} \widehat{f}(n) \exp(2\pi i n z)$. Then for n subject to $(n, N) = 1$*

$$\widehat{f}(n) \underset{N.\varepsilon}{\ll} k^{\frac{1}{2}+\varepsilon} n^{\frac{k-1}{2}+\varepsilon} \frac{(4\pi)^{\frac{k}{2}}}{(\Gamma(k))^{\frac{1}{2}}} (f, f)^{\frac{1}{2}}.$$

To prove the lemma we decompose $f(z)$ in the orthonormal basis $\{\varphi_i(z)\}$ of the space $S_k(\Gamma_0(N))$ that is obtained by orthogonalizing the Atkin-Lehner basis (see **[10]**):

$$f(z) = \sum_{i=1}^{r} \beta_i \varphi_i(z), \tag{12}$$

where $r = \dim S_k(\Gamma_0(N)) \asymp k$ (N being constant). By **[10,** Theorem 5] for n subject to $(n, N) = 1$ $(i = 1, \ldots, r)$ one has

$$\widehat{\varphi}(n) \underset{\varepsilon.N}{\ll} k^{\varepsilon} n^{\frac{k-1}{2}+\varepsilon} \frac{(4\pi)^{\frac{k}{2}}}{(\Gamma(k))^{\frac{1}{2}}}. \tag{13}$$

This is an analogue of the well-known Deligne estimate uniform with respect to the weight k. Obviously,

$$(f, f) = \sum_{i=1}^{r} |\beta_i|^2. \tag{14}$$

Lemma 3 follows directly from (12)–(14).

We pass to the proof of Theorem 2. Recall that $n = dl^2$ with a constant d and $n \equiv 1, 3, 5 \pmod 8$. As in section 2

$$r(n, B) = (B + O(\delta))r_3(n) + \sum_{m \leqslant 1/\delta^{1+\varepsilon}} \alpha_m R_m(n) + O(n^{\gamma}), \tag{15}$$

with δ to be chosen later (in the form $\delta = l^{-c}$, $c > 0$), $\gamma > 0$ a small positive number.

Let us consider the theta series with spherical function

$$\vartheta(z) = \sum_{n=1}^{\infty} n^{m/2} R_m(n) e^{2\pi i n z} = \sum_{n=1}^{\infty} c_m(n) e^{2\pi i n z};$$

it is known that

$$\vartheta(z) \in S_{3/2+m}(\Gamma_0(4)).$$

For simplicity we set $c_m(n) = c(n)$. Denote by $\vartheta_1(z)$ the cusp form of weight $k = 2m + 2$ corresponding to $\vartheta_1(z)$ by the Shimura lifting **[11]**:

$$\vartheta_1(z) = \sum_{l=1}^{\infty} a(l) e^{2\pi i l z}, \qquad \vartheta_1(z) \in S_k(\Gamma_0(2)).$$

The Fourier coefficients of ϑ and ϑ_1 are connected by

$$c(dl^2) = \sum_{t|l} \mu(t)\chi(t)t^m a(l/t),$$

$$a(l) = \sum_{t|l} \chi(t)t^m c(d(l/t)^2),$$

where $\chi = \chi_{d,m}$ is the Dirichlet character depending on d and m. By virtue of (9)

$$a(l) \ll \sum_{t|l} t^m |c(d(l/t)^2)| \ll_\varepsilon \sum_{t|l} t^m d^{m/2}(l/t)^m \Big(\frac{l^{1+\varepsilon}}{\sqrt{m}} + 1\Big) \ll_\varepsilon d^{m/2} l^{m+\varepsilon} \Big(\frac{l^{1+\varepsilon}}{\sqrt{m}} + 1\Big).$$

Thus $(k = 2m + 2)$

$$(\vartheta, \vartheta) \ll_\varepsilon d^m \int_{\sqrt{3}/2}^{\infty} y^{k-2} \sum_{l=1}^{\infty} (l^{2m+2+\varepsilon}/m + l^{2m+\varepsilon}) e^{-4\pi y l}\, dy \ll_\varepsilon d^m \frac{\Gamma(2m + 2 + \varepsilon)}{(4\pi)^{2m}};$$

therefore, using Lemma 3 one has

$$a(l) \ll_\varepsilon m^{\frac{1}{2}+\varepsilon} l^{m+\frac{1}{2}+\varepsilon} d^{\frac{m}{2}}.$$

Hence

$$c(n) = c(dl^2) \ll_\varepsilon \sum_{t|l} d^{\frac{m}{2}} m^{\frac{1}{2}+\varepsilon} t^m (l/t)^{m+\frac{1}{2}+\varepsilon} \ll_\varepsilon d^{\frac{m}{2}} m^{\frac{1}{2}+\varepsilon} l^{m+\frac{1}{2}+\varepsilon}.$$

This implies

$$(16) \qquad \sum_{x_1^2+x_2^2+x^2=n} P_m(x/\sqrt{n}) \ll_\varepsilon (dl^2)^{-\frac{m}{2}} d^{\frac{m}{2}} m^{\frac{1}{2}+\varepsilon} l^{m+\frac{1}{2}+\varepsilon} \ll_\varepsilon m^{\frac{1}{2}+\varepsilon} l^{\frac{1}{2}+\varepsilon}.$$

Taking into account (15) with $\delta = l^{-1/4}$, (16), and Lemma 2 we obtain Theorem 2.

References

1. Yu. V. Linnik, *Ergodic properties of algebraic fields*, Izdat. Leningrad. Univ., Leningrad, 1967; English transl., Springer-Verlag, Berlin and New York, 1968.
2. H. Iwaniec and C. J. Mozzochi, *On the divisor and circle problems*, J. Number Theory **29** (1988), 60–93.
3. C. Hooley, *Applications of sieve method to the theory of numbers*, Cambridge Tracts in Math., vol. 70, Cambridge Univ. Press, Cambridge, 1976.
4. H. Iwaniec, *Fourier coefficients of modular forms of half-integral weight*, Invent. Math. **87** (1987), 385–401.
5. E. P. Golubeva and O. M. Fomenko, *Asymptotic distribution of lattice points on a three-dimensional sphere*, Zap. Nauchn. Sem. Leningrad. Otdel. Mat. Inst. Steklov. (LOMI) **160** (1987), 54–71; English transl. in J. Soviet Math. **52** (1990).
6. ———, *A remark on the asymptotic distribution of lattice points on a large three-dimensional sphere*, Zap. Nauchn. Sem. Leningrad. Otdel. Mat. Inst. Steklov. (LOMI) **185** (1990), 22–28; English transl. in J. Soviet Math. **59** (1992).
7. W. Duke and R. Schulze-Pillot, *Representation of integers by positive ternary quadratic forms and equidistribution of lattice points on ellipsoids*, Invent. Math. **99** (1990), 49–57.
8. Yu. V. Linnik, *A general theorem on the representation numbers by certain ternary quadratic forms*, Izv. Akad. Nauk SSSR Ser. Mat. **3** (1939), no. 1, 87–108. (Russian)
9. W. Magnus, F. Oberhettinger, and R. P. Soni, *Formulas and theorems for the special functions of mathematical physics*, Grundlehren Math. Wiss., Band 52, Springer-Verlag, Berlin and New York, 1966.

10. O. M. Fomenko, *Estimates for scalar squares of cusp forms, and arithmetic applications*, Zap. Nauchn. Sem. Leningrad. Otdel. Mat. Inst. Steklov. (LOMI) **168** (1988), 158–179; English transl. in J. Soviet Math. **53** (1991).
11. G. Shimura, *On modular forms of half integral weight*, Ann. of Math. (2) **97** (1973), 440–481.

St. Petersburg State University for Telecommunications, 61, Mojka, St. Petersburg, 191065, Russia

Translated by B. PLAMENEVSKIĬ

Amer. Math. Soc. Transl.
(2) Vol. **159**, 1994

Local Approximations of Functions Given on Manifolds

Yu. K. Dem′yanovich

Interpolation and approximation have been studied for a long time. In the past decades smooth spline interpolations have been considered. There exist some difficulties in carrying over results of these studies to functions given on manifolds. For example, the constructions of local smooth interpolation along different paths between two points may lead to different results. However, there are papers where such difficulties have been overcome.

The present paper contains a survey of papers on local approximations for differentiable manifolds (see **[18–38]**). As a basic element of the construction we propose a procedure for transferring approximations from a plane to a manifold given by some atlas with a continual set of charts. One can easily find such an atlas for surfaces embedded smoothly in Euclidean space, for instance, a family of orthogonal projections on tangent planes. The desired plane approximations are sometimes known (the Courant approximations) or may be obtained by comparatively simple generalizations of known ones (the Zlamal, Argyris approximations and others). In practice the main problem is to construct a simplicial subdivision of the surface (see **[21]**).

Under rather general suppositions, approximations on a manifold possess the same properties of stability, smoothness, and interpolation as original plane approximations.

§1. The construction method

1. Definitions and auxiliary assertions. Let $\mathfrak{M}$ be some n-dimensional manifold (i.e., a topological space where each point possesses a neighborhood homeomorphic to the unit ball in $\mathbb{R}^n$). We suppose that there exist a family $\{U_\zeta\}_{\zeta\in\mathfrak{Z}}$ of open sets covering $\mathfrak{M}$ and homeomorphisms ψ_ζ, $\psi_\zeta\colon E_\zeta \to U_\zeta$, of open balls E_ζ in $\mathbb{R}^n$ such that the mappings

$$(1.1)\qquad \psi_\zeta^{-1}\psi_{\zeta'}\colon \psi_{\zeta'}^{-1}(U_\zeta\cap U_{\zeta'}) \to \psi_\zeta^{-1}(U_\zeta\cap U_{\zeta'}),\qquad \zeta,\zeta'\in\mathfrak{Z},$$

are continuously differentiable; $\mathfrak{Z}$ is some set of indices. The triple $\psi_\zeta\colon E_\zeta\to U_\zeta$ is called a chart and the set $\{\psi_\zeta\colon E_\zeta\to U_\zeta \mid \zeta\in\mathfrak{Z}\}$ is called an atlas representing the manifold $\mathfrak{M}$. If the mappings (1.1) are continuously differentiable k times, then $\mathfrak{M}$

1991 *Mathematics Subject Classification*. Primary 41A15, 41A65; Secondary 41A99, 41A60.

is said to be k times continuously differentiable (in this case one writes $\mathfrak{M} \in C^k$; e.g., see [**40**]).

DEFINITION 1. A function u is said to be given on $\mathfrak{M}$ if a function $F_\zeta(x)$, $x \in E_\zeta$, is defined for every $\zeta \in \mathfrak{Z}$ such that $F_\zeta(x_\zeta) = F_{\zeta'}(x_{\zeta'})$, $x_\zeta = \psi_\zeta^{-1}(\xi)$, $x_{\zeta'} = \psi_{\zeta'}^{-1}(\xi)$, $\xi \in U_\zeta \cap U_{\zeta'}$. In other words,

$$F_\zeta(\psi_\zeta^{-1}(\xi)) = F_{\zeta'}(\psi_{\zeta'}^{-1}(\xi)), \qquad \xi \in U_\zeta \cap U_{\zeta'}, \quad \zeta, \zeta' \in \mathfrak{Z}. \tag{1.2}$$

For $\xi \in U_\zeta$ one sets $u(\xi) = F_\zeta(\psi_\zeta^{-1}(\xi))$. In view of property (1.2) (which is called the identity property on intersections of charts), $u(\xi)$ is determined uniquely on $\mathfrak{M}$. The function u is said to be k times continuously differentiable if $\mathfrak{M} \in C^k$, $u \circ \psi_\zeta \in C^k(E_\zeta)$, $\zeta \in \mathfrak{Z}$.

Certainly, one can consider other spaces of differentiable functions as well. In this section we restrict ourselves to C^m, turning to Sobolev spaces in §§2, 3.

In the sequel we consider $\mathfrak{Z} = \mathfrak{M}$; besides, we suppose that the center of E_ζ coincides with 0, and that $\psi_\zeta(0) = \zeta$.

Let us give an example of coordinate systems with the above properties for an n-dimensional surface $\mathfrak{M}$ embedded smoothly in $\mathbb{R}^{n+1}$. Denote by L_ζ the tangent plane at $\zeta \in \mathfrak{M}$, and by E_ζ the unit ball of small radius ε in L_ζ whose center coincides with ζ. Then the projections of E_ζ onto $\mathfrak{M}$ may be taken as coordinate homeomorphisms. For a surface $\mathfrak{M} \subset \mathbb{R}^{n+1}$ of class C^1 represented by $G(\xi) = 0$, $\xi \in \mathbb{R}^{n+1}$ ($G \in C^1$, $\nabla G(\xi)|_{\xi \in \mathfrak{M}} \neq 0$), the mentioned mappings take the form

$$\psi_\zeta^{-1}\colon \xi \to \xi', \quad \xi' = \xi - \zeta - (\nabla G(\zeta), \xi - \zeta)\nabla G(\zeta)\|\nabla G\|^{-2}, \quad \zeta \in \mathfrak{M}. \tag{1.3}$$

From now on, parentheses for vectors in Euclidean space denote a scalar product, $\|\xi\| \overset{\text{def}}{=} (\xi, \xi)^{1/2}$, $\xi \in \mathbb{R}^{n+1}$. If $\mathfrak{M}$ is a sphere with center at the origin,

$$\mathfrak{M}\colon (\xi, \xi) = r^2, \tag{1.4}$$

then $G(\xi) = (\xi, \xi) - r^2$ and

$$\begin{aligned} \psi_\zeta\colon \xi' \to \xi, &\quad \xi = \xi' + \zeta(1 - \|\xi'\|^2 r^{-2})^{1/2}, \\ \psi_\zeta^{-1}\colon \xi \to \xi', &\quad \xi' = \xi(\xi, \zeta)\zeta r^{-2}. \end{aligned} \tag{1.5}$$

In particular, for $n = 1$, introducing an angular variable φ on $\mathfrak{M}$ one has

$$\psi_\zeta^{-1}\colon S_{\xi'} = r \sin(\varphi_\xi - \varphi_\zeta),$$

where $S_{\xi'}$ is a natural parameter on the tangent L_ζ. Transformations of this kind are called orthogonal projections onto a tangent.

2. Approximation relations. Let us consider the Banach space $X \overset{\text{def}}{=} C^{m+1}(\mathfrak{M})$ of functions given on $\mathfrak{M} \in C^{m+1}$; let X^* denote the dual space. By Definition 1, a function $u \in X$ is defined by means of $F_\zeta \in C^{m+1}(E_\zeta)$, $\zeta \in \mathfrak{M}$, $u = F_\zeta \circ \psi_\zeta^{-1}$. One has (Taylor expansion at 0)

$$F_\zeta(x) = \sum_{|\alpha|\leqslant m} \frac{1}{\alpha!} D^\alpha F_\zeta(0) x^\alpha + R_\zeta(x), \qquad x \in E_\zeta. \tag{1.6}$$

Then

$$u(\xi) = F_\zeta(\psi_\zeta^{-1}(\xi)) = \sum_{|\alpha|\leqslant m} \frac{1}{\alpha!} D^\alpha F_\zeta(0)[\psi_\zeta^{-1}(\xi)]^\alpha + R_\zeta(\psi_\zeta^{-1}(\xi)), \qquad \xi \in U_\zeta.$$

Thus, in a neighborhood U_ζ of the point ζ

$$u(\xi) = (\mathscr{P}_\zeta u)(\xi) + (\mathscr{R}_\zeta u)(\xi), \qquad \xi \in U_\zeta, \tag{1.7}$$

with

$$(\mathscr{P}_\zeta u)(\xi) = \sum_{|\alpha|\leqslant m} \frac{1}{\alpha!} D^\alpha F_\zeta(0)[\psi_\zeta^{-1}(\xi)]^\alpha, \qquad (\mathscr{R}_\zeta u)(\xi) = \mathscr{R}_\zeta(\psi_\zeta^{-1}(\xi)).$$

Let $f(\eta)$ be an abstract function on $\mathfrak{M}$ with values in X^*, where the support $\operatorname{supp} f(\eta)$ of the functional $f(\eta)$ (for a fixed η) lies in U_ζ. According to (1.7)

$$(f(\zeta), u) = (f(\zeta), \mathscr{P}_\zeta u) + (f(\zeta), \mathscr{R}_\zeta u), \qquad \zeta \in \mathfrak{M}.$$

Suppose that

$$(f(\zeta), \mathscr{P}_\zeta u) \equiv u(\zeta) \qquad \forall u \in X. \tag{1.8}$$

The expression $\widetilde{u}(\zeta) \overset{\text{def}}{=} (f(\zeta), u)$ is called an approximation of u. From (1.8) it follows that

$$\widetilde{u}(\zeta) - u(\zeta) = (f(\zeta), \mathscr{R}_\zeta u). \tag{1.9}$$

The relation (1.8) is fulfilled if and only if

$$(f(\zeta), [\psi_\zeta^{-1}(\cdot)]^\alpha) = \delta_{0.\alpha}, \qquad |\alpha| \leqslant m, \tag{1.10}$$

where $\delta_{\alpha.\beta}$ is the Kronecker symbol.

The identities (1.10) are called approximate relations.

REMARK 1. If $\mathfrak{M} \subset \mathbb{R}^{n+1}$ and

$$\psi_\zeta^{-1}(\xi) = \xi - \zeta + o(\|\xi - \zeta\|),$$

then for a small diameter of $\operatorname{supp} f(\zeta) \cup \zeta$ the relations (1.10) are close to the analogous relations on a plane (i.e., for the case $\mathfrak{M} = \mathbb{R}^n$)

$$(f(\zeta), \varphi_\alpha(\zeta)) = \delta_{0.\alpha}, \qquad |\alpha| \leqslant m, \quad \varphi_\alpha(\zeta) \overset{\text{def}}{=} (\xi - \zeta)^\alpha.$$

DEFINITION 2. Let $\{\omega_j\}$ be a family of functions given on $\mathfrak{M}$. The multiplicity of $\{\omega_j\}$ on $\mathfrak{M}$ is the number $\varkappa_\omega = \sup_{t\in\mathfrak{M}} \varkappa^t(\{\operatorname{supp}\omega_j\})$, where $\varkappa^t(\{M_j\})$ is the number of the sets M_j with $t \in M_j$.

As an example, consider a family $\{\omega_j\}$ with finite multiplicity on $\mathfrak{M}$ and set

$$f(\zeta) = \sum_j \omega_j(\zeta) f_j,$$

where $f_j \in X^*$. The conditions (1.10) take the form

$$\sum_j \omega_j(\zeta)(f_j, [\psi_\zeta^{-1}(\cdot)]^\alpha) = \delta_{0.\alpha}, \qquad |\alpha| \leqslant m. \tag{1.11}$$

If $\{\xi_j\}$ is a net on $\mathfrak{M}$ and $f_j = \delta_{\xi_j}$ (δ_ξ denotes the δ-function at $\xi \in \mathfrak{M}$), then

$$\sum_j \omega_j(\zeta)[\psi_\zeta^{-1}(\xi_j)]^\alpha = \delta_{0.\alpha}, \qquad |\alpha| \leqslant m. \tag{1.12}$$

In the case $\mathfrak{M} = \mathbb{R}^n$ one has

$$\sum_j \omega_j(\zeta)(\xi_j - \zeta)^\alpha = \delta_{0.\alpha}, \qquad |\alpha| \leqslant m. \tag{1.13}$$

If $\mathfrak{M}$ is a sphere of radius r with center at the origin, then by (1.5) we obtain

$$\sum_j \omega_j(\zeta)[\xi_j - \zeta r^{-2}(\xi_j, \zeta)]^\alpha = \delta_{0.\alpha}, \qquad |\alpha| \leqslant m. \tag{1.14}$$

As $r \to +\infty$ and $\xi_j \to \xi_j^* \in \mathbb{R}^n$ ($\mathbb{R}^n$ is a tangent plane) the relations (1.14) are transformed into (1.13) with $\xi_j = \xi_j^*$.

The operator $f(\zeta)u = \int_{\mathfrak{M}} \mathscr{K}(\zeta, \xi)u(\xi)\,d\xi$ is called a generalized averaging operator, the kernel $\mathscr{K}$ being supported near the diagonal, $\operatorname{supp}_\xi \mathscr{K}(\zeta, \xi) \subset U_\zeta$ (supp_ξ denotes a support with respect to ξ). If the condition (1.10), taking the form

$$\int_{\mathfrak{M}} \mathscr{K}(\zeta, \xi)[\psi_\zeta^{-1}(\xi)]^\alpha\,d\xi = \delta_{0.\alpha}, \qquad |\alpha| \leqslant m, \quad \zeta \in \mathfrak{M},$$

is fulfilled, then the generalized averaging operator has rank m, by definition.

3. Description of the method. To construct the approximation $\widetilde{u}$ one can use two approaches **[22]**. In the first one, the relations (1.11) are considered as a linear algebraic system (with respect to $\omega_j(\zeta)$), dependent on the parameter ζ, $\zeta \in \mathfrak{M}$. In this case a solution of (1.11) may be found directly (for example, by the Cramer rule). In the second approach, the approximations on a manifold are induced by planar (Zlamal, Argyris, and others) approximations; in complicated cases ($m \geqslant 2$ or under strong requirements of smoothness) the second approach is more convenient. Let us describe it. Suppose that a family of planar approximations is known. More precisely, we assume that for every $\zeta \in \mathfrak{M}$ an abstract function $f_\zeta(\bar{\xi}')$, $\bar{\xi}' \in E_\zeta$, is given with values in $(C^{(m+1)}(E_\zeta))^*$ such that

$$(f_\zeta(\bar{\xi}'), \varphi_\alpha(\bar{\xi}')) = \delta_{0.\alpha}, \qquad |\alpha| \leqslant m, \quad \bar{\xi}' \in E_\zeta, \quad \zeta \in \mathfrak{M}. \tag{1.15}$$

Here $\varphi_\alpha(\bar{\xi}') = (\xi' - \bar{\xi}')^\alpha$, $\xi' \in E_\zeta$.

Rewrite (1.15) in the form

$$(f_\zeta(\psi_\zeta^{-1}(\bar{\xi})), \varphi_\alpha(\psi_\zeta^{-1}(\bar{\xi}))) = \delta_{0.\alpha}, \tag{1.16}$$

$\zeta \in \mathfrak{M}$, $\bar{\xi} \in U_\zeta$, $|\alpha| \leqslant m$.

Set $\bar{\xi} = \zeta$ in (1.16). Then

$$(f_\zeta(0), \varphi_\alpha(0)) = \delta_{0.\alpha}, \qquad |\alpha| \leqslant m, \quad \zeta \in \mathfrak{M}.$$

If one defines $f(\zeta)$ by

$$(f(\zeta), u) = (f_\zeta(0), u \circ \psi_\zeta) \qquad \forall u \in X, \tag{1.17}$$

then (1.10) holds. It is easy to see that the smoothness of $(f(\zeta), u)$ is determined by the smoothness of u, $f_\zeta(0)$, and ψ_ζ (with respect to ζ). If $f_\zeta(0)$ and ψ_ζ are k times continuously differentiable and $u \in C^k(\mathfrak{M})$, $\mathfrak{M} \in C^k$, then $(f(\zeta), u) \in C^k(\mathfrak{M})$.

As an example we consider the following construction. Let ξ_j, $\xi_j \in \mathfrak{M}$, $j = 1, \dots, N$, be a net on $\mathfrak{M}$. Denote the set $\{j \mid \xi_j \in U_\zeta\}$ by $J(\zeta)$ and define in E_ζ the net $\xi'_j(\zeta) = \psi_\zeta^{-1}(\xi_j)$, $j \in J(\zeta)$. Let $\omega_{\zeta.j}(\xi')$ be a system of coordinate functions such that (1.15) holds for the abstract function

$$f_\zeta(\bar{\xi}') \overset{\text{def}}{=} \sum_{j \in J(\zeta)} \omega_{\zeta.j}(\bar{\xi}')\delta_{\xi'_j}$$

in a neighborhood V_ζ of the origin, $V_\zeta \in E_\zeta$; in other words, let

$$\sum_{j \in J(\zeta)} \omega_{\zeta.j}(\bar{\xi}')(\xi'_j(\zeta) - \bar{\xi}')^\alpha = \delta_{0.\alpha}, \qquad \bar{\xi}' \in V_\zeta, \quad \zeta \in \mathfrak{M}, \quad |\alpha| \leqslant m. \tag{1.18}$$

Defining $f(\zeta)$ by (1.17), we find

$$(f(\zeta), u) = \sum_{j \in J(\zeta)} \omega_{\zeta.j}(0) u(\xi_j).$$

Now, due to (1.18)

$$\sum_{j \in J(\zeta)} \omega_{\zeta.j}(0)[\psi_\zeta^{-1}(\xi_j)]^\alpha = \delta_{0.\alpha}, \qquad |\alpha| \leqslant m, \quad \zeta \in \mathfrak{M}.$$

The relation (1.12) follows from (1.18) with

$$\omega_j(\zeta) = \omega_{\zeta.j}(0) \text{ for } j \in J(\zeta), \quad \omega_j(\zeta) = 0 \text{ for } j \notin J(\zeta). \tag{1.19}$$

Consider (1.19) in more detail when $\mathfrak{M}$ is the sphere (1.4), $\omega_{\zeta.j}(\xi')$ are the Courant basic functions in $\mathbb{R}^n$ (see **[12**, p. 56]), and ψ_ζ^{-1} is the central projection onto L_ζ,

$$\begin{aligned} \psi_\zeta^{-1}&: \xi \to \xi', \quad \xi' = r^2\xi/(\xi, \zeta) - \zeta, \\ \psi_\zeta&: \xi \to \xi', \quad \xi = (\xi' + \zeta)r/\|\xi' + \zeta\|. \end{aligned} \tag{1.20}$$

Let $U_\zeta = \{\xi \mid (\xi, \zeta) \geqslant \varepsilon r^2, \xi \in \mathfrak{M}\}$ with a fixed $\varepsilon \in (0, 1)$.

Consider a simplicial subdivision $\mathscr{T}$ of $\mathfrak{M}$; denote the 0-skeleton by $\mathscr{T}^0 = \{\xi_j\}$ and the 1-skeleton (consisting of geodesic lines) by $\mathscr{T}^1$. Assume $\mathscr{T}$ to be so small that U_ζ contains the closure of the barycentric star for any vertex of the simplex $\bar{T}$ containing ζ, $T \in \mathscr{T}$. Evidently, $\psi_\zeta^{-1}(U_\zeta)$ is a ball E_ζ with center at the origin of L_ζ, $E_\zeta \subset L_\zeta$.

The complex Φ_ζ, which consists of the simplices lying in U_ζ, is called a fragment of $\mathcal{T}$ in U_ζ. The image Φ'_ζ of Φ under the above mapping to L_ζ is a complex consisting of rectilinear simplices. Set $\xi'_j = \psi_\zeta^{-1}\xi_j$ for $\xi_j \in \Phi_\zeta$; then

$$\xi'_j = r^2\xi_j/(\xi_j, \zeta) - \zeta. \tag{1.21}$$

By supposition, there exists a vertex $\xi'_j \in \Phi'_\zeta$, with the closure of its barycentric star $3'_j$ containing the origin of L_ζ. Let us construct the Courant function $\omega_{\zeta.j}(\xi')$, corresponding to ξ'_j. Without loss of generality we consider that $\xi' \in T'$, $T' \in 3'_j$, and that the vertices of T' are enumerated by $1, \dots, n+1$, hence $j \in \{1, \dots, n+1\}$. It is known (see **[12,** p. 56]) that

$$\omega_{\zeta.j}(\xi') = \lambda_j(\xi'), \tag{1.22}$$

where $\lambda_k(\xi')$, $k = 1, \dots, n+1$, are the barycentric coordinates of ξ' in T'. By definition, the vector $\lambda(\xi') \overset{\text{def}}{=} (\lambda_1(\xi'), \dots, \lambda_{n+1}(\xi'))$ satisfies $A\lambda(\xi') = e(\xi')$. Here A is an $(n+1)\times(n+1)$ matrix whose first row consists of 1's and whose ith row $(\xi_1'^{(i)}, \dots, \xi_{n+1}'^{(i)})$ consists of the ith components of the vertices of T', while $e(\xi')$ is the column $(1, \xi'^{(1)}, \dots, \xi'^{(n)})$, $\xi'^{(i)}$ being an ith component of ξ' in L_ζ. Therefore,

$$\lambda_i(\xi') = \Delta_{T'.i}(\xi')/\Delta_{T'}$$

(one obtains $\Delta_{T'.i}(\xi')$ by replacing the ith column in $\Delta_{T'} \overset{\text{def}}{=} \det A$ by $e(\xi')$). Thus

$$\omega_{\zeta.j}(\xi') = \begin{cases} \Delta_{T'.j}(\xi^k)/\Delta_{T'} & \text{for } \xi' \in T' \subset |3'_j|, \\ 0 & \text{for } \xi' \notin |3'_j|. \end{cases}$$

From (1.22) it follows that $\omega_{\zeta.j}$ does not depend on the choice of a coordinate system in $\mathbb{R}^{n+1}$. Writing ω_j^0 instead of ω_j one finds with the help of (1.19) that

$$\begin{aligned} &\omega_j^0(\zeta) = \Delta_{T'.j}(0)/\Delta_{t'} \quad \text{for } \zeta, \xi_j \in \bar{T}, \\ &\omega_j^0(\zeta) = 0 \quad \text{otherwise.} \end{aligned}$$

Taking into account the invariance of (1.21) under coordinate transformations, we choose basis vectors $e_1, \dots, e_n$ in $\mathbb{R}^{n+1}$ to be parallel to the vectors $e'_1, \dots, e'_n$ fixed in L_ζ, and direct e_{n+1} along the vector ζ. By projecting (1.21) onto $e_1, \dots, e_n$ we obtain

$$\xi_j'^{(k)} = r^2\xi_j^{(k)}/(\xi_j, \zeta), \qquad k = 1, \dots, n.$$

Let us substitute this in $\Delta_{T'}$; then

$$\Delta_{T'} = \zeta^{(n+1)} r^{2n} \prod_{j'=1}^{n+1} (\xi_{j'}, \zeta)^{-1} \begin{vmatrix} \xi_1^{(n+1)} & \dots & \xi_{n+1}^{(n+1)} \\ \xi_1^{(1)} & \dots & \xi_{n+1}^{(1)} \\ \vdots & & \vdots \\ \xi_1^{(n)} & \dots & \xi_{n+1}^{(n)} \end{vmatrix}.$$

Analogous transformations of $\Delta_{T'.j}(0)$ give

$$\Delta_{T'.j}(0) = r^{2n-2}\zeta^{(n+1)} \prod_{\substack{j'=1 \\ j' \neq j}}^{n+1} (\xi_{j'}, \zeta)^{-1} \begin{vmatrix} \xi_1^{(n+1)} & \dots & \zeta^{(n+1)} & \dots & \xi_{n+1}^{(n+1)} \\ \xi_1^{(1)} & \dots & \zeta^{(1)} & \dots & \xi_{n+1}^{(1)} \\ \vdots & & \vdots & & \vdots \\ \xi_1^{(n)} & \dots & \zeta^{(n)} & \dots & \xi_{n+1}^{(n)} \end{vmatrix}.$$

Thus, in invariant notation $\omega_j^0(\zeta)$ has the form

$$\omega_j^0(\zeta) = \begin{cases} r^{-2}(\xi_j, \zeta)\Delta_j^{\mathrm{T}}(\zeta)/\Delta^{\mathrm{T}}, & \zeta, \xi_j \in \bar{T} \subset |\mathfrak{Z}_j|, \\ 0, & \zeta \notin |\mathfrak{Z}_j|, \end{cases}$$

where

$$\Delta^{\mathrm{T}} = \begin{vmatrix} \xi_1^{(1)} & \dots & \xi_{n+1}^{(1)} \\ \vdots & & \vdots \\ \xi_1^{(n+1)} & \dots & \xi_{n+1}^{(n+1)} \end{vmatrix},$$

while $\Delta_j^{\mathrm{T}}(\zeta)$ is obtained by changing the jth row in Δ^{T} to $\zeta^{(1)}, \dots, \zeta^{(n+1)}$. Evidently, the choice of Cartesian coordinate system is of no importance for defining $\omega_j^0(\zeta)$. It is easy to verify that the functions $\omega_j^0(\zeta)$ are continuous, give an interpolational basis on the net $\mathscr{T}^0$, $\omega_j^0(\xi_i) = \delta_{i,j}$, and satisfy

(A) $\omega_j(\zeta) \equiv 0$ for $\zeta \in T$ and $\xi_j \notin \bar{T}$.

Replacing (1.20) by (1.5) one obtains another analog of the Courant approximations on a sphere (see **[18]**), namely:

$$\omega_j(\zeta) = \begin{cases} \Delta_j^{\mathrm{T}}(\zeta)/\Delta_*^{\mathrm{T}}(\zeta), & \text{for } \xi_j, \zeta \in \bar{T}, \\ 0 & \text{otherwise.} \end{cases} \tag{1.23}$$

Here

$$\Delta_*^{\mathrm{T}}(\zeta) = \begin{vmatrix} \xi_1^{(1)} - \xi_{n+1}^{(1)} & \dots & \xi_n^{(1)} - \xi_{n+1}^{(1)} & \zeta^{(1)} \\ \vdots & & \vdots & \vdots \\ \xi_1^{(n+1)} - \xi_{n+1}^{(n+1)} & \dots & \xi_n^{(n+1)} - \xi_{n+1}^{(n+1)} & \zeta^{(n+1)} \end{vmatrix}.$$

These functions are continuous, give an interpolational basis on the net $\{\xi_j\}$, and satisfy (A).

A piecewise-linear interpolation can serve as the simplest example of approximation on the one-dimensional sphere (circle). However, such interpolation is not convenient on a sphere of dimension > 1 because of angular coordinate singularities. The interpolation $\widetilde{u}(\zeta) = \sum_{j \in J} u(\xi_j)\omega_j(\zeta)$ is not influenced by this fact, $\omega_j(\zeta)$ being given by (1.23). Let us consider (1.23) for the one-dimensional case. Introduce in $\mathbb{R}^2$ Cartesian coordinates (x, y); denote by $\{\zeta_j\}$ a net on $\mathfrak{M}$: $x^2 + y^2 = r^2$, ordered by the increase of the (polar) angle, $\zeta_j = (x_j, y_j)$. From (1.23) one finds for points $\zeta = (x, y) \in \mathfrak{M}$ lying between ξ_j and ξ_{j+1}

$$\omega_j(\zeta) = (yx_{j+1} - xy_{j+1})/((x_{j+1} - x_j)y - (y_{j+1} - y_j)x),$$
$$\omega_{j+1}(\zeta) = (xy_j - yx_j)/((x_{j+1} - x_j)y - (y_{j+1} - y_j)x).$$

In polar coordinates $x = r\cos\varphi$, $y = r\sin\varphi$ for $\widetilde{\omega}_j(\varphi) \stackrel{\text{def}}{=} \omega_j(r\cos\varphi, r\sin\varphi)$ we obtain

$$\widetilde{\omega}_j(\varphi) = \begin{cases} \dfrac{\sin(\varphi_{j-1} - \varphi)}{2\sin\frac{\varphi_{j-1}-\varphi_j}{2}\cos(\varphi - \frac{\varphi_{j-1}+\varphi_j}{2})} & \text{for } \varphi \in [\varphi_{j-1}, \varphi_j), \\ \dfrac{\sin(\varphi - \varphi_{j+1})}{2\sin\frac{\varphi_j-\varphi_{j+1}}{2}\cos(\varphi - \frac{\varphi_j+\varphi_{j+1}}{2})} & \text{for } \varphi \in [\varphi_j, \varphi_{j+1}), \\ 0 & \text{for } \varphi \notin [\varphi_{j-1}, \varphi_{j+1}). \end{cases}$$

§2. Approximation estimates

1. Estimates in the uniform metric. In this section the estimates of $\widetilde{u} - u$ are given in the spaces C and L_p. Write the remainder in the Taylor formula (1.6) in integral form

$$\mathscr{R}_\zeta(\xi) = R_\zeta(\psi_\zeta^{-1}(\xi)) = \frac{1}{m!}\int_0^1 (1-\tau)^m \sum_{|\alpha|=m+1} \frac{1}{\alpha!} D^\alpha F_\zeta(\tau\xi')\xi'^\alpha \, d\tau,$$

where $\xi' = \psi_\zeta^{-1}(\xi)$, $\xi \in \operatorname{supp} f(\zeta)$. Setting

$$\mathscr{S}(\zeta) = \operatorname{supp} f(\zeta), \quad \mathscr{S}'(\zeta) = \psi_\zeta^{-1}(\mathscr{S}(\zeta)), \quad h(\zeta) = \operatorname{diam} \mathscr{S}'(\zeta),$$

for $\xi \in \mathscr{S}(\zeta)$ we obtain

$$\|\mathscr{R}_\zeta\|_{C(\mathscr{S}(\zeta))} \leqslant C_0[h(\zeta)]^{m+1} \max_{|\alpha|=m+1} \|D^\alpha F_\zeta\|_{C(\mathscr{S}'(\zeta))} \tag{2.1}$$

with $C_0 =$ constant independent of ζ and F_ζ.

If $f(\zeta)$ is a trajectory in $C^*(\mathfrak{M})$, then from (1.9) it follows that

$$|u(\zeta) - \widetilde{u}(\zeta)| \leqslant \|f(\zeta)\|_{C^*(\mathfrak{M})} \|\mathscr{R}_\zeta\|_{C(\mathscr{S}(\zeta))}.$$

Using (2.1) one has

$$|u(\zeta) - \widetilde{u}(\zeta)| \leqslant C_0[h(\zeta)]^{m+1} \|f(\zeta)\|_{C^*(\mathfrak{M})} \max_{|\alpha|=m+1} \|D^\alpha F_\zeta\|_{C(E_\zeta)}.$$

Introducing the notation

$$h = \sup_{\zeta\in\mathfrak{M}} h(\zeta), \quad K_0 = \sup_{\zeta\in\mathfrak{M}} \|f(\zeta)\|_{C^*(\mathfrak{M})}, \quad \|u\|_{\check{C}^k(\mathfrak{M})} = \sup_{\zeta\in\mathfrak{M}} \|F_\zeta\|_{C^k(E_\zeta)}, \tag{2.2}$$

we obtain the following result.

THEOREM 1. *Let $f(\zeta)$ be a continuous function on a compact manifold $\mathfrak{M}$ with values in $C^*(\mathfrak{M})$. Then*

$$\|u - \widetilde{u}\|_{C(\mathfrak{M})} \leqslant C_0 K_0 h^{m+1} \|u\|_{\check{C}^{m+1}(\mathfrak{M})}. \tag{2.3}$$

Recall that an atlas $A = \{\chi_i\colon E_i \to M_i \mid i = 1, \dots, i_0\}$ is admissible for a differentiable manifold of class C^k if the homeomorphisms

$$\chi_i^{-1}\chi_{i'}\colon \chi_{i'}^{-1}(M_i \cap M_{i'}) \to \chi_i^{-1}(M_i \cap M_{i'})$$

are continuously differentiable k times. We shall consider that E_i and E_ζ are balls in $\mathbb{R}^n$ with centers at the origin.

Under certain conditions the norm $\|u\|_{\check{C}^k(\mathfrak{M})}$ is equivalent to the norm defined in the following way. Suppose that $\mathfrak{M} \in C^k$, the atlas $A = \{\chi_i\colon E_i \to M_i \mid i = 1, \dots, i_0\}$ is admissible and a function $u(\zeta)$, $\zeta \in \mathfrak{M}$, is represented by functions Φ_i in the atlas A. By definition, $u \in \widehat{C}^k(\mathfrak{M})$ if the norm

$$\|u\|_{\widehat{C}^k(\mathfrak{M})} = \max_{i=1,\dots,i_0} \|\Phi_i\|_{C^k(\bar{E}_i)} \tag{2.4}$$

is finite. It is not difficult to see that $\widehat{C}^*(\mathfrak{M})$ is a Banach space and that for different admissible atlases the corresponding norms are equivalent.

We introduce the following condition.

(B) On $\mathfrak{M}$ there exist an admissible atlas $A = \{\chi_i : E_i \to M_i \mid i = 1, \dots, i_0\}$ and a finite covering $\{\mathfrak{M}_i \mid i = 1, \dots, i_0\}$ consisting of open sets $\mathfrak{M}_i$ such that for any $i \in \{1, \dots, i_0\}$ the mappings

$$\chi_i^{-1}\psi_\zeta : E_\zeta \cap \psi_\zeta^{-1}\chi_i E_i \to \chi_i^{-1}\psi_\zeta E_\zeta \cap E_i$$

are continuously differentiable k times and their derivatives are uniformly bounded with respect to $\zeta \in \mathfrak{M}_i$. Besides, we assume that for any finite covering $\{U_{\zeta_i} \mid i = 1, \dots, i_0\}$ there exists an admissible atlas $U \overset{\text{def}}{=} \{\psi_{\zeta_i} : E_{\zeta_i} \to U_{\zeta_i} \mid i = 1, \dots, i_1\}$.

Note that (B) is fulfilled in the case of a manifold $\mathfrak{M} \in C^{k+2}$ embedded smoothly in $\mathbb{R}^{n+1}$, ψ_ζ^{-1} being the orthogonal projection onto the tangent plane L_ζ at ζ (see §1.1).

THEOREM 2. *Under the condition* (B) *for a compact manifold* $\mathfrak{M} \in C^k$ *the norms* (2.2), (2.4) *are equivalent.*

2. Estimates in L_p. We restrict ourselves to the case where $f(\zeta)$ is given by (1.17).

Let $A = \{\chi_i : E_i \to M_i \mid i = 1, \dots, i_0\}$ be an admissible atlas on $\mathfrak{M}$, $\mathfrak{M}_0$ a subset in $\mathfrak{M}$ with measurable $M'_{i.0} \overset{\text{def}}{=} \chi_i^{-1}(M_i \cap \mathfrak{M}_0)$, $M_i \cap \mathfrak{M}_0 \neq \varnothing$. Consider a function $w(\zeta)$ with a representation in A consisting of measurable functions $\Phi_i(\xi')$ on $M'_{i.0}$, $i = 1, \dots, i_0$. The sum

$$\sum_{\substack{i=1 \\ M'_{i.0} \neq \Lambda}}^{i_0} \int_{M'_{i.0}} \Phi_i(\xi')\, d\xi'$$

is called an A-integral of w over $\mathfrak{M}_0$ and is denoted by

$$(A) \int_{\mathfrak{M}_0} w(\zeta)\, d\zeta.$$

Evidently, the A-integral depends on A and does not coincide with an ordinary integral (if the latter may be defined) because M_i and $M_{i'}$ can be overlapping. It is convenient to use the A-integral in definitions of the integral norms (see **[41**, p. 377])

$$\|w\|_{L_p(\mathfrak{M}_0)} = \left((A) \int_{\mathfrak{M}_0} |w(\zeta)|^p\, d\zeta \right)^{1/p}, \qquad p \geqslant 1.$$

If $A' = \{\chi'_j : E'_j \to M'_j \mid j = 1, \dots, j_0\}$ is another admissible atlas in $\mathfrak{M} \in C^k$, $k \geqslant 1$, then the norm

$$\|w\|'_{L_p(\mathfrak{M}_0)} \overset{\text{def}}{=} \left((A') \int_{\mathfrak{M}_0} |w(\zeta)|^p\, d\zeta \right)^{1/p}$$

is equivalent to $\|w\|_{L_p(\mathfrak{M}_0)}$. Analogously, for admissible atlases the norms

$$\|w\|_{\widehat{W}_p^s(\mathfrak{M}_0)} \overset{\text{def}}{=} \left(\sum_{\substack{i=1 \\ M_i \cap \mathfrak{M}_0 \neq \Lambda}}^{i_0} \int_{M'_{i.0}} \sum_{|\alpha| \leqslant s} |D^\alpha \Phi_i(\xi')|^p\, d\xi' \right)^{1/p}, \tag{2.5}$$

$s \leqslant k$, turn out to be equivalent. Set

$$K_1 = \sup_{\zeta \in \mathfrak{M}} \|f_\zeta(0)\|_{L_{p'}}, \qquad p' = (1-p^{-1})^{-1},$$

$$\|u\|_{\check{W}_p^s(\mathfrak{M})} = \max_{i=1,\dots,i_0} \sup_{\xi' \in B_{h_i}(0)} \left(\sum_{|\alpha| \leqslant s} \|D^\alpha(u \circ \psi_\zeta)(\xi')\|^p_{L_p(\zeta \in \mathfrak{M}_i)} \right)^{1/p}. \tag{2.6}$$

THEOREM 3 (see **[37]**). *The following estimate is valid:*

$$\|u - \widetilde{u}\|_{L_p(\mathfrak{M})} \leqslant K_1 h^{m+1} \|u\|_{\check{W}_p^{m+1}(\mathfrak{M})}. \tag{2.7}$$

However, the norm (2.6) differs from $\|u\|_{\hat{W}_p^{m+1}(\mathfrak{M})}$ in (2.5). Under certain conditions these norms can be shown to be equivalent (see **[37]**). Here we state conditions only for majorizing (2.6) by (2.5). We shall sometimes write $\psi(\zeta, \xi')$ instead of $\psi_\zeta(\xi')$; set also $\widetilde{\mathfrak{M}}_j(\xi') = \psi(\mathfrak{M}_j, \xi')$. Let P_j, Q_j be neighborhoods on $\mathfrak{M}$ such that

$$P_j \subset \bigcap_{\xi' \in B_{h_j}(0)} \widetilde{\mathfrak{M}}_j(\xi'), \qquad \bigcup_{\xi' \in B_{h_j}(0)} \widetilde{\mathfrak{M}}_j(\xi') \subset Q_j.$$

(C) Suppose that

1) there exists an admissible atlas $A = \{\chi_i : E_i \to M_i \mid i = 1, \dots, i_0\}$ such that for $j \in \{1, \dots, j_0\}$ one can find $i = i(j)$, $i \in \{1, \dots, i_0\}$, with $Q_j \subset M_i$ while the Jacobian $\mathscr{J}_i(\widetilde{\xi}'', \xi')$ of the mapping

$$\widetilde{\xi}'' = \chi_i^{-1} \psi(\chi_i(\xi''), \xi'), \quad \xi'' \in \chi_i^{-1}(Q_j), \quad \xi' \in B_{h_j}(0), \tag{2.8}$$

satisfies

$$0 < c_0 \leqslant |\mathscr{J}_i(\widetilde{\xi}'', \xi')| \leqslant c_1 < +\infty \tag{2.9}$$

with some constants c_0, c_1 independent of ξ', $\widetilde{\xi}''$;

2) the atlas $Q = \{\chi_{i(j)} : E_j^Q \to Q_j\}$, $E_j^Q \stackrel{\text{def}}{=} \chi_{i(j)}^{-1}(Q_j)$, is admissible;

3) the γ-order derivatives, $|\gamma| \leqslant s$, with respect to ξ' of the functions $\varphi_i(\xi'', \xi') \stackrel{\text{def}}{=} \chi_i^{-1} \psi(\chi_i(\xi''), \xi')$, $i = i(j)$, are bounded by $L_0 < +\infty$ uniformly on the set $\{(\xi', \xi'') : \xi' \in B_{h_i}(0), \xi'' \in E_i\}$.

Under the condition (C) one has

$$\begin{aligned}(A) \int_{\mathfrak{M}_j} |u(\psi(\zeta, \xi'))|^p \, d\zeta &= \int_{\chi_i^{-1}(\mathfrak{M}_j)} |u(\psi(\chi_i(\widetilde{\xi}''), \xi'))|^p \, d\widetilde{\xi}'' \\ &= \int_{\chi_i^{-1}(\widetilde{\mathfrak{M}}_j(\xi'))} |u(\chi_i(\widetilde{\xi}''))|^p |\mathscr{J}_i(\widetilde{\xi}'', \xi')| \, d\widetilde{\xi}''.\end{aligned}$$

Due to (2.9)

$$(A) \int_{\mathfrak{M}_j} |u(\psi(\zeta, \xi'))|^p \, d\zeta \leqslant c_1 \int_{\chi_i^{-1}(\widetilde{\mathfrak{M}}_j(\xi'))} |u(\chi_i(\xi''))|^p \, d\xi'',$$

whence

$$(A) \int_{\mathfrak{M}_j} |u(\psi(\zeta, \xi'))|^p \, d\zeta \leqslant (A) \int_{Q_j} |u(\zeta)|^p \, d\zeta \cdot c_1. \tag{2.10}$$

Using 2) in (C) and (2.10) we obtain

$$(2.11)\qquad (A)\int_{\mathfrak{M}_j} |u(\psi(\zeta,\xi'))|^p\, d\zeta \leqslant (Q)\int_{Q_j} |u(\zeta)|^p\, d\zeta \cdot c_1, \qquad \xi' \in B_{h_j}(0).$$

THEOREM 4. *If* $\mathfrak{M} \in C^k$, $0 \leqslant s \leqslant k$, *and mappings* $\{\psi_\zeta\}_{\zeta\in\mathfrak{M}}$ *satisfy* (C), *then*

$$\|u\|_{\check{W}_p^s(\mathfrak{M})} \leqslant c_0 \|u\|_{\widehat{W}_p^s(\mathfrak{M})},$$

where c_0 *does not depend on* u.

PROOF. Our assertion follows from (2.11) in the case $s = 0$. For $s > 0$ the argument is somewhat more complicated (see **[37]**) □

COROLLARY 1. *Under the conditions of Theorems* 3 *and* 4

$$\|\widetilde{u} - u\|_{L_p(\mathfrak{M})} \leqslant c' h^{m+1} \|u\|_{\widehat{W}_p^{m+1}(\mathfrak{M})}$$

with constants c' *and* h *independent of* u.

REMARK 1. Under certain conditions the constants K_0 and K_1 in (2.3) and (2.7) do not depend on h; such conditions are given in §3 (see Theorem 5 there).

REMARK 2. The cases where the finiteness of $\|f(\zeta)\|_{(C^k(\mathfrak{M}))^*}$ $(\|f_\zeta(0)\|_{(L_p^k)^*})$ is assumed, instead of that of $\|f(\zeta)\|_{C^*(\mathfrak{M})}$ $(\|f_\zeta(0)\|_{L_{p'}})$, are treated similarly.

§3. Stability

1. Representation of basis functions with the help of an atlas. We consider the approximations

$$(3.1)\qquad \widetilde{u}(\xi) = \sum_{j\in J} v_j \omega_j(\xi), \qquad \xi \in \mathfrak{M},$$

where $\{\omega_j(\xi)\}$ is some system of compactly supported functions dependent on a positive parameter h (J may depend on h as well). The diameters of the supports of these functions tend to zero as $h \to 0$. By stability of approximation (3.1) one means the equivalence (with constants independent of h) of $\|\widetilde{u}\|_{L_p(\mathfrak{M})}$, $p \geqslant 1$, and $\|v\|_p$ for a net function $v = (v_j)_{j\in J}$ given on a net $\{\xi_j\}_{j\in J} \subset \mathfrak{M}$ (see below for a description of $\|v\|_p$). For a more general formulation of the problem (with other norms and approximations) see **[19]**.

Let $A = \{\chi_i \colon E_i \to M_i \mid i = 1, \dots, k\}$ be an admissible atlas on $\mathfrak{M}$. We define the admissible atlas A^{N+1} for $\mathfrak{M}^{N+1} = \mathfrak{M} \times \cdots \times \mathfrak{M}$, N being the number of elements in J, by

$$A^{N+1} \stackrel{\text{def}}{=} \{\chi_I \colon E_I \to M_I\},$$

where

$$\begin{aligned}
I &= (i^{(0)}, \dots, i^{(N)}), & E_I &= E_{i^{(0)}} \times E_{I_0},\\
I_0 &= (i^{(1)}, \dots, i^{(N)}), & E_{I_0} &= E_{i^{(1)}} \times \cdots \times E_{i^{(N)}},\\
M_I &= M_{i^{(0)}} \times M_{I_0}, & M_{I_0} &= M_{i^{(1)}} \times \cdots \times M_{i^{(N)}},\\
\chi_I &= (\chi_{i^{(0)}}, \chi_{I_0}), & \chi_{I_0} &= (\chi_{i^{(1)}}, \dots, \chi_{i^{(N)}}),\\
1 &\leqslant i^{(l)} \leqslant k, & l &= 0, 1, \dots, N.
\end{aligned}$$

Consider functions $\Omega_j(\xi', \Xi')$ on a net $\Xi' = \{\xi'_{j'}\}_{j' \in J}$ in $\mathbb{R}^n$ such that $\Omega_j(\xi', \Xi') \equiv 0$ for $\xi'_j \notin B_{\mathrm{ch}}(\xi')$. We suppose that $\Omega_j(\xi', \Xi')$ depends only on those ξ'_j that belong to $B_{\mathrm{ch}}(\xi')$. We shall sometimes write

$$\Omega_j(\xi', \Xi') = \Omega_j(\xi', \{\xi'_j \mid \xi'_j \in B_{\mathrm{ch}}(\xi')\})$$

to underscore this fact.

Let $\Xi_0 \stackrel{\mathrm{def}}{=} \{\xi_j \mid j \in J, \xi_j \in \mathfrak{M}\}$ be a net on $\mathfrak{M}$, dependent on a parameter h (condensing as $h \to 0$), where J is an ordered set of indices (dependent on h). Denoting the mapping $\psi_\zeta^{-1}(\xi)$ by $\psi^{-1}(\zeta, \xi)$, we set

$$\begin{aligned} \xi'_j &= \psi^{-1}(\zeta, \xi_j) \qquad \xi_j \in U_\zeta, \quad \Xi = (\zeta, \Xi_0), \\ J_{(\zeta,\mathrm{ch})} &= \{j' \mid \xi'_{j'} = \psi^{-1}(\zeta, \xi_{j'}) \in B_{\mathrm{ch}}(0)\}, \\ \Xi_{(\zeta,\mathrm{ch})} &= \{\xi_j \mid j \in J_{(\zeta,\mathrm{ch})}\}, \end{aligned}$$

and $\Xi'_{(\zeta,\mathrm{ch})} = \psi^{-1}(\zeta, \Xi_{(\zeta,\mathrm{ch})})$, so that $\Xi'_{(\zeta,\mathrm{ch})} = \{\xi'_{j'} \mid j' \in J_{(\zeta,\mathrm{ch})}\}$. We assume the above sets to be ordered in accordance with the ordering of J.

Define $\omega_j(\zeta, \Xi_0)$ by

$$\omega_j(\zeta, \Xi_0) = \begin{cases} \Omega_j(0, \{\psi^{-1}(\zeta, \xi_{j'})\}_{j' \in J(\zeta,\mathrm{ch})}) & \xi_j \in U_\zeta, \\ 0 & \xi_j \notin U_\zeta. \end{cases} \tag{3.2}$$

The mapping $\psi^{-1}(\zeta, \xi)$ is defined only for $\xi \in U_\zeta$, $\zeta \in \mathfrak{M}$; let us extend it in some way to the whole of $\mathfrak{M}$.

Assume that in the atlas

$$A_2 \stackrel{\mathrm{def}}{=} \{(\chi_i, \chi_j) \colon E_i \times E_j \to M_i \times M_j \mid i, j = 1, \dots, k\}$$

the mapping $\psi^{-1}(\zeta, \xi)$ is represented as a system of the functions $\{F_{(\psi^{-1})ij}\}$ so that

$$\xi' = \psi^{-1}(\zeta, \xi) = F_{(\psi^{-1})ij}(\chi_i^{-1}(\zeta), \chi_j^{-1}(\xi)), \tag{3.3}$$

while

$$F_{(\psi^{-1})ij}(\chi_i^{-1}(\zeta), \chi_j^{-1}(\xi)) \equiv F_{(\psi^{-1})i'j'}(\chi_{i'}^{-1}(\zeta), \chi_{j'}^{-1}(\xi))$$

for any $i, i', j, j' \in \{1, \dots, k\}$, $\zeta \in M_i \cap M_{i'}$, $\xi \in M_j \cap M_{j'}$.

Now denote by $F_{(\omega_j)I}$ the representation of ω_j in A^{N+1}. Then (3.2) and (3.3) imply

$$\begin{aligned} \omega_j(\Xi) &= F_{(\omega_j)I}(\chi_I^{-1}(\Xi)) \\ &= \begin{cases} \Omega_j(0, \{F_{(\psi^{-1})il_{j'}}(\chi_i^{-1}(\zeta), \chi_{l_{j'}}^{-1}(\xi_{j'}))\}_{j' \in J(\zeta,\mathrm{ch})}) & \xi_j \in U_\zeta, \\ 0 & \xi_j \notin U_\zeta \end{cases} \end{aligned}$$

if $\zeta \in M_i$, $\xi'_j \in M_{l_{j'}}$; here $i, l_{j'} \in \{1, \dots, k\}$, $I = (i, \{l_{j'}\}_{j' \in J(\zeta,\mathrm{ch})})$. It is clear that $\omega_j(\Xi)$ depends on $(\zeta, \{\xi_{j'}\}_{j' \in J(\zeta,\mathrm{ch})})$ only.

Thus, the representation of ω_j in A^{N+1} may be written by means of the set $\{F_{(\omega_j)I}\}$,

$$F_{(\omega_j)I}(\Xi'') = \begin{cases} \Omega_j(0, \{F_{\psi^{-1}il_{j'}}(\zeta'', \xi''_{j'})\}_{\xi''_{j'} \in B_{\mathrm{ch}}(0)}) & \chi_i(\zeta'') \in U_\zeta, \\ 0 & \chi_i(\zeta'') \notin U_\zeta \end{cases} \tag{3.4}$$

with $\Xi'' = (\zeta'', \{\xi''_{j'}\}_{\xi''_{j'} \in B_{\mathrm{ch}}(0)})$, $I = (i, \{l_{j'}\}_{j' \in J(\zeta, \mathrm{ch})})$.

2. Conditions for the set of generators. In a certain sense the functions ω_j may be reduced to affine transformations of the variables in some standard functions which are called generators (see **[20]**). We introduce the generators and state some conditions, valid in the case of a smooth manifold.

Let us consider a function $\Omega^{(0)}(\bar{\Xi})$, $\bar{\Xi} = (\bar{\xi}, \bar{\xi}_1, \dots, \bar{\xi}_k)$, $\bar{\xi}, \bar{\xi}_j \in \mathbb{R}^n$. Suppose that

1) $\Omega^{(0)}$ is defined on $\mathscr{D} = \mathbb{R}^n \times \bar{\mathscr{D}}_0$, where $\mathscr{D}_0$ is a subdomain of $(\mathbb{R}^n)^k$, $\xi \in \mathbb{R}^n$, $\bar{\Xi}_0 \overset{\mathrm{def}}{=} (\bar{\xi}_1, \dots, \bar{\xi}_k) \in \mathscr{D}_0$;

2) for any fixed $\bar{\Xi}_0 \in \mathscr{D}_0$ the support $\operatorname{supp}\varphi$, $\varphi(\bar{\xi}) \overset{\mathrm{def}}{=} \Omega^{(0)}(\bar{\xi}, \bar{\Xi}_0)$, is a compact set in $\mathbb{R}^n$, $\operatorname{supp}\varphi = \bar{s}_1 \cup \dots \cup \bar{s}_r$, where the s_j are disjoint domains and r is a finite number independent of $\bar{\Xi}_0$; $\varphi \in C^l(\bar{s}_p)$, $p = 1, \dots, r$, and $\varphi \in C^{\mathcal{S}}(\mathbb{R}^n)$, where l and $\mathcal{S}$ are fixed, $l = 0, 1, \dots$, $\mathcal{S} = -1, 0, 1, \dots$; $C^{-1}(\mathbb{R}^n)$ is the space of piecewise continuous bounded functions.[1]

We introduce the notation

$$\bar{\Xi}_0 - \bar{\xi} \overset{\mathrm{def}}{=} (\bar{\xi}_1 - \bar{\xi}, \dots, \bar{\xi}_k - \bar{\xi}), \quad c\bar{\Xi}_0 \overset{\mathrm{def}}{=} (c\bar{\xi}_1, \dots, c\bar{\xi}_k), \qquad c \in \mathbb{R}_1,$$

$$B^k_\rho(\bar{\xi}) \overset{\mathrm{def}}{=} \{\bar{\Xi}_0 \mid \bar{\Xi}_0 = (\bar{\xi}_1, \dots, \bar{\xi}_k), \sum_{i=1}^k \|\bar{\xi}_i - \bar{\xi}\|^2 \leqslant \rho^2\}, \qquad \rho > 0,$$

and the additional suppositions

3) if $(\bar{\xi}, \bar{\Xi}_0) \in \mathscr{D}$, then $(0, \bar{\Xi}_0 - \bar{\xi}) \in \mathscr{D}$, $(c\bar{\xi}, c\bar{\Xi}_0) \in \mathscr{D}$ for any $c \neq 0$ and

$$\Omega^{(0)}(\bar{\xi}, \bar{\Xi}_0) = \Omega^{(0)}(0, \bar{\Xi}_0 - \bar{\xi}), \quad \Omega^{(0)}(c\bar{\xi}, c\bar{\Xi}_0) = \Omega^{(0)}(\bar{\xi}, \bar{\Xi}_0);$$

4) for some positive ε and R there exists a domain, the ε-neighborhood of which lies in $\mathscr{D}_0 \cap B^k_R(0)$; denote this domain by $\mathscr{E}^k_{R,\varepsilon}$.

REMARK 1. As a rule, $\mathscr{D}_0$ does not contain the ball $B^k_\rho(0)$, whatever $\rho > 0$. The point is that $B^k_\rho(0)$ intersects the surfaces where $\Omega^{(0)}$ takes "infinite values" (for example, the Courant function takes such values on the planes $\bar{\xi}_{i'} = \bar{\xi}_{i''} = \bar{\xi}_{i'''}$).

At last we introduce the following condition:

5) the function $\Omega^{(0)}(\Xi)$ is continuous in the ε-neighborhood of $\mathbb{R}^n \times \mathscr{E}^k_{R,\varepsilon}$.

We assume that the conditions 1)–5) are fulfilled for a fixed k. The set of functions

$$\{\varphi_{\bar{\Xi}_0}(\bar{\xi})\}, \quad \varphi_{\bar{\Xi}_0}(\bar{\xi}) \overset{\mathrm{def}}{=} \Omega^{(0)}(\bar{\xi}, \bar{\Xi}_0), \quad \bar{\Xi}_0 \in \mathscr{E}^k_{R,\varepsilon}$$

will be denoted by $\mathscr{K}(\mathscr{E}^k_{R,\varepsilon})$.

---

[1] The smoothness conditions may be relaxed, for example, if $\varphi \in W^l_{q_1}(s_p) \cap W^{\mathcal{S}}_{q_2}(\mathbb{R}^n)$ for fixed $l, s \geqslant 0$, $q_1, q_2 \in [1, +\infty]$.

Set

$$\mathscr{B}_K = \bigcup_{i=1}^{t} \mathscr{K}(\mathscr{E}^{k_i}_{R_i,\varepsilon_i}),$$

where $K = \{k_1, \dots, k_t\}$, k_i is a positive integer, $0 < \varepsilon_i < R_i < +\infty$. The elements of $\mathscr{B}_K$ are called generators. Denote by G_j the open set such that $\bar{G}_j = \operatorname{supp}\omega_j$, $j \in J$. We consider that $\mathfrak{M} = \bigcup_{j\in J} \bar{G}_j$ and

$$\sup_{t\in\mathfrak{M}} \varkappa^{(t)}(\{G_j\}) \leqslant \varkappa$$

($\varkappa$ does not depend on h). The nonempty intersection

$$\tau = \bigcap_{l=1,\dots,s} G_{j_l}, \qquad j_l \in J,$$

of the sets G_j is called elementary if $\tau \cap G_j = \varnothing$ for any $j \in J$, $j \notin \{j_1, \dots, j_s\}$.

Consider all elementary intersections τ_k of the family $\{G_j\}_{j\in J}$, $k \in \mathfrak{K}_0$, $\mathfrak{K}_0$ being the corresponding set of indices. Evidently, $\mathscr{T} \overset{\text{def}}{=} \{\tau_k\}$ forms a subdivision of $\mathfrak{M}$, so that

$$\mathfrak{M} = \bigcup_{k\in\mathfrak{K}_0} \bar{\tau}_k, \quad \tau_k \cap \tau_{k'} = \varnothing \text{ for } k \neq k'.$$

Set $J_\tau = \{j \mid \tau \subset \operatorname{supp}\omega_j, j \in J\}$. If $\operatorname{supp}\omega_j \subset M_i$ for $j \in J_\tau$, then $\tau \subset M_i$ and $\tau'' = \chi_i^{-1}(\tau)$ is defined. According to (3.4) the relation

$$\widetilde{u}(\zeta) = \sum_{j\in J_\tau} v_j\omega_j(\zeta), \qquad \zeta \in \tau,$$

takes the form

$$\widetilde{U}(\zeta'') \overset{\text{def}}{=} \widetilde{u}(\chi_i(\zeta'')) = \sum_{j\in J_\tau} v_j\Omega_j\left(0, \{F_{(\psi^{-1})ii_{j'}}(\zeta'', \xi''_{j'})\}_{\xi''_{j'}\in B_{\text{ch}}(0)}\right),$$

where $\zeta'' \in \tau''$.

Suppose that the following condition is fulfilled:

(D) For any $\tau \in \mathscr{T}$ there exists an affine transformation $\mathscr{C}_\tau$ of $\mathbb{R}^n$ and linearly independent functions $\Omega_i^{(0)} \in \mathscr{B}_K$, $i = i(j)$, $j \in J_\tau$, on $\tau'' \overset{\text{def}}{=} \mathscr{C}_\tau\tau'$ such that

$$\Omega_j(\xi'', \{\xi''_{j'}\}) = \Omega_i^{(0)}(\mathscr{C}_\tau(\xi''), \{\mathscr{C}_\tau(\xi''_{j'})\}), \qquad i = i(j);$$

here

$$\Xi''_0 \overset{\text{def}}{=} (\mathscr{C}_\tau(\xi''_{j'})) \in \mathscr{E}^{k_i}_{R_i,\varepsilon_i},$$

$\mathscr{C}_\tau(\xi'') = B_\tau\xi'' + b_\tau$, $b_\tau \in \mathbb{R}^n$, B_τ is an $n\times n$ matrix whose maximal and minimal singular numbers ν_τ and μ_τ are subject to $0 < c_0h^{-1} \leqslant \nu_\tau \leqslant \mu_\tau \leqslant c_1h^{-1}$ with constants c_0 and c_1 independent of h and $\tau \in \mathscr{T}$.

The condition (D) means that for any $\tau \in \mathscr{T}$ the representation $\widetilde{U}(\xi'') \overset{\text{def}}{=} \sum_{j\in J_\tau} v_j\Omega_j(\xi'')$, $\xi'' \in \tau''$, may be written as a linear combination (with the same coefficients) of generators whose arguments are shifted and stretched by a factor h^{-1} in all directions (uniformly in $\tau \in \mathscr{T}$).

For finite-element approximations, conditions 1)–5) and (D) are fulfilled (see **[12, 20]**).

3. Equivalence of L_p-norms. As to $\psi^{-1}(\zeta, \xi)$, we suppose that

(E) the functions $F_{(\psi^{-1})ii_{j'}}$ representing ψ^{-1} in the atlas A_2 have the asymptotics (as $\xi''_{j'} - \zeta'' \to 0$)

$$F_{(\psi^{-1})ii_{j'}}(\zeta'', \xi''_{j'}) = (\xi''_{j'} - \zeta'')\varphi_{ii_{j'}}(\zeta'') + o_{\xi''_{j'}\zeta''}(\xi''_{j'} - \zeta'')$$

for $0 < c_0 \leqslant |\varphi_{i.i_{j'}}(\zeta'')| \leqslant c_1 < +\infty$, $\|o(\eta)\| \leqslant c_2\|\eta\|^{1+\delta}$ for $\|\eta\| \leqslant c_3$, and the positive constants c_k $(k = 0, 1, 2, 3)$ and δ do not depend on i, $i_{j'}$, j', $\xi''_{j'}$, and ζ''.

In view of the identity on the intersections of charts and of the smallness of $\xi''_{j'} - \zeta''$ it suffices to verify (E) for $i = i_{j'}$.

Let $\{\mathfrak{M}_j\}_{j=1,\dots,\bar{k}}$ be a covering of $\mathfrak{M}$ independent of h. Set

$$\mathfrak{M}_{j.(h)} = \overline{\bigcup_{\substack{\tau \cap \mathfrak{M}_j \neq \Lambda, \\ \tau \in \mathscr{T}}} \tau}, \qquad \mathfrak{M}_j^{(h)} = \overline{\bigcup_{\substack{\tau \subset \mathfrak{M}_j, \\ \tau \in \mathscr{T}}} \tau}.$$

Evidently,

$$\mathfrak{M}_j^{(h)} \subset \mathfrak{M}_j \subset \mathfrak{M}_{j.(h)},$$

where $\{\mathfrak{M}_{j.(h)}\}$ is a covering of $\mathfrak{M}$ for each fixed h.

(F) There exists h_0 such that the set $\{\mathfrak{M}_{j.(h)}\}_{j=1,\dots,\bar{k}}$ forms a covering of $\mathfrak{M}$ for each $h \in (0, h_0]$. We suppose that the maximal multiplicity of this covering does not exceed a number $\varkappa_0$ independent of h. We assume also that an admissible atlas $A = \{\chi_i : E_i \to M_i \mid i = 1, \dots, k\}$ exists such that for any $j \in \{1, \dots, \bar{k}\}$ one can find $i = i(j) \in \{1, \dots, k\}$ so that $\mathfrak{M}_{j.(h)} \subset M_i$.

It is clear that (E) and (F) hold for a smooth surface embedded in $\mathbb{R}^{n+1}$.

THEOREM 5 (**[4]**). *Under the conditions* 1)–5), (D)–(F) *the inequality*

$$c_0\|v\|_p \leqslant \|\widetilde{u}\|_{L_p(\mathfrak{M})} \leqslant c_1\|v\|_p$$

holds with c_0, c_1 *independent of* $\widetilde{u}$, v, h, *and*

$$\|v\|_p \overset{\text{def}}{=} \Big(\sum_{\tau \in \mathscr{T}} \sum_{j \in J_\tau} |v_j|^p h^n\Big)^{1/p}.$$

§4. Realization of the approximations

1. Approximation in a domain. Let E_0 be a domain homeomorphic to a ball in $\mathbb{R}^n$, T' its regular simplicial subdivision (curvilinear, generally speaking). Assume that a function u given on E_0 belongs to a Banach space X. In the finite element method an approximation of u is defined by

$$\widetilde{u}(\xi') = \sum_{T' \in \mathscr{T}'} \chi_{T'}(\xi') \sum_{s \in \mathscr{S}(T')} p_s^{T'}(\lambda_{T'}(\xi'))(f_s, u), \tag{4.1}$$

where $f_s \in X^*$, $\lambda_{T'}(\xi')$ are coordinates (as a rule, barycentric) of ξ' in the simplex T', $p_s^{T'}(\lambda)$ are given functions (the finite element basis functions on T'), and $\chi_{T'}(\xi')$ is the characteristic function of T'.

The function

$$\Omega_s(\xi') = \sum_{\substack{T'',T'\in\mathcal{T}'\\ s\in\mathcal{S}(T')}} \chi_{T'} p_s^{T'}(\lambda_{T'}(\xi')), \qquad s\in\mathcal{S}, \quad \xi'\in E_0,$$

is called a coordinate function corresponding to the functional (f_s, u). It may be written as

$$\Omega_s(\xi') = \begin{cases} p_s^{T'}(\lambda_{T'}(\xi')) & \xi'\in T'\in \mathfrak{Z}'_s, \\ 0 & \text{otherwise} \end{cases} \tag{4.2}$$

with $\mathfrak{Z}'_s = \{T' \mid s\in\mathcal{S}(T')\}$. In the simplest case of piecewise-linear Courant functions the set $\mathfrak{Z}'_s$ consists of simplices of a barycentric star for the vertex with number s in $\mathcal{T}'$.

2. Construction of simplicial subdivisions of charts. Let $\mathfrak{M}$ be a manifold endowed with a simplicial subdivision $\mathcal{T}$, $\dim\mathfrak{M} = n$. To every point $\zeta\in\mathfrak{M}$ we associate its neighborhood U_ζ and diffeomorphism

$$\psi_\zeta\colon E_\zeta\to\mathfrak{M}, \quad E_\zeta\subset\mathbb{R}^n, \quad \psi_\zeta(E_\zeta) = U_\zeta, \quad \psi_\zeta(0) = \zeta.$$

A sufficiently small subdivision $\mathcal{T}$ and the ψ_ζ give rise to a regular subdivision $\mathcal{T}'_\zeta$ of E_ζ dependent on ζ. (A special homeomorphism of E_ζ makes such a subdivision into a rectilinear one.)

Note that one can do otherwise. Having a topological structure (i.e., incidence matrices) of some simplicial subdivision $\mathcal{T}$ on $\mathfrak{M}$ and 0-skeleton $\mathcal{T}^0 = \{\xi_j\}_{j\in J}$, consider $\xi'_j = \psi_\zeta^{-1}(\xi_j)$ for $\xi_j\subset U_\zeta$. We connect ξ'_{j_1} and ξ'_{j_2} by the straight line segment $l'(\xi'_{j_1},\xi'_{j_2})$ if $\xi_{j_1} = \psi_\zeta(\xi'_{j_1})$ and $\xi_{j_2} = \psi_\zeta(\xi'_{j_2})$ are to be connected in $\mathcal{T}$. If $\psi_\zeta(\xi')$ is differentiable with respect to ζ and ξ', then

$$l(\xi_{j_1},\xi_{j_2}) = \{\zeta \mid \zeta\in\mathfrak{M}, 0\in l'(\psi_\zeta^{-1}(\xi_{j_1}), \psi_\zeta^{-1}(\xi_{j_2}))\}$$

is a simple curve on $\mathfrak{M}$ connecting ξ_{j_1} and ξ_{j_2} (we consider ξ_{j_1} and ξ_{j_2} to be close). All such intervals $l(\xi_{j_1},\xi_{j_2})$ form a 1-skeleton $\mathcal{T}^1$ of $\mathcal{T}$; $\mathcal{T}^k$ may be constructed in a similar way, $k = 2,\dots,n$. Note that in the sequel we need neither exact methods nor approximate ones to construct curvilinear simplices from $\mathcal{T}$.

Suppose the subdivision $\mathcal{T}$ to be so fine that for any $\zeta\in\mathfrak{M}$ the simplex $\bar{T}$, containing ζ, is the image of some simplex $\bar{T}'$ in E_ζ under the mapping ψ_ζ.

3. Coordinate functions on a manifold. Let us construct coordinate functions of the form (4.2) on the subdivision $\mathcal{T}'_\zeta$ for any $\zeta\in\mathfrak{M}$; we shall denote them by $\omega_{\zeta,s}(\xi')$, $s\in\mathcal{S}$. With $\mathcal{S}$ assumed to be independent of ζ, we suppose this set to be defined by means of a subdivision $\mathcal{T}$ on $\mathfrak{M}$. In the sequel the mapping $\mathcal{S}\to\omega_\zeta \overset{\text{def}}{=} \{\omega_{\zeta,s}\}$ determines the identification of coordinate functions for different ζ. Thus

$$\omega_{\zeta,s}(\xi') = \begin{cases} p_{\zeta,s}^{T'}(\lambda_{T'}(\xi')) & \text{for } \xi'\in T'\in\mathfrak{Z}'_s, \\ 0 & \text{otherwise;} \end{cases}$$

here $T' = \psi_\zeta^{-1}(T)$ and $p_{\zeta,s}^{T'}(\lambda)$ are finite element basis functions on T' (they are now dependent on $\zeta\in\mathfrak{M}$, in general).

In analogy to (4.1) one has

$$\widetilde{u}(\xi') = \sum_{s \in \mathcal{S}(T')} \omega_{\zeta,s}(\xi')(f_{\zeta,s}, u), \qquad \xi' \in T',$$

where $f_{\zeta,s} \in X_\zeta^*$, $\omega_{\zeta,s} \in X_\zeta$, X_ζ being a Banach space of functions u on E_ζ. The representation (1.7) may be rewritten as $\widetilde{u}(\xi') = (f_\zeta(\xi'), u)$ with

$$f_\zeta(\xi') = \sum_{s \in \mathcal{S}(T')} \omega_{\zeta,s}(\xi') f_{\zeta,s}, \qquad \xi' \in T'.$$

Consider a Banach space $X = X(\mathfrak{M})$ of functions on $\mathfrak{M}$ endowed with the norm

$$\|w\|_{X(\mathfrak{M})} = \sup_{\zeta \in \mathfrak{M}} \|w \circ \psi_\zeta\|_{X_\zeta}.$$

For every $\zeta \in \mathfrak{M}$ we define a functional $f(\zeta)$ by

$$(f(\zeta), w) = (f_\zeta(0), w \circ \psi_\zeta), \qquad w \in X(\mathfrak{M}).$$

Evidently,

$$|(f(\zeta), w)| \leqslant \|f_\zeta(0)\|_{X_\zeta^*} \|w\|_{X(\mathfrak{M})},$$

whence

$$\|f(\zeta)\|_{X^*} \leqslant \|f_\zeta(0)\|_{X_\zeta^*}. \tag{4.3}$$

Thus, $f(\zeta)$ is a trajectory in X^*, $\zeta \in \mathfrak{M}$. In the following, we suppose that

$$(f(\zeta), w) \in X, \quad w \in X. \tag{4.4}$$

If X is a space with absolute norm (for example, $L_p(\mathfrak{M})$), then due to (4.3) the relation $\|f_\zeta(0)\|_{X_\zeta^*} \in X$ implies (4.4). In the case of Banach spaces of differentiable functions the inclusion (4.4) may be provided by the corresponding smoothness of $f_\zeta(0)$ and ψ_ζ with respect to ζ.

We find

$$(f(\zeta), w) = \sum_{s \in \mathcal{S}} \omega_{\zeta,s}(0)(f_{\zeta,s}, w \circ \psi_\zeta). \tag{4.5}$$

Definition 1. A family of functionals $\{\varphi_\zeta \mid \varphi_\zeta \in X_\zeta^*, \zeta \in \mathfrak{M}\}$ is called decomposable on $\mathfrak{M}$ with respect to ψ_ζ in functions $\varkappa_l(\zeta)$, $l = 1, \dots, L$, if

$$(\varphi_\zeta, w \circ \psi_\zeta) = \sum_{l=1}^{L} \varkappa_l(\zeta)(F_l, w),$$

where $F_l \in X^*$ does not depend on ζ.

Supposing the families $\{f_{\zeta,s}\}$ in (4.5) to be decomposable with respect to ψ_ζ,

$$(f_{\zeta,s}, w \circ \psi_\zeta) = \sum_{j \in J_s} \varkappa_{(s,j)}(\zeta)(F_{(s,j)}, w),$$

one has

$$(f(\zeta), w) = \sum_{s \in \mathcal{S}} \sum_{j \in J_s} \omega_{\zeta,s}(0) \varkappa_{(s,j)}(\zeta)(F_{(s,j)}, w)$$

with some finite sets J_s.

We order the set $l \stackrel{\text{def}}{=} (s,j)$ of indices and denote it by $\mathscr{L}$. We introduce

$$\omega_{(s,j)}(\zeta) = \omega_{\zeta,s}(0)\varkappa_{(s,j)}(\zeta)$$

and set $F_l = F_{(s,j)}$, $\varkappa_l = \varkappa_{(s,j)}$, $\omega_l = \omega_{(s,j)}$.

Suppose that $\omega_l \in X$. From (4.6) it follows that

$$(f(\zeta), w) = \sum_{l\in\mathscr{L}} \omega_l(\zeta)(F_l, w).$$

DEFINITION 2. A family of functionals $\{\varphi_\zeta \mid \varphi_\zeta \in X_\zeta^*, \zeta \in \mathfrak{M}\}$ is called point-differential if $\varphi_\zeta = \sum_{\alpha,j} a_{\alpha,j}(\zeta)\delta_{\eta'_j}^{(\alpha)}$, where $\zeta \in \mathfrak{M}$, $\delta_{\eta'_j}^{(\alpha)}$ is the order α derivative of the Dirac function at the point $\eta'_j = \psi_\zeta^{-1}(\eta_j)$, and $a_{\alpha,j}$ are measurable bounded functions.

The above family turns out to be decomposable on $\mathfrak{M}$ with respect to ψ_ζ. Indeed,

$$\delta_{\eta'_j}^{(\alpha)}(w \circ \psi_\zeta) = \sum_\beta p_\beta(\zeta)(D_\eta^\beta w)(\psi_\zeta(\eta'_j)) = \sum_\beta p_\beta(\zeta)(D_\eta^\beta w(\eta'_j)),$$

$p_\beta(\zeta)$ being defined by ψ_ζ.

§5. Some applications

1. Third-order Lagrange approximations. Let a_i be the vertices of the rectilinear triangle $T' \subset \mathbb{R}^2$; set $a_{kl} = \sigma_{kl}a_k + \sigma_{lk}a_l$, $\sigma_{kl} = 1 - \sigma_{lk}$, $\sigma_{kl} \in (0,1)$, $k < l$. It is not difficult to check that the function

$$\begin{aligned}\tilde{u}(\xi') = &\sum_{i=1}^{3} \lambda_i(\xi')((\lambda_i(\xi') - 1)q_i/2 + u(a_i)) \\ &+ \sum_{\substack{k<l\\ k,l\in\{1,2,3\}}} \lambda_k(\xi')\lambda_l(\xi')\sigma_{kl}^{-1}\sigma_{lk}^{-1}u(a_{kl}), \qquad \xi' \in T',\end{aligned} \tag{5.1}$$

interpolates $u \in C(\bar{T})$ at a_i, $i = 1,2,3$, and at a_{kl}, $k < l$, $k,l \in \{1,2,3\}$, if

$$\begin{aligned}
q_1 &= u(a_1)(\sigma_{21}^{-1} + \sigma_{31}^{-1}) + u(a_2)(\sigma_{12}^{-1} - \sigma_{32}^{-1}) + u(a_3)(\sigma_{13}^{-1} - \sigma_{23}^{-1}),\\
q_2 &= u(a_1)(\sigma_{21}^{-1} - \sigma_{31}^{-1}) + u(a_2)(\sigma_{12}^{-1} + \sigma_{32}^{-1}) + u(a_3)(\sigma_{23}^{-1} - \sigma_{13}^{-1}),\\
q_3 &= u(a_1)(\sigma_{31}^{-1} - \sigma_{21}^{-1}) + u(a_2)(\sigma_{32}^{-1} - \sigma_{12}^{-1}) + u(a_3)(\sigma_{13}^{-1} + \sigma_{23}^{-1}).
\end{aligned}$$

For $\sigma_{kl} = 1/2$, $k < l$, $k,l \in \{1,2,3\}$, the approximation (5.1) coincides with the Zlamal element.

THEOREM 6 (see **[22]**). *The approximate equality $\tilde{u}(\xi') \approx u(\xi')$ is exact on second-degree polynomials.*

Let a triangulation $\mathscr{T}'$ be constructed on $\mathbb{R}^2$ with $\mathscr{T}^0 \stackrel{\text{def}}{=} \{\xi'_j\}_{j\in J}$, J being an ordered set. To every pair of vertices ξ'_k, ξ'_l, $k < l$, $k,l \in J$, we associate a number $\sigma_{kl} \in (0,1)$ and set $\xi'_{kl} = \sigma_{kl}\xi'_k + \sigma_{lk}\xi'_l$, where $\sigma_{lk} = 1 - \sigma_{kl}$. For each triangle $T' \in \mathscr{T}'$

we construct the approximation (5.1). Denote by $\Omega_i^z(\xi')$, $\Omega_{kl}^z(\xi')$ the coordinate functions corresponding to the functionals $u(\xi_i')$, $u(\xi_{kl}')$ respectively,

$$\Omega_i^z(\xi') = \begin{cases} \lambda_i(\xi')(1 - \sum\limits_{\substack{j\neq i \\ \xi_j' \in \bar{T}'}} \lambda_j(\xi')\sigma_{ji}^{-1}) & \text{for } \xi', \xi_i' \in \bar{T}', \\ 0 & \text{otherwise}, \end{cases} \tag{5.2}$$

$$\Omega_{kl}^z(\xi') = \begin{cases} \lambda_k(\xi')\lambda_l(\xi')\sigma_{kl}^{-1}\sigma_{lk}^{-1} & \text{for } \xi', \xi_{kl}' \in \bar{T}', \\ 0 & \text{otherwise} \end{cases} \tag{5.3}$$

with $i, k, l \in J$, $k < l$.

The functions (5.2), (5.3) are continuous on $\mathbb{R}^2$, $\operatorname{supp}\Omega_i^z(\xi')$ is the closure of the barycentric star of ξ_i', $\operatorname{supp}\Omega_{kl}^z(\xi') = \bar{T}_1' \cup \bar{T}_2'$ with triangles $\bar{T}_1'$, $\bar{T}_2'$ such that $\xi_k', \xi_l' \in \bar{T}_1' \cap \bar{T}_2'$, $T_s' \in \mathscr{T}$, $s = 1, 2$. Besides, the coordinate functions satisfy $\Omega_i^z(\xi_j') = \delta_{i,j}$, $\Omega_{kl}^z(\xi_j') = 0$, $\Omega_i^z(\xi_{kl}') = 0$, $\Omega_{kl}^z(\xi_{k'l'}') = \delta_{k,k'}\delta_{l,l'}$, where $k < l$, $k' < l'$, and i, j, k, l, k', l' lie in J.

Now we construct coordinate functions on the sphere (1.4), which will be denoted by $\mathfrak{M}$. Suppose $\mathfrak{M}$ to be triangulated in accordance with §1, subsection 3. Apart from the vertices ξ_i of $\mathscr{T}$, $i \in J$, we introduce the point $\xi_{kl} = (\xi_k + \xi_l)\|\xi_k + \xi_l\|^{-1} \in \mathfrak{M}$ for a pair of vertices ξ_k, ξ_l in every triangle $T \in \mathscr{T}$, $k < l$. Setting $U_\zeta \overset{\text{def}}{=} \{\xi \mid \varepsilon r^2 \leqslant (\xi, \zeta)\}$ we consider the mapping (1.20), $\varepsilon \in (0, 1)$, $\varepsilon = \text{constant}$. Assume the triangulation to be so small that $\zeta \in \bar{T}$ implies $\bar{T} \subset U_\zeta$ for $T \in \mathscr{T}$. Let ξ_i, $\xi_{kl} \in \bar{T}$, and $\xi_i' \overset{\text{def}}{=} \psi_\zeta^{-1}(\xi_i) \in L_\zeta$, $\xi_{kl}' \overset{\text{def}}{=} \psi_\zeta^{-1}(\xi_{kl}) \in L_\zeta$. Denote by $\mathscr{T}'$ the triangle in L_ζ with the vertices ξ_i'. Evidently, $0 \in \bar{T}'$. In $\bar{T}'$ we define $\Omega_i^z(\xi')$ and $\Omega_{kl}^z(\xi')$ by (5.2) and (5.3); these functions depend on ζ owing to the dependence of ξ_i', ξ_{kl}' on ζ. In what follows, $\omega_{\zeta,i}^z(\xi')$ and $\omega_{\zeta,kl}^z(\xi')$ stand for $\Omega_i^z(\xi')$ and $\Omega_{kl}^z(\xi')$.

Note that

$$\sigma_{ij} = (\zeta, \xi_i)/(\zeta, \xi_i + \xi_j), \qquad i, j \in J. \tag{5.4}$$

To obtain $\omega_{\zeta,i}^z(0)$ and $\omega_{\zeta,kl}^z(0)$ (see (5.2), (5.3)) one should find the barycentric coordinates of the point 0 in T' depending on $\zeta \in \mathfrak{M}$. If ξ_1, ξ_2, ξ_3 are the vertices of the corresponding triangle $T \subset \mathscr{T}$, then

$$\lambda_i(0) = \frac{(\xi_i, \zeta)\Delta_i^{\mathrm{T}}(\zeta)}{r^2\Delta^{\mathrm{T}}}, \qquad i = 1, 2, 3, \tag{5.5}$$

where Δ^{T} is the third-order determinant, whose columns are the vectors ξ_1, ξ_2, ξ_3, and the determinant Δ_i^{T} is obtained by replacing the ith column in Δ^{T} by the vector ζ.

Using (1.19), (5.2)–(5.5), we find basis functions of the Zlamal approximations on $\mathfrak{M}$, namely,

$$\omega_i^z(\zeta) = \begin{cases} \dfrac{(\xi_i, \zeta)^2\Delta_i^{\mathrm{T}}(\zeta)}{r^4(\Delta^{\mathrm{T}})^2}\Big[\Delta_i^{\mathrm{T}}(\zeta) - \sum\limits_{j, j\neq i} \Delta_j^{\mathrm{T}}(\zeta)\Big] & \text{for } \zeta, \xi_i \in \bar{T}, \\ 0 \quad \text{otherwise}, \end{cases} \tag{5.6}$$

$$\omega_{kl}^z(\zeta) = \begin{cases} \dfrac{(\zeta, \xi_k + \xi_l)^2\Delta_k^{\mathrm{T}}(\zeta)\Delta_l^{\mathrm{T}}(\zeta)}{r^4(\Delta^{\mathrm{T}})^2} & \text{for } \xi_k, \xi_l, \zeta \in \bar{T}, \\ 0 \quad \text{otherwise}. \end{cases} \tag{5.7}$$

As are the Zlamal planar functions, these functions are continuous and satisfy the following interpolation relations:

$$\omega_i^z(\xi_j) = \delta_{i,j}, \quad \omega_{kl}^z(\xi_j) = \omega_i^z(\xi_{kl}) = 0, \quad \omega_{kl}^z(\xi_{k'l'}) = \delta_{k,k'}\delta_{l,l'},$$

with $k < l$, $k' < l'$, and $i, j, k, l, k', l' \in J$.

The set $\operatorname{supp}\omega_i^z$ is the closure of the barycentric star of ξ_i, and $\operatorname{supp}\omega_{kl}^z$ coincides with the union of two closed triangles with vertices ξ_k and ξ_l, $k < l$. From the construction of the functions (5.6), (5.7) and from Theorems 1, 2 it follows that

$$\widetilde{u}(\xi) = \sum_{i\in J} u(\xi_i)\omega_i^z(\xi) + \sum_{\substack{k<l\\ k,l\in J}} u(\xi_{kl})\omega_{kl}^z(\xi)$$

approximates $u \in C^3(\mathfrak{M})$ with error $O(h^3)$ in $C(\mathfrak{M})$ where

$$h \stackrel{\text{def}}{=} \max_{T\in\mathcal{T}} \max_{\xi_i,\xi_j\in\bar{T}} \|\xi_i - \xi_j\|. \tag{5.8}$$

If $u \in W_q^3(\mathfrak{M})$, then one obtains the same approximation order in $L_q(\mathfrak{M})$. Analogous results for a star surface are given in **[23]**.

2. Sixth-order approximations. First we consider the duality of approximate and interpolational relations. In the Banach space X of functions u given on $\mathbb{R}^n$ we introduce the approximation

$$\widetilde{U}(\xi') = \sum_j v_j\Omega_j(\xi'), \qquad \xi' \in \mathbb{R}^n,$$

where $\{\Omega_j(\xi')\}$ is a family of functions in X with a finite multiplicity on $\mathbb{R}^n$. Let the condition

$$\sum_{j\in J_{T'}} (F_j, \varphi_\alpha)\Omega_j(\xi') = \xi'^\alpha, \qquad \xi' \in T', \quad |\alpha| \leqslant m, \quad \varphi_\alpha = \eta'^\alpha, \tag{5.9}$$

be fulfilled in a domain $T' \subset \mathbb{R}^n$. Suppose that X contains all polynomials of degree $\leqslant m$, $F_j \in X^*$, and $J_{T'} \stackrel{\text{def}}{=} \{j \mid \operatorname{supp}\Omega_j \cap T' \neq \varnothing\}$. With the sets $J_{T'}$, $\{\alpha \mid |\alpha| \leqslant m\}$ strictly ordered, we rewrite (5.9) in the form

$$W_{T'}\Omega(\xi') = e(\xi'), \qquad \xi' \in T', \tag{5.10}$$

where $W_{T'}$ is the matrix with entries (F_j, φ_α), $j \in J_{T'}$, $\alpha \leqslant m$, Ω and e are vectors, $\Omega(\xi') = \{\Omega_j(\xi')\}_{j\in J_{T'}}$, $e(\xi') = \{\xi'^\alpha\}_{|\alpha|\leqslant m}$. Further,

$$\widetilde{U}(\xi') = (V, \Omega(\xi')), \qquad \xi' \in T'; \tag{5.11}$$

here $V = (v_j)_{j\in J_{T'}}$, and $(\cdot,\cdot)$ is the inner product.

Consider a polynomial $\mathscr{P}(\xi') = (p, l(\xi'))$ with the coefficients $p \stackrel{\text{def}}{=} (p_\alpha)_{|\alpha| \leqslant m}$ subject to

$$(F_j, \mathscr{P}) = v_j, \qquad j \in J_{T'}. \tag{5.12}$$

One has

$$W^*_{T'} p = V. \tag{5.13}$$

From (5.10)–(5.12) it follows that

$$\widetilde{U}(\xi') = (p, W_{T'}\Omega(\xi')) = (p, e(\xi')), \qquad \xi' \in T',$$

whence $\widetilde{U}(\xi') \equiv \mathscr{P}(\xi')$ for $\xi' \in T'$.

Suppose that the sets $J_{T'}$ and $\{\alpha \mid |\alpha| \leqslant m\}$ have the same number of elements. Then unique solvability of the problem (5.13) is equivalent to solvability (with respect to Ω) of the linear algebraic system (5.10), and

$$\Omega(\xi') = W^{-1}_{T'} e(\xi'), \qquad \xi' \in T'. \tag{5.14}$$

Now we turn to constructing the generalized Argyris element on $\mathbb{R}^2$. Set $n = 2$, $m = 5$ and take for T' the triangle with vertices a_i, $i = 1, 2, 3$. As in §1, introduce the points a_{kl}. Let (x', y') be components of ξ'. As the functionals (F_j, u), $j = J_{T'}$, we take the δ-function, its first and second derivatives at a_i, $i = 1, 2, 3$, and its normal derivatives at the points a_{kl}, $k < l$. Thus, the number of elements in every $J_{T'}$ is equal to 21 and

$$J_{T'} = \{(\beta, i) \mid |\beta| \leqslant 2, \ i = 1, 2, 3\} \cup \{(1, 2), (2, 3), (1, 3)\}.$$

The functionals F_j, $j \in J_{T'}$, are given by

$$\begin{aligned} (F_{(\beta,i)}, U) &= U^{(\beta)}(a_i), && |\beta| \leqslant 2, \quad i = 1, 2, 3, \\ (F_{(k,l)}, U) &= \frac{\partial}{\partial \nu_{kl}} U(a_{kl}), && k < l, \quad k, l \in \{1, 2, 3\}, \end{aligned}$$

where ν_{kl} is the normal to the side (a_k, a_l).

It is not difficult to prove unique solvability of the interpolational problem (5.12) (in analogy to [**12**, p. 77]). Thereby, the matrix $W_{T'}$ is invertible and (5.14) is valid for the coordinate functions Ω_j. From (5.14) it follows that $\Omega_j(\xi')$ in T' is a fifth-degree polynomial whose coefficients coincide with the components of the jth column in $W^{-1}_{T'}$.

We can argue as in §1 up to (5.4). So we have the triangulation $\mathscr{T}$ on $\mathfrak{M}$, the points ξ_{kl} and the mapping ψ^{-1}_ζ. On the triangle $T' \subset L_\zeta$, containing the point 0, we construct the generalized Argyris element in accordance with (5.11), where $V_j = (F_j, U)$, $U \in C^2(E_\zeta)$.

One can consider a function $u \in C^2(\mathfrak{M})$ to be extended to an ε-neighborhood $\mathfrak{M}_\varepsilon$ of $\mathfrak{M}$ in $\mathbb{R}^3$ so that $u(\tau\zeta) \equiv u(\zeta)$ for $\tau \in (1-\varepsilon/r, 1+\varepsilon/r)$, $\zeta \in \mathfrak{M}$. We aim at clarifying the structure of the functionals $(F_j, u \circ \psi_\zeta)$. Obviously,

$$(F_{(1,0,i)}, u \circ \psi_\zeta) = (\nabla u(\xi_i), \frac{\partial \psi_\zeta}{\partial x'}(\psi_\zeta^{-1}(\xi_i))), \tag{5.15}$$

$$(F_{(0,1,i)}, u \circ \psi_\zeta) = (\nabla u(\xi_i), \frac{\partial \psi_\zeta}{\partial y'}(\psi_\zeta^{-1}(\xi_i))). \tag{5.16}$$

Further,

$$(F_{(2,0,i)}, u \circ \psi_\zeta) = (\nabla^2 u(\xi_i), \frac{\partial \psi_\zeta}{\partial x'}, \frac{\partial \psi_\zeta}{\partial x'}) + (\nabla u(\xi_i), \frac{\partial^2 \psi_\zeta}{\partial x'^2}), \tag{5.17}$$

where $(\nabla^2 u(\xi_i), x, y)$ is a bilinear form corresponding to the second-order differential of u at the point ξ_i, $\partial\psi_\zeta/\partial x'$ and $\partial^2\psi_\zeta/\partial x'^2$ being calculated at $\xi_i' = \psi_\zeta^{-1}(\xi_i)$. We obtain an analogous representation for $(F_{(1,1,i)}, u \circ \psi_\zeta)$ and $(F_{(0,2,i)}, u \circ \psi_\zeta)$.

One has

$$\begin{aligned}(F_{(k,l)}, u \circ \psi_\zeta) =& \frac{x_k' - x_l'}{\rho}(\nabla u(\xi_{kl}), \frac{\partial \psi_\zeta}{\partial x'}(\psi_\zeta^{-1}(\xi_{kl})))\\ &+ \frac{y_l' - y_k'}{\rho}(\nabla u(\xi_{kl}), \frac{\partial \psi_\zeta}{\partial y'}(\psi_\zeta^{-1}(\xi_{kl}))), \qquad \rho \stackrel{\text{def}}{=} \|\xi_k' - \xi_l'\|,\end{aligned} \tag{5.18}$$

for $k < l$, $k, l \in \{1,2,3\}$.

Owing to (5.15)–(5.18) the functionals $f_{\zeta,j}$,

$$\begin{aligned}(f_{\zeta,(\beta,i)}, u) &= (F_{(\beta,i)}, u \circ \psi_\zeta), \qquad |\beta| \leqslant 2, \quad i = 1,2,3,\\ (f_{\zeta,(k,l)}, u) &= (F_{(k,l)}, u \circ \psi_\zeta), \qquad k,l \in \{1,2,3\}, \quad k < l,\end{aligned}$$

are point-differential. Therefore, the approximations

$$\widetilde{u}(\zeta) = \sum_j (f_{\zeta,j}, u)\omega_{\zeta,j}(0),$$

obtained in accordance with (1.17), can be represented as

$$\widetilde{u}(\zeta) = \sum_{j'} (f_{j'}, u)\omega_j(\zeta), \tag{5.19}$$

$\omega_{j'}(\zeta)$ being defined by (4.5), and the functionals $f_{j'}$ do not depend on ζ. The error of approximating $u \in C^6(\mathfrak{M})$ by the expressions (5.19) is $O(h^6)$ in $C(\mathfrak{M})$, h being determined by (5.8). The error of approximating $u \in W_q^6(\mathfrak{M})$ is $O(h^6)$ in $L_q(\mathfrak{M})$.

REMARK 1. The Bell element (see **[12]**) differs from the Argyris element in that the Bell element does not contain normal derivatives. Analogous constructions are valid for the Bell element, but the error of approximating is $O(h^5)$.

To illustrate practical possibilities some calculations were implemented by means of BESM-6 using the analytical calculation system CABAΓ worked out earlier by the author. Three sets of functions given on a sphere, torus, and sphere with two handles have been considered. Errors of the above approximations were calculated at control points for a small enough triangulation (also obtained with the help of the computer); these errors corresponded to the theoretical estimates.

References

1. I. J. Schoenberg, *Contributions to the problem of approximation of equidistant data by analytic functions. Part A. On the problem of smoothing or graduation. A first class of analytic approximation formulae*, Quart. Appl. Math. **4** (1946), 45–99; *Part B. On the problem of osculatory interpolation. A second class of analytic approximation formulae*, Quart. Appl. Math. **4** (1946), 112–141.
2. J. H. Ahlberg, E. N. Nilson, and J. L. Walsh, *The theory of splines and their applications*, Academic Press, New York and London, 1967.
3. S. B. Stechkin and Yu. N. Subbotin, *Splines in numerical analysis mathematics*, "Nauka", Moscow, 1976. (Russian)
4. Yu. S. Zavyalov, B. I. Kvasov, and V. L. Miroshnichenko, *Methods of spline-functions*, "Nauka", Moscow, 1980. (Russian)
5. A. I. Grebennikov, *The method of splines and solution of ill-posed problems of approximation theory*, Moskov. Gos. Univ., Moscow, 1983. (Russian)
6. N. P. Korneĭchuk, *Splines in approximation theory*, "Nauka", Moscow, 1984. (Russian)
7. V. S. Ryabenkiĭ, *On the stability of finite difference equations*, Thesis, Moscow, 1952. (Russian)
8. V. S. Ryabenkiĭ and A. F. Filippov, *On stability of difference equations*, GITTL, Moscow, 1956. (Russian)
9. V. S. Ryabenkiĭ, *The local formulas of smooth fulfillment and interpolation of functions along their values at points of an irregular rectangular grid*, Preprint, Inst. Prikl. Mat. Akad. Nauk SSSR, Moscow, 1974. (Russian)
10. ———, *The method of difference potentials for some problems of continuum mechanics*, "Nauka", Moscow, 1987. (Russian)
11. S. G. Mikhlin, *Variational net-approximation. Numerical methods and automatic programming*, Zap. Nauchn. Sem. Leningrad. Otdel. Mat. Inst. Steklov. (LOMI) **48** (1974), 32–188; English transl. in J. Soviet Math. **10** (1978).
12. P. G. Ciarlet, *The finite element method for elliptic problems*, North-Holland, Amsterdam, 1978.
13. V. G. Korneev, *Schemes of the finite element method for high orders of accuracy*, Leningrad Univ., Leningrad, 1977. (Russian)
14. L. A. Oganesyan and L. A. Rukhovets, *Variation-difference methods for solving elliptic equations*, Akad. Nauk Armyan. SSR, Erevan, 1979. (Russian)
15. V. N. Malozemov and A. B. Pevnyĭ, *Polynomial splines*, Leningrad. Univ., Leningrad, 1986. (Russian)
16. P. I. Lizorkin and S. M. Nikolskiĭ, *A theorem concerning approximation on the sphere*, Anal. Math. **9** (1983), 207–221.
17. V. A. Vasilenko, *Spline functions: theory, algorithms, programs*, "Nauka", Novosibirsk, 1983. (Russian)
18. Yu. K. Dem′yanovich, *Construction of spaces of local functions on manifolds*, Metody Vychisl. **1985**, no. 14, 100–109. (Russian)
19. ———, *On the construction and stability of local approximations*, Preprint No. 1586-V87, VINITI, 1987. (Russian)
20. ———, *Approximation by spaces of local functions*, Vestnik Leningrad. Univ. Mat. Mekh. Astronom. **1977**, no. 1, 35–41; English transl. in Vestnik Leningrad Univ. Math. **10** (1977).
21. ———, *On the numerical realization of Courant approximation on a simply connected surface*, Preprint No. 7605-V, VINITI, 1985. (Russian)
22. ———, *The construction of local construction spaces on manifolds*, Preprint No. 2100, VINITI, 1984. (Russian)
23. ———, *On spline approximations on surfaces without boundary*, Boundary Value Problems and Automation of Their Solution, Kharkov, 1985, pp. 95–100. (Russian)
24. ———, *On the construction of the homogeneous spaces of local functions and inverse theorems of approximation*, Zap. Nauchn. Sem. Leningrad. Otdel. Mat. Inst. Steklov. (LOMI) **90** (1979), 5–23; English transl. in J. Soviet Math. **20** (1982).
25. I. G. Burova and Yu. K. Dem′yanovich, *Hermitian minimal splines of a variable defect*, Preprint No. 6907-84, VINITI, Leningrad, 1984. (Russian)
26. ———, *On the construction of smoothed splines with minimal support*, Vestnik Leningrad. Univ. Mat. Mekh. Astronom. **1983**, no. 3, 10–15; English transl. in Vestnik Leningrad Univ. Math. **16** (1984).
27. Yu. K. Dem′yanovich, *The stability of approximation by minimal splines on an irregular grid*, Preprint No. 250-V, VINITI, 1986. (Russian)
28. ———, *The local approximation of functions with increasing derivatives*, Preprint No. 6908-84, VINITI, 1984. (Russian)

29. ———, *Approximation and interpolation by local functions on a nonuniform grid*, Vestnik Leningrad. Univ. Mat. Mekh. Astronom. **1982**, no. 13, 15–19; English transl. in Vestnik Leningrad Univ. Math. **15** (1982).
30. I. G. Burova and Yu. K. Dem′yanovich, *Local approximations of variable height*, Vestnik Leningrad. Univ. Mat. Mekh. Astronom. **1986**, no. 13, 46–51; English transl. in Vestnik Leningrad Univ. Math. **19** (1986).
31. Yu. K. Dem′yanovich, *On the construction of spaces of local functions with given smoothness*, Preprint No. 7606-V, VINITI, 1985. (Russian)
32. I. G. Burova, *Approximation by complex splines*, Vestnik Leningrad. Univ. Mat. Mekh. Astronom. **1986**, no. 2, 3–9; English transl. in Vestnik Leningrad Univ. Math. **19** (1986).
33. Yu. K. Dem′yanovich, *On the construction of embedded spaces of local functions*, Preprint No. 280-V, VINITI, 1987. (Russian)
34. ———, *The local approximations on manifolds and weight estimates*, Zap. Nauchn. Sem. Leningrad. Otdel. Mat. Inst. Steklov. (LOMI) **139** (1984), 125–138; English transl. in J. Soviet Math. **36** (1987).
35. ———, *Simplicial extensions of functions on nets*, Metody Vychisl. **8** (1973), 32–50. (Russian)
36. ———, *On the construction of local functions on a nonuniform mesh*, Zap. Nauchn. Sem. Leningrad. Otdel. Mat. Inst. Steklov. (LOMI) **124** (1983), 140–153; English transl. in J. Soviet Math. **29** (1985).
37. ———, *Estimates of approximation by local function spaces on manifolds*, Preprint No. 1328-V88, VINITI, 1988. (Russian)
38. ———, *The stability of local approximations on manifolds*, Preprint No. 1329-V88, VINITI, 1988. (Russian)
39. ———, *On the system of analytical computations in system "Dubna"*, Programmirovanie **1981**, no. 4, 57–63; English transl. in Programming and Computer Software.
40. M. M. Postnikov, *Smooth manifolds, Lectures in geometry: Semester* III, "Nauka", Moscow, 1987; English transl., "Mir", Moscow, 1989.
41. O. V. Besov, V. P. Il'in, and S. M. Nikolskiĭ, *Integral representations of functions and embedding theorems*, "Nauka", Moscow, 1975; English transl., Wiley, New York, 1978.
42. Yu. K. Dem′yanovich, *Splines on manifolds and their applications*, Vychisl. Mekh. **1990**, no. 1, 108–128. (Russian)

Department of Mathematics and Mechanics, St. Petersburg State University, 2, Bibliotechnaya pl., Stary Petergof, St. Petersburg, 198904, Russia

Translated by B. PLAMENEVSKIĬ

Amer. Math. Soc. Transl.
(2) Vol. **159**. 1994

Seminorms and Higher-Order Moduli of Continuity

V. V. Zhuk

§1. Introduction

1. Notation. Hereafter $\mathbb{C}$, $\mathbb{R}$, $\mathbb{R}_+$, $\mathbb{Z}$, $\mathbb{Z}_+$, $\mathbb{N}$ are, respectively, the sets of complex, real, nonnegative, integer, nonnegative integer, and natural numbers; $[a]$ for $a \in \mathbb{R}$ is the integer part of a. The notation $k = \overline{a, b}$ with $a, b \in \mathbb{R} \cup \{-\infty, +\infty\}$ means that k runs through all the integers between a and b including a and b provided they are integers. Functions of a real variable having a removable discontinuity at some point are redefined there by continuity; in other cases the symbol $0/0$ is understood as 0; the symbol $a/0$ with $a > 0$ means $+\infty$; $+\infty \cdot 0 = 0 \cdot (+\infty)$ means $+\infty$. In the sequel the notation introduced by the sign $:=$ (equality by definition) is used without additional comments. If a function $U(f)$ is put into correspondence with the function f, then, strictly speaking, the value of $U(f)$ at x must be denoted by $U(f)(x)$. However, we shall often write $U(f, x)$ instead of $U(f)(x)$ and having defined $U(f, x)$ we shall use the symbol $U(f)$ without additional comments. Functions having the period 2π are simply called periodic. The space of continuous periodic functions $f : \mathbb{R} \to \mathbb{C}$ with norm $\|f\| = \max_{x\in\mathbb{R}} |f(x)|$ is denoted by C; $C^{(r)} = \{f \in C;\ \exists f^{(r)} \in C\}$; L_p for $1 \leqslant p < \infty$ is the space of measurable periodic functions $f : \mathbb{R} \to \mathbb{C}$ subject to $\int_{-\pi}^{\pi} |f(t)|^p\,dt < \infty$ with norm $\|f\|_p = (\int_{-\pi}^{\pi} |f(t)|^p\,dt)^{1/p}$; L_∞ is the space of periodic a.e. bounded on $\mathbb{R}$ functions $f : \mathbb{R} \to \mathbb{C}$ with norm $\|f\|_\infty = \operatorname{ess\,sup}_{x\in\mathbb{R}} |f(x)|$. If $r \in \mathbb{N}$, $1 \leqslant p \leqslant \infty$, then $W_p^{(r)}$ is the set of $f \in C$ having an absolutely continuous derivative of order $(r-1)$ (here $f^{(0)} = f$) and $f^{(r)} \in L_p$. Let the function $f : \mathbb{R} \to \mathbb{C}$ be integrable over all finite intervals, $h > 0$, $r - 1 \in \mathbb{N}$. The first-order Steklov function for f with step h is the function $\mathcal{S}_{h.1}(f)$ defined by

$$\mathcal{S}_{h.1}(f, x) = (1/h) \int_{-h/2}^{h/2} f(x + t)\,dt.$$

The order r Steklov function with step h is $\mathcal{S}_{h.r}(f) = \mathcal{S}_{h.1}(\mathcal{S}_{h.r-1}(f))$.

1991 *Mathematics Subject Classification*. Primary 26A15, 26A24; Secondary 26D15, 26D99.

Let $r \in \mathbb{Z}_+$, $t \in \mathbb{R}$. The quantities

$$\Delta_t^r(f,x) = \sum_{k=0}^{r}(-1)^{r-k}\binom{r}{k} f(x+kt),$$

$$\delta_t^r(f,x) = \sum_{k=0}^{r}(-1)^{k}\binom{r}{k} f(x+(r-2k)t/2)$$

are called, respectively, order r difference and central difference of f at x with step t. For $r \in \mathbb{Z}_+$, $h \geqslant 0$, $f \in L_\infty$ we set $\omega_r(f,h)_\infty = \sup_{|t|\leqslant h} \|\Delta_t^r(f)\|_\infty$. The function $\omega_r(f)_\infty$ is called the order r modulus of continuity of f in L_∞. Next, $H_n = \{\sum_{k=-n}^{n} c_k e^{ikx}; c_k \in \mathbb{C}\}$ is the set of trigonometric polynomials of order no greater than n. For $n \in \mathbb{Z}_+$, $f \in L_\infty$ we set $E_n(f)_\infty = \inf_{T\in H_n} \|f - T\|_\infty$; this quantity is called the best nth-order approximation of f in L_∞. By $\mathfrak{M}$ we denote the space of functionals $\Phi\colon C \to \mathbb{R}_+$ such that $\Phi(f+g) \leqslant \Phi(f) + \Phi(g)$ for all $f, g \in C$. If $\Phi \in \mathfrak{M}$ we write $m_l(\Phi)_\infty = \sup_{f\in C^{(l)}} \{\Phi(f)/\|f^{(l)}\|\}$.

2. Statement of the problems. In the approximation theory of periodic functions, structural properties of a function are usually characterized by moduli of continuity of the function itself and of its derivatives. A question often arises in this theory on estimating some seminorm defined on the set of periodic functions in terms of their moduli of continuity. Such questions appear most commonly (but not always) in studying estimates of deviations of approximation methods. For problems of this kind there exists a general and at the same time efficient (from the applicational point of view) method. We describe it here as applied to the second-order moduli of continuity.

First of all, note (see, e.g., **[1, pp. 224–226]**) that if $f \in C$, then $\mathcal{S}_{h,2}(f) \in C^{(2)}$ and

$$\|f - \mathcal{S}_{h,2}(f)\| \leqslant (1/2)\omega_2(f,h)_\infty, \tag{1}$$

$$\|\mathcal{S}_{h,2}''(f)\| \leqslant h^{-2}\omega_2(f,h)_\infty. \tag{2}$$

Let $\Phi \in \mathfrak{M}$, $f \in C$, $h > 0$. Using (1), (2) we obtain that

$$\begin{aligned}\Phi(f) &\leqslant \Phi(f - \mathcal{S}_{h,2}(f)) + \Phi(\mathcal{S}_{h,2}(f)) \\ &\leqslant m_0(\Phi)_\infty \|f - \mathcal{S}_{h,2}(f)\| + m_2(\Phi)_\infty \|\mathcal{S}_{h,2}''(f)\| \\ &\leqslant m_0(\Phi)_\infty (1/2)\omega_2(f,h)_\infty + m_2(\Phi)_\infty h^{-2}\omega_2(f,h)_\infty.\end{aligned}$$

Thus we have proved the following assertion: if $\Phi \in \mathfrak{M}$, $f \in C$, $h > 0$, then

$$\Phi(f) \leqslant \{m_0(\Phi)_\infty/2 + h^{-2} m_2(\Phi)_\infty\}\omega_2(f,h)_\infty. \tag{3}$$

In order to apply (3) to some concrete functionals it is sufficient to know upper estimates of $m_l(\Phi)_\infty$ for $l = 2$ and $l = 0$. The analysis of $m_l(\Phi)_\infty$ is an independent problem; it has been extensively studied before and keeps evoking interest even now. Some information on the results regarding this problem was given in **[2–5]**. In particular, if $\Phi(f) = E_n(f)_\infty$, then (see, e.g., **[1, pp. 241, 242]**) $m_0(\Phi)_\infty \leqslant 1$, $m_2(\Phi)_\infty \leqslant \pi^2/(8(n+1)^2)$. With regard for these estimates, setting $h = \pi\gamma/(2n+2)$

with $\gamma > 0$ and applying (3) to the functional $\Phi(f) = E_n(f)_\infty$, we obtain that for every $f \in C$

$$E_n(f)_\infty \leqslant \frac{1}{2}\left(1+\frac{1}{\gamma^2}\right)\omega_2\left(f, \frac{\gamma\pi}{2(n+1)}\right)_\infty. \tag{4}$$

In particular, it follows from (4) that

$$\begin{aligned} E_n(f)_\infty &\leqslant 1\cdot\omega_2\left(f, \frac{\pi}{2(n+1)}\right)_\infty, \\ E_n(f)_\infty &\leqslant \frac{5}{8}\omega_2\left(f, \frac{\pi}{n+1}\right)_\infty. \end{aligned} \tag{5}$$

The estimate (5) is precise in the sense ([**6**], see [**7**, pp. 321, 322]) that

$$\sup_{n\in\mathbb{N}}\sup_{f\in C}\left\{E_n(f)_\infty \Big/ \omega_2\left(f, \frac{\pi}{2n+2}\right)_\infty\right\} = 1.$$

It is still unknown whether it is possible to replace the constant $(1+\gamma^{-2})/2$ in (4) (with γ fixed, $\gamma \neq 1$) by a smaller one. If one takes $\omega_2(f, \lambda h)$ as $\Phi(f)$ with $\lambda, h \geqslant 0$, then in this case $m_0(\Phi)_\infty \leqslant 4$, $m_2(\Phi)_\infty \leqslant \lambda^2 h^2$ and, applying (3) to this functional, we get the following result: If $f \in C$, $\lambda, h \geqslant 0$, then

$$\omega_2(f, \lambda h)_\infty \leqslant (\lambda^2+2)\omega_2(f,h)_\infty. \tag{6}$$

The constant 2 in (6) may not be replaced by a smaller one [**8**]. It would be easy to give a lot of other examples of application of (3) to specific functionals but we shall not dwell on it. To sum up, we stress that the inequality (3) often produces inequalities with almost best possible constants, which is important for specific applications. Other methods fail sometimes not only to improve these constants but even to reproduce them. Note that Steklov functions were originally used in the above situation by N. I. Akhiezer [**1**].

Intending to extend the inequality (3) to the case of higher-order moduli of continuity and to use the same method of proof (to the author's knowledge there is no other method) we have to construct analogs of the function $\mathcal{S}_{h,2}(f)$. This means that one has to find the functions $U_{h,r}(f)$, subject to the inclusion $U_{h,r}(f) \in C^{(r)}$ for $f \in C$ (in fact, one can restrict oneself to the weaker condition $U_{h,r}(f) \in W_\infty^{(r)}$; in this case one uses $\sup_{f\in W_\infty^{(r)}}\{\Phi(f)/\|f^{(r)}\|_\infty\}$ instead of $m_r(\Phi)_\infty$ in the version of (3)) and

$$\|f - U_{h,r}(f)\| \leqslant A(r)\omega_r(f,h)_\infty, \tag{1'}$$

$$\|U_{h,r}^{(r)}(f)\| \leqslant B(r)h^{-r}\omega_r(f,h)_\infty, \tag{2'}$$

where $A(r)$, $B(r)$ are some constants, $h > 0$. Provided such functions are constructed, for $\Phi \in \mathfrak{M}$, $f \in C$, $h > 0$, we have

$$\begin{aligned} \Phi(f) &\leqslant \Phi(f - U_{h,r}(f)) + \Phi(U_{h,r}(f)) \\ &\leqslant m_0(\Phi)_\infty\|f - U_{h,r}(f)\| + m_r(\Phi)_\infty\|U_{h,r}^{(r)}(f)\| \\ &\leqslant m_0(\Phi)_\infty A(r)\omega_r(f,h)_\infty + m_r(\Phi)_\infty B(r)h^{-r}\omega_r(f,h)_\infty. \end{aligned}$$

This is the way to establish the following assertion: if $\Phi \in \mathfrak{M}$, $r-2 \in \mathbb{N}$, $h>0$, $f \in C$, then

$$\Phi(f) \leqslant (A(r)m_0(\Phi)_\infty + B(r)m_r(\Phi)_\infty h^{-r})\omega_r(f,h)_\infty. \tag{3'}$$

One is tempted to choose $\mathcal{S}_{h,r}(f)$ as $U_{h,r}(f)$, since for all $r \in \mathbb{N}$ we have $\mathcal{S}_{h,r}(f) \in C^{(r)}$ and $\|\mathcal{S}^r_{h,r}(f)\| \leqslant h^{-r}\omega_r(f,h)_\infty$ for $f \in C$. This, however, is impossible when $r \geqslant 3$, since for each nonconstant $f \in C^{(2)}$

$$\varliminf_{h\to 0+} \frac{\|f-\mathcal{S}_{h,r}(f)\|}{\omega_r(f,h)_\infty} \geqslant \varliminf_{h\to 0+} \frac{\|f-\mathcal{S}_{h,r}(f)\|}{2^{r-3}\omega_3(f,h)_\infty} = +\infty,$$

and thus (1′) fails to hold. The required functions $U_{h,r}(f)$ can be constructed, and in various ways at that (see, e.g., **[9, 10]**, **[1,** pp. 51, 52], **[7,** Ch. 5], **[12]**). However, if we try to use the currently known functions $U_{h,r}(f)$ for proving inequalities similar to (3′), then in the most interesting cases (including the case $\Phi(f) = E_n(f'))$ we obtain considerably overstated constants. These constants, being too large, are not even given in final form in the literature. Thus it remains necessary to construct new functionals $U_{h,r}$ enabling one to establish inequalities of the form (3′) with "accurate" constants and to develop a technique for obtaining such inequalities. The aim of this paper is to contribute to the study of these problems. We have chosen the constants in the generalized Jackson inequality (see, e.g., **[2,** p. 274])

$$E_n(f)_\infty \leqslant C(r,\gamma)\omega_r\left(f, \frac{\pi\gamma}{2(n+1)}\right)_\infty,$$

giving the estimate for the best approximation in terms of moduli of continuity for testing our progress in specific questions.

3. Review of the contents of the paper. In §2 we set forth a number of results of approximation theory for periodic functions and of interpolation theory, which we use systematically in the sequel. Some basic notation is also introduced.

The constants of importance (e.g., $\mathcal{T}_k^{(\gamma)}(0)$, $\mathcal{Q}_k^{(\gamma)}(0)$, $\mathcal{A}(p,\gamma)$, etc.) are closely connected with the so-called central factorial numbers (c. f. n.). These numbers are somewhat similar to the Stirling numbers of the first kind. In this connection we give in §2 a number of facts about c. f. n. and Stirling numbers. Some of these facts are not used in the paper, but they should be born in mind when studying the above constants in more detail than is done in the present paper.

Usually, one resorts to numerical differentiation in the cases when the function is given by means of a table or has a complicated analytical expression. In such cases the following method is usually applied: one constructs an interpolational polynomial of the function to be differentiated and the derivative of order γ of this polynomial is taken in place of the derivative of f of order γ.

The standard material on the numerical differentiation formulas based on this approach is presented in **[13,** Ch. 2, §§15, 16], **[14,** Ch. 3, §§1, 2], **[15,** Ch. 3, §1]. In §3 of the paper we give some new representations of the remainder terms of the aforementioned numerical differentiation formulas.

In §4 we study the dependence of numerical differentiation formulas (n. d. f.) as functions of the point of differentiation.

In §5 mean functions of the type of Stirling and Bessel n. d. f. are constructed. It is these functions that play the part of the aforementioned functions $U_{h,r}(f)$ in obtaining equalities of the type (3′). They turn out to be natural generalizations of the Steklov functions (which also may be treated as n. d. f.) and they are meant for working with moduli of continuity of arbitrary order.

In §§6 and 7 we develop a technique for estimating functionals defined on sets of periodic functions. It is based on the results presented in §§3–5. In establishing general theorems we face the situation when there exist several ways of reasoning, and it depends on specific interrelations of the parameters which way is the best. An attempt to describe all the cases in a unified manner would lead to very complicated formulations. Therefore we have chosen the following method of presentation: first (§6) we consider general theorems for the moduli of continuity of not very high order and, when possible, we attempt to give final form to these results. In the most general situation (§7) we do not strive for such completeness.

The statements of the same type (definitions, lemmas, theorems, remarks, formulas) have their own enumeration within each section. When referring to statements from another section, statement x from section y is denoted by $y.x$.

§2. Preliminary material

The space C_q. Let a seminorm q be defined in C. We say that q belongs to the class A if the following conditions are fulfilled: 1) there exists a constant M not depending on f such that $q(f) \leqslant M\|f\|$ for all $f \in C$; 2) the seminorm q is invariant with respect to shift, i.e., for each $f \in C$ and arbitrary $h \in \mathbb{R}$, we have $q(f(\cdot + h)) = q(f)$.

LEMMA A (see [**16**, pp. 21, 22]). *Let $a, b \in \mathbb{R}$, $a < b$, $\Omega = \mathbb{R} \times [a, b]$ and let the function $K\colon \Omega \to \mathbb{C}$ be a continuous function, $K(x + 2\pi, t) = K(x, t)$ for all $(x, t) \in \Omega$. Then*

$$q\left(\int_a^b K(\cdot, t)\, dt\right) \leqslant \int_a^b q(K(\cdot, t))\, dt,$$

provided the seminorm q is in A.

The space C equipped with a seminorm $q \in A$ will be called the space C_q. By $C_q^{(r)}$ we denote the set $C^{(r)}$ considered as part of C_q. Let $r \in \mathbb{Z}_+$, $f \in C_q$. The function $\omega_r(f)$ defined on $\mathbb{R}_+$ by the equality $\omega_r(f, h) = \sup_{|t| \leqslant h} q(\Delta_t^r(f))$ is called the rth-order modulus of continuity of f in the space C_q. The modified rth-order modulus of continuity of f in C_q is the function $\omega_r^*(f, h) = \sup_{|t| \leqslant h} q(2^{-1}\Delta_t^r(f + f(\cdot + t)))$. It is clear that $\omega_r^*(f, h) \leqslant \omega_r(f, h)$ for $f \in C_q$.

Let $f \in C_q$, $r \in \mathbb{Z}_+$, $h > 0$, $0 \leqslant \alpha \leqslant r$. In this case we set

$$\Omega_{r,\alpha}(f, h) = \sup_{0 < t \leqslant h} t^{-\alpha}\omega_r(f, t).$$

We note some properties of the moduli of continuity (see, e.g., [**7**, pp. 96–106]).

M.1. *For r and h fixed, the mapping associating to $f \in C_q$ the number $\omega_r(f, h)$ is a class A seminorm in C.*

M.2. *If $r \in \mathbb{N}$, then $\omega_r(f, 0) = 0$.*

M.3. *The function $\omega_r(f)$ is uniformly continuous on $\mathbb{R}_+$.*

M.4. *The function $\omega_r(f)$ increases on $\mathbb{R}_+$.*

M.5. *If* $n \in \mathbb{N}$, *then* $\omega_r(f, nh) \leqslant n^r \omega_r(f, h)$; *if* $\lambda \in \mathbb{R}_+$, *then*

$$\omega_r(f, \lambda h) \leqslant ([\lambda] + 1)^r \omega_r(f, h).$$

M.6. *If* $l \leqslant r$, *then* $\omega_r(f, h) \leqslant 2^{r-l} \omega_l(f, h)$; *in particular,* $\omega_r(f, h) \leqslant 2^r q(f)$.

M.7. *Let* $l < r$, $f \in C_q^{(r-l)}$. *Then* $\omega_r(f, h) \leqslant h^{r-l} \omega_l(f^{(r-l)}, h)$; *in particular, for* $l = 0$ *we have* $\omega_r(f, h) \leqslant h^r q(f^{(r)})$.

If we replace ω by ω^* in M.1–M.7 then all these statements (with the exception of M.5) remain valid. In what follows the statement obtained from M.j by changing ω to ω^* is referred to as the property M*.j. Set

$$\overline{\omega}_r(f, h) = \begin{cases} \omega_r^*(f, h), & \text{for } (r+1)/2 \in \mathbb{N}, \\ \omega_r(f, h), & \text{for } r/2 \in \mathbb{N}; \end{cases}$$

$$\underline{\omega}_r(f, h) = \begin{cases} \omega_r(f, h), & \text{for } (r+1)/2 \in \mathbb{N}, \\ \omega_r^*(f, h), & \text{for } r/2 \in \mathbb{N}. \end{cases}$$

Let $f \in C_q$, $n \in \mathbb{Z}_+$. The best approximation of order n of f in C_q is $E_n(f) = \inf_{T \in H_n} q(f - T)$. The polynomial $T_n(f) \in H_n$ for which $q(f - T_n(f)) = E_n(f)$ is called the polynomial of the best approximation of order n of f in C_q. The following properties of best approximations hold (see, e.g., [**16**, pp. 23–25, 50, 51]).

N.1. *If* $f \in C_q$, $n \in \mathbb{Z}_+$, *then there exists a polynomial* $T \in H_n$ *such that* $q(f - T) = E_n(f)$.

N.2. *For* $n \in \mathbb{Z}_+$ *fixed, the mapping associating* $E_n(f)$ *to* $f \in C_q$ *is a class A seminorm in* C.

In approximation theory an important role is played by the quantities

$$K_r := \frac{4}{\pi} \sum_{l=0}^{\infty} \frac{(-1)^{(r+1)l}}{(2l+1)^{r+1}},$$

which are often called Favard constants in the literature. In order to find the numerical values of K_r one can use the formulas (see, e.g., [**17**, p. 21])

$$K_{2r-1} = \frac{2(2^{2r} - 1)\pi^{2r-1}}{(2r)!} |B_{2r}| \qquad (r \in \mathbb{N}),$$

$$K_{2r} = \frac{\pi^{2r}}{2^{2r}((2r)!)} |\mathscr{E}_{2r}| \qquad (r \in \mathbb{Z}_+),$$

where B_k and $\mathscr{E}_k$ are the Bernoulli and Euler numbers, respectively (see, e.g., [**17**, pp. 1093, 1094], [**18**, Chapter 23]). Note that usually the kth Euler number is denoted by E_k. We abandon this tradition, since the symbol E_k is connected here with the best approximations. It follows immediately from the definition of K_r that $K_1 > K_3 > K_5 > \cdots > (4/\pi)$ and $K_0 < K_2 < K_4 < \cdots < (4/\pi)$.

We present here the first twelve K_r: $K_0 = 1$, $K_1 = \pi/2$, $K_2 = \pi^2/8$, $K_3 = \pi^3/24$, $K_4 = 5\pi^4/384$, $K_5 = \pi^5/240$, $K_6 = \frac{61\pi^6}{6!2^6} = \frac{61\pi^6}{46080}$, $K_7 = \frac{17\pi^7}{8!} = \frac{17\pi^7}{40320}$, $K_8 = \frac{1385\pi^8}{8!2^8} = \frac{277\pi^8}{2064384}$, $K_9 = \frac{155\pi^9}{10!} = \frac{31\pi^9}{725760}$, $K_{10} = \frac{50521\pi^{10}}{10!2^{10}}$, $K_{11} = \frac{2073\pi^{11}}{12!}$.

The series $\sum_{k=-\infty}^{\infty} c_k$, $c_k \in \mathbb{C}$, is $c_0 + \sum_{k=1}^{\infty}(c_k + c_{-k})$ by definition. Let $f \in L_1$, $k \in \mathbb{Z}$, $x \in \mathbb{R}$. Then

$$a_k(f) = \frac{1}{\pi}\int_{-\pi}^{\pi} f(t)\cos kt\,dt, \quad b_k(f) = \frac{1}{\pi}\int_{-\pi}^{\pi} f(t)\sin kt\,dt,$$

$$c_k(f) = \frac{1}{2\pi}\int_{-\pi}^{\pi} f(t)e^{-ikt}\,dt \text{ are the Fourier coefficients of } f;$$

$$A_k(f,x) = \begin{cases} a_0(f)/2, & \text{for } k = 0, \\ a_k(f)\cos kx + b_k(f)\sin kx, & \text{for } k \in \mathbb{N}; \end{cases}$$

$$\sigma(f,x) = \sum_{k=0}^{\infty} A_k(f,x) = \sum_{k=-\infty}^{\infty} c_k(f)e^{ikx} \text{ is the Fourier series of } f.$$

For $f \in L_1$, $n \in \mathbb{Z}_+$, $r \in \mathbb{N}$, $x \in \mathbb{R}$, the sums

$$X_{n.r}(f,x) = \sum_{k=0}^{n} \varphi_r\left(\frac{\pi k}{2(n+1)}\right) A_k(f,x),$$

where

$$\varphi_r(t) = \begin{cases} \dfrac{t^r}{(r-1)!}(\cos t/\sin^2 t)^{(r-2)} & \text{for } \dfrac{r}{2} \in \mathbb{N}, \\ \dfrac{t^r}{(r-1)!}(\cot t)^{(r-1)} & \text{for } \dfrac{r+1}{2} \in \mathbb{N} \end{cases}$$

are called Akhiezer–Kreĭn–Favard sums.

THEOREM A (see, e.g., [**7,** p. 148]). *Let $n \in \mathbb{Z}_+$, $r \in \mathbb{N}$, $f \in C_q^{(r)}$. Then*

$$q(f - X_{n.r}(f)) \leqslant K_r(n+1)^{-r} q(f^{(r)}).$$

COROLLARY A. *Let $n \in \mathbb{Z}_+$, $r \in \mathbb{N}$, $f \in C_q^{(r)}$. Then*

$$E_n(f) \leqslant K_r(n+1)^{-r} E_n(f^{(r)}).$$

COROLLARY B. *Let $n, r \in \mathbb{Z}_+$. Then*

$$\sup_{f \in C_q^{(r)}} \{E_n(f)/q(f^{(r)})\} \leqslant K_r(n+1)^{-r}.$$

For $r \in \mathbb{N}$, $f \in C$, by $f^{(-r)}$ we denote the function in $C^{(r)}$ such that $(f^{(-r)})^{(r)} = f - c_0(f)$, $c_0(f^{(-r)}) = 0$.

2. Interpolation by algebraic polynomials. Henceforth

$$\mathbb{P}_n = \left\{\sum_{k=0}^{n} c_k x^k; c_k \in \mathbb{C}\right\}$$

is the set of algebraic polynomials of order no greater than n; in the case $b < a$ the product $\prod_a^b$ is considered equal to 1 and the sum $\sum_a^b$ equal to 0.

We recall some elementary facts in the theory of interpolation by polynomials (see [**13,** Ch. 2], [**14,** Ch. 2], [**15,** pp. 179–196], [**19,** pp. 120–135]).

Let $x_k \in \mathbb{R}$, $y_k \in \mathbb{C}$ $(k = \overline{0,n})$, and let all x_k be distinct. Set

$$\omega(x) = \prod_{k=0}^{n}(x - x_k), \quad l_k(x) = \frac{\omega(x)}{\omega'(x_k)(x - x_k)}.$$

The function

$$L(x) = \sum_{k=0}^{n} y_k l_k(x) \tag{1}$$

is the only polynomial in $\mathbb{P}_n$ assuming the values y_k at the nodes x_k. Such a polynomial is called interpolational and its representation (1) the interpolational formula of Lagrange. In the case when the nodes x_k are equally spaced, i.e., $x_k = x_0 + kh$, $h \neq 0$, $k = \overline{0,n}$, then the polynomial (1) can be written in Newton form

$$L(x) = y_0 + \sum_{k=1}^{n} \frac{\Delta^k y_0}{k!h^k} \prod_{j=0}^{k-1}(x - x_j), \tag{2}$$

where $\Delta^k y_0 = \sum_{m=0}^{k}(-1)^{k+m}\binom{k}{m} y_m$. By setting

$$\binom{t}{k} = \frac{t(t-1)\cdots(t-k+1)}{k!} \qquad (k \in \mathbb{N}), \quad \binom{t}{0} = 1,$$

one can rewrite (2) as

$$L(x_0 + th) = \sum_{k=0}^{n} \binom{t}{k} \Delta^k y_0. \tag{3}$$

We introduce the notation

$$\mathscr{T}_k(t) = (2k+1)!\binom{t+k}{2k+1} = t\prod_{l=1}^{k}(t^2 - l^2) \qquad (k \in \mathbb{Z}_+),$$

$$Q_k(t) = t\mathscr{T}_{k-1}(t) \qquad (k \in \mathbb{N}), \quad Q_0(t) = 1,$$

$$M_k(t) = (2k)!t\binom{t + (k - 1/2)}{2k} = t\prod_{l=1}^{k}(t^2 - (l - 1/2)^2) \quad \text{for } k \in \mathbb{N},$$

$$M_0(t) = t,$$

$$\Lambda_k(t) = M_k(t)/t = \prod_{l=1}^{k}(t^2 - (l - 1/2)^2) \qquad (k \in \mathbb{Z}_+).$$

Let $p \in \mathbb{Z}_+$, $h > 0$, $Q_{h,p} := [-ph, ph]$, $f : Q_{h,p} \to \mathbb{C}$. Then we denote by $\mathscr{L}_{h,p}(f)$ the polynomial in $\mathbb{P}_{2p}$ such that $\mathscr{L}_{h,p}(f, kh) = f(kh)$ for $k = \overline{-p,p}$. Let $p \in \mathbb{N}$, $h > 0$,

$$W_{h,p} := [(-p + 1/2)h, (p - 1/2)h], \qquad f : W_{h,p} \to \mathbb{C}.$$

Then by $\mathscr{R}_{h,p}(f)$ we denote the polynomial in $\mathbb{P}_{2p-1}$ such that

$$\mathscr{R}_{h,p}(f, (k + 1/2)h) = f((k + 1/2)h) \qquad \text{for } k = \overline{-p, p-1}.$$

By setting $z = ht$ in (1), we get

$$\mathcal{L}_{h,p}(f,z) = \frac{\mathcal{T}_p(t)}{(2p)!}\sum_{k=-p}^{p} f(kh)\frac{(-1)^{k+p}\binom{2p}{k+p}}{t-k},$$

$$\mathcal{R}_{h,p}(f,z) = \frac{\Lambda_p(t)}{(2p-1)!}\sum_{k=-p}^{p-1} f((k+1/2)h)\frac{(-1)^{p+1+k}\binom{2p-1}{k+p}}{t-k-1/2}.$$

Now it is not hard to verify, starting with (2) (or (3), which is the same), that

$$\begin{aligned}\mathcal{L}_{h,p}(f,z) =& f(0) + \sum_{k=0}^{p-1}\frac{\mathcal{T}_k(t)}{(2k+1)!}\frac{\Delta_h^{2k+1}(f,-kh)+\Delta_h^{2k+1}(f,-(k+1)h)}{2}\\ &+\sum_{k=1}^{p}\frac{Q_k(t)}{(2k)!}\Delta_h^{2k}(f,-kh),\end{aligned} \tag{4}$$

$$\begin{aligned}\mathcal{R}_{h,p}(f,z) =& \sum_{k=0}^{p-1}\frac{\Lambda_k(t)}{(2k)!}\frac{\Delta_h^{2k}(f,-(k+1/2)h)+\Delta_h^{2k}(f,-(k-1/2)h)}{2}\\ &+\sum_{k=0}^{p-1}\frac{M_k(t)}{(2k+1)!}\Delta_h^{2k+1}(f,-(k+1/2)h).\end{aligned} \tag{5}$$

Equalities (4) and (5) are called formulas of Stirling and Bessel, respectively.

The Stirling numbers of the first kind $S(n,k)$ are defined by

$$n!\binom{x}{n} = \sum_{k=0}^{\infty} S(n,k)x^k \qquad (n \in \mathbb{Z}_+).$$

Note some properties of the Stirling numbers.

S.1. $S(n+1,k) = S(n,k-1) - nS(n,k) \quad (n \in \mathbb{Z}_+,\ k \in \mathbb{N})$.

S.2. $(-1)^{n+k}S(n,k) \geqslant 0 \quad (n,k \in \mathbb{Z}_+)$.

S.3. $S(n,k) = (\ln^k(1+x))^{(n)}_{x=0}/(k!) \quad (n,k \in \mathbb{Z}_+)$.

S.4. *If* $t \in (-1,1)$, $x \in \mathbb{R}$, *then*

$$(1+t)^x = \sum_{n=0}^{\infty}\left(\sum_{k=0}^{n} S(n,k)x^k\right)\frac{t^n}{n!}.$$

For $k = o(\ln n)$ $(n \to \infty)$ the Jordan formula holds:

$$|S(n,k)| = \frac{(n-1)!}{(k-1)!}(\ln n + C)^{k-1}(1+o(1)), \tag{6}$$

where $C = 0.55772\ldots$ is the Euler constant. When $n,k \to \infty$ while $k \leqslant n - O(n^\alpha)$, $0 < \alpha < 1$, it was shown by Moser and Wyman **[20]** that

$$S(n,k) = (-1)^{n+k}\frac{\Gamma(n+R)}{R^k\Gamma(R)\sqrt{2\pi H}}(1+o(1)), \tag{7}$$

where R is the only solution of the equation $\sum_{l=0}^{n-1} R/(R+l) = k$, $H = k -$

$\sum_{l=0}^{n-1} R^2/(R+l)^2$, $\Gamma(\cdot)$ is the gamma function. If, however, $n - o(\sqrt{n}) \leqslant k \leqslant n$, Moser and Wyman established that then

$$S(n,k) = (-1)^{n+k}\binom{n}{k}(k/2)^{n-k}(1+o(1)). \tag{8}$$

For properties of the Stirling numbers see, e.g., **[21,** pp. 38–40]. Formulas (6)–(8) together with their proofs can be found in **[22,** pp. 149–161]. In **[23,** p. 48] a table of the Stirling numbers for $n = \overline{1,8}$ is given. Set

$$T_n(x) = \begin{cases} Q_{n/2}(x), & \text{for } n/2 \in \mathbb{Z}_+, \\ M_{(n-1)/2}(x), & \text{for } (n-1)/2 \in \mathbb{Z}_+. \end{cases}$$

It is not difficult to see that $T_0(x) = 1$, $T_1(x) = x$,

$$T_{n+2}(x) = (x^2 - (n/2)^2)T_n(x) \qquad (n \in \mathbb{Z}_+).$$

The central factorial numbers $t(n,k)$, $k \in \mathbb{Z}$ (this is not a conventional notation; we follow **[24,** p. 209]) are defined by

$$T_n(x) = \sum_{k=-\infty}^{\infty} t(n,k)x^k.$$

It is clear that

$$Q_n(x) = \sum_{k=0}^{\infty} t(2n,k)x^k, \quad M_n(x) = \sum_{k=0}^{\infty} t(2n+1,k)x^k.$$

We note here some properties of $t(n,k)$.

T.1. $t(n+2,k) = t(n,k-2) - (n/2)^2 t(n,k) \quad (n \in \mathbb{Z}_+,\ k \in \mathbb{Z})$.

T.2. $t(n,k) = 0$ *if* $(n+k)/2 \notin \mathbb{Z}$ *or* $-k \in \mathbb{N}$ *or* $k > n$.

T.3. $i^{n-k}t(n,k) \geqslant 0 \quad (n \in \mathbb{Z}_+,\ k \in \mathbb{Z})$.

T.4. *If* $n, k \in \mathbb{Z}_+$ *then*

$$t(2n+2,k+1) = \sum_{l=k}^{\infty}\binom{l}{k} n^{l-k} S(2n+1,l), \tag{9}$$

$$t(2n+1,k+1) = \sum_{l=k}^{\infty}\binom{l}{k}(n-1/2)^{l-k} S(2n,l). \tag{10}$$

Proof. We establish (9). We have

$$\sum_{k=0}^{\infty} t(2n+2,k)x^k = Q_{n+1}(x) = x\mathcal{T}_n(x) = x(2n+1)!\binom{x+n}{2n+1}$$

$$= x\sum_{k=0}^{\infty} S(2n+1,k)(x+n)^k = x\sum_{k=0}^{\infty} S(2n+1,k)\sum_{l=0}^{k}\binom{k}{l}x^l n^{k-l}$$

$$= x\sum_{l=0}^{\infty} x^l \sum_{k=l}^{\infty}\binom{k}{l} n^{k-l} S(2n+1,k) = \sum_{k=0}^{\infty} x^{k+1}\sum_{l=k}^{\infty}\binom{l}{k} n^{l-k} S(2n+1,l).$$

Now compare the coefficients of like powers of x in the starting and final parts of this equality to obtain (9). Formula (10) is derived in a similar way. □

T.5. *If* $n \in \mathbb{N}$, $k \in \mathbb{Z}_+$, *then*

$$t(n, k+1) = \left((1+x)^{n/2-1} \frac{\ln^k(1+x)}{k!} \right)^{(n-1)}_{x=0}.$$

PROOF. Let $n = 2r + 2$ where $r \in \mathbb{Z}_+$. Applying (9) and the property S.3 of the Stirling numbers, we get

$$\begin{aligned} k!t(n, k+1) &= k! \sum_{l=k}^{\infty} \binom{l}{k} r^{l-k} S(2r+1, l) \\ &= \left(\sum_{l=k}^{\infty} \frac{r^{l-k}}{(l-k)!} \ln^l(1+x) \right)^{(2r+1)}_{x=0} = \left(\sum_{l=0}^{\infty} \frac{r^l}{l!} \ln^{l+k}(1+x) \right)^{(2r+1)}_{x=0} \\ &= \left(\ln^k(1+x) e^{\ln(1+x)^r} \right)^{(2r+1)}_{x=0} = \left((1+x)^{n/2-1} \ln^k(1+x) \right)^{(n-1)}_{x=0}. \end{aligned}$$

The case $n = 2r + 1$ is dealt with in a similar way with the help of (10). □

T.6. *The following equalities hold*: $t(n, n) = 1$ $(n \in \mathbb{Z}_+)$,

$$t(n, n-2) = -\frac{1}{4}\binom{n}{3} \qquad (n-2 \in \mathbb{N}),$$

$$16t(n, n-4) = 9\binom{n}{5} + 10\binom{n}{6} \qquad (n-5 \in \mathbb{N}),$$

$$-64t(n, n-6) = 225\binom{n}{7} + 504\binom{n}{8} + 280\binom{n}{9} \qquad (n-8 \in \mathbb{N}),$$

$$\begin{aligned} 256t(n, n-8) &= 11025\binom{n}{9} + 37206\binom{n}{10} + 41580\binom{n}{11} \\ &\quad + 15400\binom{n}{12} \qquad (n-11 \in \mathbb{N}). \end{aligned}$$

T.7. *If* $n, k \in \mathbb{Z}_+$, *then*

$$\mathcal{T}_n^{(k)}(0) = k!t(2n+2, k+1), \tag{11}$$

$$\Lambda_n^{(k)}(0) = k!t(2n+1, k+1). \tag{12}$$

PROOF. Both relations (11) and (12) follow immediately from

$$\mathcal{T}_n(x) = Q_{n+1}(x)/x = \sum_{k=1}^{\infty} t(2n+2, k)x^{k-1} = \sum_{k=0}^{\infty} t(2n+2, k+1)x^k,$$

$$\Lambda_n(x) = M_n(x)/x = \sum_{k=1}^{\infty} t(2n+1, k)x^{k-1} = \sum_{k=0}^{\infty} t(2n+1, k+1)x^k. \qquad \square$$

In connection with c. f. n. see [**24**, pp. 209–211, 225, 226]. In [**24**, p. 249] a table for $t(n, k)$, $n = \overline{0, 13}$ is given.

3. S. N. Bernstein polynomials. If $X \subset \mathbb{R}$, then $C(X)$ denotes the set of functions $f\colon X \to \mathbb{C}$ continuous on X; $C^{(r)}(X) = \{f \in C(X); \exists f^{(r)} \in C(X)\}$, $\|f|X\| = \max_{x\in X}|f(x)|$. Let $f\colon [a,b] \to \mathbb{C}$, $n \in \mathbb{N}$. The polynomial

$$B_n(f,[a,b],x) = \sum_{k=0}^{n} f\left(a + \frac{(b-a)k}{n}\right)\binom{n}{k}\left(\frac{x-a}{b-a}\right)^k\left(\frac{b-x}{b-a}\right)^{n-k}$$

is called the Bernstein polynomial of order n for f.

THEOREM B (see, e.g., [**11**, p. 15]). *Let $k, r \in \mathbb{Z}_+$, $k \leqslant r$, $f \in C^{(r)}([a,b])$. Then*

$$\lim_{n\to\infty} \|f^{(k)} - B_n^{(k)}(f,[a,b])|[a,b]\| = 0.$$

§3. Interpolation and numerical differentiation

1. The remainder term of interpolation.

LEMMA 1. *Let $f \in C^{(1)}([\alpha,\beta])$, $m \in \mathbb{Z}_+$, $\{x_k\}_{k=0}^m \subset [\alpha,\beta]$, $x_k \neq x_i$ for $k \neq i$, $\omega(x) = \prod_{k=0}^m (x - x_k)$,*

$$l_k(x) = \frac{\omega(x)}{(x-x_k)\omega'(x_k)}, \quad L_m(f,x) = \sum_{k=0}^{m} f(x_k)l_k(x). \tag{1}$$

Then for $x \in [\alpha,\beta]$

$$f(x) = L_m(f,x) + \omega(x)\int_0^1 \sum_{k=0}^{m} \frac{f'((x_k - x)u + x)}{\omega'(x_k)}\,du. \tag{2}$$

PROOF. It is clear that

$$\begin{aligned} f(x) - L_m(f,x) &= \sum_{k=0}^{m} (f(x) - f(x_k))l_k(x) \\ &= \sum_{k=0}^{m} l_k(x)(x - x_k)\int_0^1 f'((x_k - x)u + x)\,du \\ &= \omega(x)\int_0^1 \sum_{k=0}^{m} (\omega'(x_k))^{-1} f'((x_k - x)u + x)\,du. \end{aligned}$$ □

COROLLARY 1 (see [**25**, pp. 10, 11]). *Let $f \in C^{(1)}([\alpha,\beta])$, $m \in \mathbb{N}$, $[a,b] \subset [\alpha,\beta]$, $h = (b-a)/m$, $x_k = a + kh$, $\omega(x) = \prod_{k=0}^m (x - x_k)$, and let $l_k(x)$ and $L_m(f,x)$ be defined by* (1). *Then for $x \in [\alpha,\beta]$*

$$f(x) = L_m(f,x) + \frac{\omega(x)}{m!h^m}\int_0^1 \Delta_{uh}^m(f', au + x(1-u))\,du.$$

The proof of Corollary 1 is obtained by comparing the equality

$$\omega'(x_k) = (-1)^{m-k}(m!)h^m\left(\binom{m}{k}\right)^{-1}$$

with (2).

COROLLARY 2. *Let* $f \in C^{(1)}([\alpha,\beta])$, $p \in \mathbb{Z}_+$, $h > 0$, $Q_{h.p} \subset [\alpha,\beta]$, $x \in [\alpha,\beta]$. *Then*

$$f(x) = \mathscr{L}_{h.p}(f,x) + \frac{x}{(2p)!h^{2p}} \times \prod_{k=1}^{p}(x^2-(kh)^2)\int_0^1 \delta_{uh}^{2p}(f',x(1-u))\,du.$$

COROLLARY 3. *Let* $f \in C^{(1)}([\alpha,\beta])$, $p \in \mathbb{N}$, $h > 0$, $W_{h.p} \subset [\alpha,\beta]$, $x \in [\alpha,\beta]$. *Then*

$$f(x) = \mathscr{R}_{h.p}(f,x) + \frac{1}{(2p-1)!h^{2p-1}} \times \prod_{k=1}^{p}(x^2-(k-1/2)^2h^2)\int_0^1 \delta_{uh}^{2p-1}(f',x(1-u))\,du.$$

These two corollaries follow easily from Corollary 1.

2. Remainder terms of the numerical differentiation formulas.

THEOREM 1. *Let* $p \in \mathbb{N}$, $\gamma \in \mathbb{Z}_+$, $h > 0$.

(i) *If* $f \in C^{(\gamma)}(Q_{h.p})$, *then*

$$f^{(\gamma)}(0) = \mathscr{L}_{h.p}^{(\gamma)}(f,0) + \frac{1}{(2p)!} \times \sum_{l=1}^{\gamma}\binom{\gamma}{l}\left(\int_0^1 \delta_{uh}^{2p}(f^{(\gamma+1-l)},0)(1-u)^{\gamma-l}\,du\right)h^{1-l}\mathscr{T}_p^{(l)}(0). \tag{3}$$

(ii) *If* $f \in C^{(\gamma+1)}(W_{h.p})$, *then*

$$f^{(\gamma)}(0) = \mathscr{R}_{h.p}^{(\gamma)}(f,0) + \frac{1}{(2p-1)!} \times \sum_{l=0}^{\gamma}\binom{\gamma}{l}\left(\int_0^1 \delta_{uh}^{2p-1}(f^{(\gamma+1-l)},0)(1-u)^{\gamma-l}\,du\right)h^{1-l}\Lambda_p^{(l)}(0). \tag{4}$$

PROOF. We establish (3). One can assume that $\gamma \in \mathbb{N}$, since for $\gamma = 0$ (3) is obvious. Set $\lambda(x) = x\prod_{k=1}^{p}(x^2-(kh)^2)$. Suppose first that $f \in C^{(\gamma+1)}(Q_{h.p})$. Apply Leibniz's formula to find the derivatives of the product and note that $\lambda(0) = 0$ to get

$$\begin{aligned}\left(\lambda(x)\int_0^1 \delta_{uh}^{2p}(f',x(1-u))\,du\right)^{(\gamma)}_{x=0} &= \sum_{l=0}^{\gamma}\binom{\gamma}{l}\left(\int_0^1 \delta_{uh}^{2p}(f',x(1-u))\,du\right)^{(\gamma-l)}_{x=0}\lambda^{(l)}(0)\\ &= \sum_{l=1}^{\gamma}\binom{\gamma}{l}\left(\int_0^1 \delta_{uh}^{2p}(f^{(\gamma+1-l)},0)(1-u)^{\gamma-l}\,du\right)\lambda^{(l)}(0).\end{aligned} \tag{5}$$

Obviously,

$$\lambda^{(l)}(0) = h^{2p+1-l}\mathscr{T}_p^{(l)}(0). \tag{6}$$

Comparing Corollary 2, formulas (5) and (6) we obtain (3) for the case $f \in C^{(\gamma+1)}(Q_{h.p})$. Now let $f \in C^{(\gamma)}(Q_{h.p})$. By virtue of A. A. Markov's inequality for the derivative of a polynomial (see, e.g., **[26,** p. 179]) we have for $P \in \mathbb{P}_n$

$$\|P^{(\gamma)}|[a,b]\| \leqslant C(\gamma,n,b-a)\|P|[a,b]\|,$$

where $C(\gamma,n,b-a)$ depends only on the arguments specified. Thus the last inequality gives

$$\begin{aligned}\|\mathscr{L}_{h.p}^{(\gamma)}(f) - \mathscr{L}_{h.p}^{(\gamma)}(B_n(f,Q_{h.p}))|Q_{h.p}\| \\ \leqslant C(\gamma,2p,2ph)\|\mathscr{L}_{h.p}(f - B_n(f,Q_{h.p}))|Q_{h.p}\| \\ \leqslant C(\gamma,2p,2ph)\|\mathscr{L}_{h.p}\| \cdot \|f - B_n(f,Q_{h.p})|Q_{h.p}\|,\end{aligned}$$

where $\|\mathscr{L}_{h.p}\| = \sup_{f\in C(Q_{h.p})}\{\|\mathscr{L}_{h.p}(f)|Q_{h.p}\|/\|f|Q_{h.p}\|\} < \infty$. Applying here Theorem 2.B (for $r = 0$) we obtain

$$\lim_{n\to\infty} \|\mathscr{L}_{h.p}^{(\gamma)}(f) - \mathscr{L}_{h.p}^{(\gamma)}(B_n(f,Q_{h.p}))|Q_{h.p}\| = 0. \tag{7}$$

Since (3) is already proved for $f \in C^{(\gamma+1)}(Q_{h.p})$, we can apply it to the polynomial $B_n(f,Q_{h.p})$ for each $n \in \mathbb{N}$. Passing to the limit as $n \to \infty$ in the resulting inequality (here (7) and Theorem 2.B are used) we obtain (3) for the general case. Now we prove (4). Set $\beta(x) = \prod_{k=1}^{p}(x^2 - (k-1/2)^2h^2)$,

$$\begin{aligned}&\left(\beta(x)\int_0^1 \delta_{uh}^{2p-1}(f',x(1-u))\,du\right)_{x=0}^{(\gamma)} \\ &\quad= \sum_{l=0}^{\gamma}\binom{\gamma}{l}\left(\int_0^1 \delta_{uh}^{2p-1}(f^{(\gamma+1-l)},0)(1-u)^{\gamma-l}\,du\right)\beta^{(l)}(0).\end{aligned} \tag{8}$$

It is clear that

$$\beta^{(l)}(0) = h^{2p-l}\Lambda_p^{(l)}(0). \tag{9}$$

Comparing (8), (9), and Corollary 3 we get (4). □

COROLLARY 4. *Let* $p \in \mathbb{N}$, $h > 0$, $f \in C^{(2p+1)}(Q_{h.p})$. *Then*

$$\begin{aligned}f^{(2p+1)}(0) &= \frac{1}{(2p)!}\sum_{l=1}^{2p+1}\binom{2p+1}{l} \\ &\quad\times\left(\int_0^1 \delta_{uh}^{2p}(f^{(2p+2-l)},0)(1-u)^{2p+1-l}\,du\right)h^{1-l}\mathscr{T}_p^{(l)}(0).\end{aligned} \tag{10}$$

To prove Corollary 4 it is sufficient to set $\gamma = 2p+1$ in (3) and note that $\mathscr{L}_{h.p}^{(2p+1)}(f,0) = 0$, since $\mathscr{L}_{h.p}(f) \in \mathbb{P}_{2p}$.

COROLLARY 5. *Let* $p \in \mathbb{N}$, $h > 0$, $f \in C^{(2p+1)}(W_{h,p})$. *Then*

$$
\begin{aligned}
f^{(2p)}(0) = \frac{1}{(2p-1)!} \sum_{l=0}^{2p} \binom{2p}{l} \\
\times \left(\int_0^1 \delta_{uh}^{2p-1}(f^{(2p+1-l)}, 0)(1-u)^{2p-l}\, du \right) h^{1-l} \Lambda_p^{(l)}(0).
\end{aligned} \tag{11}
$$

To prove Corollary 5 we set $\gamma = 2p$ in (4) and note that $\mathscr{R}_{h,p}^{(2p)}(f, 0) = 0$, since $\mathscr{R}_{h,p}(f) \in \mathbb{P}_{2p-1}$.

COROLLARY 4′. *Let* $p \in \mathbb{N}$, $h > 0$, $f \in C^{(2p+1)}(Q_{h,p})$. *Then*

$$
\begin{aligned}
f^{(2p+1)}(0) = \frac{2p+1}{h^{2p+1}} \sum_{l=1}^{2p+1} \frac{t(2p+2, 2p+3-l)}{(l-1)!} \\
\times \left(\int_0^1 \delta_{uh}^{2p}(f^{(l)}, 0)(1-u)^{l-1}\, du \right) h^l.
\end{aligned} \tag{10′}
$$

PROOF. Compare (10) and (2.11) to get

$$
\begin{aligned}
f^{(2p+1)}(0) &= (2p+1) \sum_{l=1}^{2p+1} \frac{t(2p+2, l+1) h^{1-l}}{(2p+1-l)!} \\
&\quad \times \int_0^1 \delta_{uh}^{2p}(f^{(2p+2-l)}, 0)(1-u)^{2p+1-l}\, du \\
&= (2p+1) \sum_{l=1}^{2p+1} \frac{t(2p+2, 2p+3-l)}{(l-1)!} h^{-2p-1+l} \\
&\quad \times \int_0^1 \delta_{uh}^{2p}(f^{(l)}, 0)(1-u)^{l-1}\, du.
\end{aligned}
$$

□

COROLLARY 5′. *Let* $p \in \mathbb{N}$, $h > 0$, $f \in C^{(2p+1)}(W_{h,p})$. *Then*

$$
\begin{aligned}
f^{(2p)}(0) = \frac{2p}{h^{2p-1}} \sum_{l=0}^{2p} \frac{t(2p+1, 2p+1-l)}{l!} \\
\times \left(\int_0^1 \delta_{uh}^{2p-1}(f^{(l+1)}, 0)(1-u)^l\, du \right) h^l.
\end{aligned} \tag{11′}
$$

PROOF. Compare (11) and (2.12):

$$f^{(2p)}(0) = 2p \sum_{l=0}^{2p} \frac{t(2p+1,l+1)h^{1-l}}{(2p-l)!} \times \int_0^1 \delta_{uh}^{2p-1}(f^{(2p+1-l)},0)(1-u)^{2p-l}\,du$$
$$= 2p \sum_{l=0}^{2p} \frac{t(2p+1,2p+1-l)}{l!} h^{-2p+l+1} \times \int_0^1 \delta_{uh}^{2p-1}(f^{(l+1)},0)(1-u)^{l}\,du. \qquad \square$$

Set

$$\mathscr{K}(p,k,\gamma) := \frac{(-1)^{k+p}}{(2p)!}\binom{2p}{k+p}\sum_{l=1}^{\gamma}\frac{\mathscr{T}_p^{(l)}(0)}{l!}k^{l-1} = \frac{(-1)^{k+p}}{(2p)!}\binom{2p}{k+p}\sum_{l=1}^{\gamma} t(2p+2,l+1)k^{l-1},$$

$$\mathfrak{L}(p,k,\gamma) := \frac{(-1)^{k+p+1}}{(2p-1)!}\binom{2p-1}{k+p}\sum_{l=0}^{\gamma}\frac{\Lambda_p^{(l)}(0)}{l!}(k+1/2)^{l} = \frac{(-1)^{k+p+1}}{(2p-1)!}\binom{2p-1}{k+p}\sum_{l=0}^{\gamma} t(2p+1,l+1)(k+1/2)^{l}.$$

THEOREM 2. *Let* $p,\gamma \in \mathbb{N}$, $h>0$.
(i) *If* $\gamma \leqslant 2p+1$, $f \in C^{(\gamma)}(Q_{h.p})$, *then*

$$f^{(\gamma)}(0) = \mathscr{L}_{h.p}^{(\gamma)}(f,0) + \gamma\int_0^1 \left\{\sum_{k=-p}^{p}\mathscr{K}(p,k,\gamma)f^{(\gamma)}(khu)\right\}(1-u)^{\gamma-1}\,du. \tag{12}$$

(ii) *If* $\gamma \leqslant 2p-1$, $f \in C^{(\gamma+1)}(W_{h.p})$ *then*

$$f^{(\gamma)}(0) = \mathscr{R}_{h.p}^{(\gamma)}(f,0) + h\int_0^1 \left\{\sum_{k=-p}^{p}\mathfrak{L}(p,k,\gamma)f^{(\gamma+1)}((k+1/2)hu)\right\}(1-u)^{\gamma}\,du. \tag{13}$$

PROOF. Note first that if $n \in [0,t)\cap\mathbb{Z}_+$, $a \in \mathbb{R}$, then

$$\sum_{m=0}^{t}(-1)^m\binom{t}{m}(m-a)^n = 0. \tag{14}$$

We prove (12). Integrating by parts $l-1$ times ($l \leqslant \gamma$), taking (14) into account and using the equality

$$\delta_{uh}^{2p}(f^{(\gamma+1-l)},0) = \sum_{m=0}^{2p}(-1)^m\binom{2p}{m}f^{(\gamma+1-l)}(uh(m-p)),$$

we come to

$$\int_0^1 \delta_{uh}^{2p}(f^{(\gamma+1-l)},0)(1-u)^{\gamma-l}\,du$$
$$= \int_0^1 \sum_{m=0}^{2p}(-1)^m \binom{2p}{m} f^{(\gamma+1-l)}(uh(m-p))(1-u)^{\gamma-l}\,du$$
$$= \frac{(\gamma-l)!}{(\gamma-1)!}h^{l-1}\int_0^1\left(\sum_{m=0}^{2p}(-1)^m\binom{2p}{m}\right.$$
$$\left.\times f^{(\gamma)}(uh(m-p))(m-p)^{l-1}\right)(1-u)^{\gamma-1}\,du.$$

Now apply (3) to obtain

$$f^{(\gamma)}(0) - \mathscr{L}_{h.p}^{(\gamma)}(f,0) = \frac{\gamma}{(2p)!}\sum_{l=1}^{\gamma}\frac{\mathscr{T}_p^{(l)}(0)}{l!}$$
$$\times \int_0^1\left(\sum_{m=0}^{2p}(-1)^m\binom{2p}{m} f^{(\gamma)}(uh(m-p))(m-p)^{l-1}\right)(1-u)^{\gamma-1}\,du$$
$$= \gamma\int_0^1\sum_{m=0}^{2p} f^{(\gamma)}(uh(m-p))$$
$$\times\left\{\frac{(-1)^m}{(2p)!}\binom{2p}{m}\sum_{l=1}^{\gamma}\frac{\mathscr{T}_p^{(l)}(0)}{l!}(m-p)^{l-1}\right\}(1-u)^{\gamma-1}\,du$$
$$= \gamma\int_0^1\left\{\sum_{k=-p}^{p} f^{(\gamma)}(uhk)\mathscr{K}(p,k,\gamma)\right\}(1-u)^{\gamma-1}\,du.$$

In order to prove (13) integrate by parts l times ($l \leqslant \gamma$) and use (14) and the equality

$$\delta_{uh}^{2p-1}(f^{(\gamma+1-l)},0) = \sum_{m=0}^{2p-1}(-1)^{m+1}\binom{2p-1}{m} f^{(\gamma+1-l)}(uh(m-p+1/2))$$

to get

$$\int_0^1 \delta_{uh}^{2p-1}(f^{(\gamma+1-l)},0)(1-u)^{\gamma-l}\,du = \frac{(\gamma-l)!}{\gamma!}h^l$$
$$\times\int_0^1\left(\sum_{m=0}^{2p-1}(-1)^{m+1}\binom{2p-1}{m}\right.$$
$$\left.\times f^{(\gamma+1)}(uh(m-p+1/2))(m-p+1/2)^l\right)(1-u)^{\gamma}\,du.$$

Finally, apply (4):

$$f^{(\gamma)}(0)-\mathscr{R}_{h,p}^{(\gamma)}(f,0)=\frac{h}{(2p-1)!}\sum_{l=0}^{\gamma}\frac{\Lambda_p^{(l)}(0)}{l!}$$

$$\times\int_0^1\left(\sum_{m=0}^{2p-1}(-1)^{m+1}\binom{2p-1}{m}\right.$$

$$\left.\times f^{(\gamma+1)}(uh(m-p+1/2))(m-p+1/2)^l\right)(1-u)^\gamma\,du.$$

$$=h\int_0^1\sum_{m=0}^{2p-1}f^{(\gamma+1)}(uh(m-p+1/2))$$

$$\times\left\{\frac{(-1)^{m+1}}{(2p-1)!}\binom{2p-1}{m}\sum_{l=0}^{\gamma}\frac{\Lambda_p^{(l)}(0)}{l!}(m-p+1/2)^l\right\}(1-u)^\gamma\,du$$

$$=h\int_0^1\left\{\sum_{k=-p}^{p-1}f^{(\gamma+1)}(uh(k+1/2))\mathfrak{L}(p,k,\gamma)\right\}(1-u)^\gamma\,du.\qquad\square$$

Corollary 6. *Let* $p,\gamma\in\mathbb{N}$. *Then*

$$\sum_{k=-p}^{p}\mathscr{K}(p,k,\gamma)=\begin{cases}0, & \text{if } \gamma\leqslant 2p,\\ 1, & \text{if } \gamma=2p+1.\end{cases}$$

To verify Corollary 6 set $f(x)=x^\gamma/\gamma!$ in (12).

Corollary 7. *Let* $p,\gamma\in\mathbb{N}$, $\gamma\leqslant 2p+1$, $h>0$, $f\in C^{(\gamma)}(Q_{h,p})$. *Then*

$$\mathscr{L}_{h,p}^{(\gamma)}(f,0)=-\gamma\int_0^1\left\{\sum_{k=1}^{p}\mathscr{K}(p,k,\gamma)(f^{(\gamma)}(khu)+f^{(\gamma)}(-khu))\right\}(1-u)^{\gamma-1}\,du.$$

To prove Corollary 7 it is sufficient to note that $\mathscr{K}(p,k,\gamma)=\mathscr{K}(p,-k,\gamma)$ for $k=\overline{1,p}$, $\mathscr{K}(p,0,\gamma)=1$, and to use (12).

§4. Numerical differentiation formulas as functions of the differentiation point

Definition 1. Let $f\in C(\mathbb{R})$, $h>0$, $p\in\mathbb{Z}_+$, $\gamma\in\mathbb{N}$, $x\in\mathbb{R}$. Then we set

$$U_{h,p,\gamma}(f,x)=(\mathscr{L}_{h,p}(f(\cdot+x),z))_{z=0}^{(\gamma)},$$

where the derivative is taken with respect to the z variable.

DEFINITION 2. Let $f \in C(\mathbb{R})$, $h > 0$, $p \in \mathbb{N}$, $\gamma \in \mathbb{Z}_+$, $x \in \mathbb{R}$. Then we set

$$V_{h,p,\gamma}(f,x) = (\mathscr{R}_{h,p}(f(\cdot + x), z))^{(\gamma)}_{z=0},$$

where the derivative is taken with respect to the z variable.

Introduce

$$\mathscr{A}(p,\gamma,\alpha) := \frac{1}{(2p)!}\sum_{l=1}^{\gamma}\binom{\gamma}{l}2^{2[(\gamma+1)/2]-l-1}|\mathscr{T}_p^{(l)}(0)|B(l+\alpha,\gamma-l+1),$$

$$\mathscr{B}(p,\gamma,\alpha) := \frac{1}{(2p-1)!}\sum_{l=0}^{\gamma}\binom{\gamma}{l}2^{2[\gamma/2]-l}|\Lambda_p^{(l)}(0)|B(l+\alpha+1,\gamma-l+1),$$

where $B(a,b) = \int_0^1 x^{a-1}(1-x)^{b-1}\,dx$ is the Beta function;

$$\mathscr{A}(p,\gamma) := \mathscr{A}(p,\gamma,0) = \frac{1}{(2p)!}\sum_{l=1}^{\gamma}2^{2[(\gamma+1)/2]-l-1}\frac{|\mathscr{T}_p^{(l)}(0)|}{l},$$

$$\mathscr{B}(p,\gamma) := \mathscr{B}(p,\gamma,0) = \frac{1}{(2p-1)!(\gamma+1)}\sum_{l=0}^{\gamma}2^{2[\gamma/2]-l}|\Lambda_p^{(l)}(0)|,$$

$$\mathscr{D}(p,\gamma) := \begin{cases} \displaystyle\sum_{k=\gamma/2}^{p}\frac{2^{2k-\gamma}}{(2k)!}|Q_k^{(\gamma)}(0)|, & \text{for } \dfrac{\gamma}{2}\in\mathbb{N}, \\ \displaystyle\sum_{k=(\gamma-1)/2}^{p-1}\frac{2^{2k+1-\gamma}}{(2k+1)!}|\mathscr{T}_k^{(\gamma)}(0)|, & \text{for } \dfrac{\gamma+1}{2}\in\mathbb{N}; \end{cases}$$

$$\mathscr{E}(p,\gamma) := \begin{cases} \displaystyle\sum_{k=\gamma/2}^{p-1}\frac{2^{2k-\gamma}}{(2k)!}|\Lambda_k^{(\gamma)}(0)|, & \text{for } \dfrac{\gamma}{2}\in\mathbb{Z}_+, \\ \displaystyle\sum_{k=(\gamma-1)/2}^{p-1}\frac{2^{2k+1-\gamma}}{(2k+1)!}|M_k^{(\gamma)}(0)|, & \text{for } \dfrac{\gamma+1}{2}\in\mathbb{N}. \end{cases}$$

THEOREM 1. *Let $p \in \mathbb{Z}_+$, $\gamma \in \mathbb{N}$, $k = 2[(\gamma+1)/2] \leqslant 2p+2$, $0 \leqslant \alpha \leqslant 2p+2-k$, $h > 0$, $f \in C_q^{(\gamma)}$. Then*

$$q(f^{(\gamma)} - U_{h,p,\gamma}(f)) \leqslant \mathscr{A}(p,\gamma,\alpha)h^{\alpha}\Omega_{2p+2-k,\alpha}(f^{(\gamma)},h).$$

PROOF. By virtue of (3.3) (for $p = 0$, (1) is obvious) we have

(1)
$$f^{(\gamma)}(x) - U_{h,p,\gamma}(f,x) = \frac{1}{(2p)!}\sum_{l=1}^{\gamma}\binom{\gamma}{l} \times \left(\int_0^1 \delta_{uh}^{2p}(f^{(\gamma+1-l)},x)(1-u)^{\gamma-l}\,du\right)h^{1-l}\mathscr{T}_p^{(l)}(0).$$

Setting $d_l = \binom{\gamma}{l}|\mathscr{T}_p^{(l)}(0)|/((2p)!)$ and applying Lemma 2.A we find

$$
\begin{aligned}
q(f^{(\gamma)} - U_{h,p,\gamma}(f)) &\leqslant \sum_{l=1}^{\gamma} d_l h^{1-l} q\left(\int_0^1 \delta_{uh}^{2p}(f^{(\gamma+1-l)})(1-u)^{\gamma-l}\,du\right) \\
&\leqslant \sum_{l=1}^{\gamma} d_l h^{1-l} \int_0^1 q(\delta_{uh}^{2p}(f^{(\gamma+1-l)}))(1-u)^{\gamma-l}\,du \qquad (2)\\
&\leqslant \sum_{l=1}^{\gamma} d_l h^{1-l} \int_0^1 \omega_{2p}(f^{(\gamma+1-l)}, uh)(1-u)^{\gamma-l}\,du.
\end{aligned}
$$

Since (cf. 2.M.7)

$$\omega_{2p}(f^{(\gamma+1-l)}, uh) \leqslant (uh)^{l-1}\omega_{2p+1-l}(f^{(\gamma)}, uh),$$

it follows from (2) that

$$q(f^{(\gamma)} - U_{h,p,\gamma}(f)) \leqslant \sum_{l=1}^{\gamma} d_l \int_0^1 \omega_{2p+1-l}(f^{(\gamma)}, uh)u^{l-1}(1-u)^{\gamma-l}\,du. \qquad (3)$$

First let $\gamma/2 \in \mathbb{N}$. Since (see property 2.M.6)

$$
\begin{aligned}
\omega_{2p+1-l}(f^{(\gamma)}, uh) &\leqslant 2^{\gamma-l-1}\omega_{2p+2-\gamma}(f^{(\gamma)}, uh) \\
&\leqslant 2^{\gamma-l-1}(uh)^{\alpha}\Omega_{2p+2-\gamma,\alpha}(f^{(\gamma)}, uh)
\end{aligned}
$$

and $\mathscr{T}_p^{(\gamma)}(0) = 0$, it follows from (3) that

$$
\begin{aligned}
&q(f^{(\gamma)} - U_{h,p,\gamma}(f)) \\
&\quad\leqslant h^{\alpha} \sum_{l=1}^{\gamma-1} d_l 2^{\gamma-l-1} \int_0^1 \Omega_{2p+2-\gamma,\alpha}(f^{(\gamma)}, uh)u^{l+\alpha-1}(1-u)^{\gamma-l}\,du \\
&\quad\leqslant \Omega_{2p+2-\gamma,\alpha}(f^{(\gamma)}, h)h^{\alpha} \sum_{l=1}^{\gamma-1} d_l 2^{\gamma-l-1} B(l+\alpha, \gamma-l+1) \\
&\quad= \mathscr{A}(p,\gamma,\alpha)h^{\alpha}\Omega_{2p+2-\gamma,\alpha}(f^{(\gamma)}, h).
\end{aligned}
$$

When $(\gamma+1)/2 \in \mathbb{N}$, similar reasoning leads to

$$
\begin{aligned}
&q(f^{(\gamma)} - U_{h,p,\gamma}(f)) \\
&\quad\leqslant h^{\alpha} \sum_{l=1}^{\gamma} d_l 2^{\gamma-l} \int_0^1 \Omega_{2p+1-\gamma,\alpha}(f^{(\gamma)}, uh)u^{l+\alpha-1}(1-u)^{\gamma-l}\,du \\
&\quad\leqslant \Omega_{2p+1-\gamma,\alpha}(f^{(\gamma)}, h)h^{\alpha} \sum_{l=1}^{\gamma} d_l 2^{\gamma-l} B(l+\alpha, \gamma-l+1) \\
&\quad= \mathscr{A}(p,\gamma,\alpha)h^{\alpha}\Omega_{2p+1-\gamma,\alpha}(f^{(\gamma)}, h).
\end{aligned}
$$

□

THEOREM 2. *Let* $p \in \mathbb{N}$, $\gamma \in \mathbb{Z}_+$, $k = 2[\gamma/2] \leqslant 2p-1$, $0 \leqslant \alpha \leqslant 2p-1-k$, $h > 0$, $f \in C_q^{(\gamma+1)}$. *Then*

$$q(f^{(\gamma)} - V_{h,p,\gamma}(f)) \leqslant \mathscr{B}(p,\gamma,\alpha)h^{1+\alpha}\Omega_{2p-1-k,\alpha}(f^{(\gamma+1)}, h).$$

PROOF. According to (3.4) we have

$$f^{(\gamma)}(x) - V_{h,p,\gamma}(f,x) \tag{4}$$
$$= \frac{1}{(2p-1)!}\sum_{l=0}^{\gamma}\binom{\gamma}{l}\left(\int_0^1 \delta_{uh}^{2p-1}(f^{(\gamma+1-l)},x)(1-u)^{\gamma-l}\,du\right)h^{1-l}\Lambda_p^{(l)}(0).$$

Set here $d_l = \binom{\gamma}{l}|\Lambda_p^{(l)}(0)|/((2p-1)!)$ and apply Lemma 2.A to get

$$q(f^{(\gamma)} - V_{h,p,\gamma}(f)) \leqslant \sum_{l=0}^{\gamma} d_l h^{1-l}\int_0^1 \omega_{2p-1}(f^{(\gamma+1-l)},uh)(1-u)^{\gamma-l}\,du. \tag{5}$$

Since $\omega_{2p-1}(f^{(\gamma+1-l)},uh) \leqslant (uh)^l\omega_{2p-l-1}(f^{(\gamma+1)},uh)$, (5) implies

$$q(f^{(\gamma)} - V_{h,p,\gamma}(f)) \leqslant h\sum_{l=0}^{\gamma} d_l \int_0^1 \omega_{2p-l-1}(f^{(\gamma+1)},uh)u^l(1-u)^{\gamma-l}\,du. \tag{6}$$

First let $(\gamma+1)/2 \in \mathbb{N}$. Since

$$\begin{aligned}\omega_{2p-l-1}(f^{(\gamma+1)},uh) &\leqslant 2^{\gamma-l-1}\omega_{2p-\gamma}(f^{(\gamma+1)},uh)\\ &\leqslant 2^{\gamma-l-1}(uh)^{\alpha}\Omega_{2p-\gamma,\alpha}(f^{(\gamma+1)},uh)\end{aligned}$$

and $\Lambda_p^{(\gamma)}(0) = 0$, on the basis of (6) we have

$$\begin{aligned}&q(f^{(\gamma)} - V_{h,p,\gamma}(f))\\ &\quad\leqslant h^{1+\alpha}\sum_{l=0}^{\gamma} d_l 2^{\gamma-l-1}\int_0^1 \Omega_{2p-\gamma,\alpha}(f^{(\gamma+1)},uh)u^{l+\alpha}(1-u)^{\gamma-l}\,du\\ &\quad\leqslant \Omega_{2p-\gamma,\alpha}(f^{(\gamma+1)},h)h^{1+\alpha}\sum_{l=0}^{\gamma} d_l 2^{\gamma-l-1}B(l+\alpha+1,\gamma-l+1)\\ &\quad= \mathscr{B}(p,\gamma,\alpha)h^{1+\alpha}\Omega_{2p-\gamma,\alpha}(f^{(\gamma+1)},h).\end{aligned}$$

For $\gamma/2 \in \mathbb{Z}_+$ similar reasoning gives

$$\begin{aligned}&q(f^{(\gamma)} - V_{h,p,\gamma}(f))\\ &\quad\leqslant h^{1+\alpha}\sum_{l=0}^{\gamma} d_l 2^{\gamma-l}\int_0^1 \Omega_{2p-1-\gamma,\alpha}(f^{(\gamma+1)},uh)u^{l+\alpha}(1-u)^{\gamma-l}\,du\\ &\quad\leqslant \Omega_{2p-1-\gamma,\alpha}(f^{(\gamma+1)},h)h^{1+\alpha}\sum_{l=0}^{\gamma} d_l 2^{\gamma-l}B(l+\alpha+1,\gamma-l+1)\\ &\quad= \mathscr{B}(p,\gamma,\alpha)h^{1+\alpha}\Omega_{2p-1-\gamma,\alpha}(f^{(\gamma+1)},h).\end{aligned}$$

□

COROLLARY 1. *Let* $p \in \mathbb{Z}_+$, $\gamma \in \mathbb{N}$, $k = 2[(\gamma+1)/2] \leqslant 2p+2$, $h > 0$, $f \in C_q^{(\gamma)}$. *Then*

$$q(f^{(\gamma)} - U_{h,p,\gamma}(f)) \leqslant \mathscr{A}(p,\gamma)\omega_{2p+2-k}(f^{(\gamma)},h).$$

COROLLARY 2. *Let* $p \in \mathbb{N}$, $\gamma \in \mathbb{Z}_+$, $k = 2[\gamma/2] \leqslant 2p-1$, $h > 0$, $f \in C_q^{(\gamma+1)}$. *Then*

$$q(f^{(\gamma)} - V_{h,p,\gamma}(f)) \leqslant \mathscr{B}(p,\gamma) h \omega_{2p-1-k}(f^{(\gamma+1)}, h).$$

Both corollaries are obtained, respectively, from Theorems 1, 2 by setting $\alpha = 0$.
Let $f \in C(\mathbb{R})$, $p \in \mathbb{Z}_+$, $h > 0$, $x \in \mathbb{R}$. It follows from (2.4) that

$$U_{h,p,\gamma}(f,x) = h^{-\gamma} \sum_{k=\gamma/2}^{p} \frac{Q_k^{(\gamma)}(0)}{(2k)!} \Delta_h^{2k}(f, x - kh) \tag{7}$$

if $\gamma/2 \in \mathbb{N}$, and

$$\begin{aligned} U_{h,p,\gamma}(f,x) = h^{-\gamma} &\sum_{k=(\gamma-1)/2}^{p-1} \frac{\mathscr{T}_k^{(\gamma)}(0)}{(2k+1)!} \\ &\times \tfrac{1}{2}\{\Delta_h^{2k+1}(f, x-kh) + \Delta_h^{2k+1}(f, x-(k+1)h)\} \end{aligned} \tag{8}$$

if $(\gamma+1)/2 \in \mathbb{N}$. Also, if $f \in C(\mathbb{R})$, $p \in \mathbb{N}$, $h > 0$, $x \in \mathbb{R}$ it follows from (2.5) that

$$\begin{aligned} V_{h,p,\gamma}(f,x) = h^{-\gamma} &\sum_{k=\gamma/2}^{p-1} \frac{\Lambda_k^{(\gamma)}(0)}{(2k)!} \\ &\times \tfrac{1}{2}\{\Delta_h^{2k}(f, x-(k+1/2)h) + \Delta_h^{2k}(f, x-(k-1/2)h)\} \end{aligned} \tag{9}$$

for $\gamma/2 \in \mathbb{Z}_+$, and

$$V_{h,p,\gamma}(f,x) = h^{-\gamma} \sum_{k=(\gamma-1)/2}^{p-1} \frac{M_k^{(\gamma)}(0)}{(2k+1)!} \Delta_h^{2k+1}(f, x-(k+1/2)h) \tag{10}$$

for $(\gamma+1)/2 \in \mathbb{N}$.
Relying on (7)–(10) one can easily verify the following Theorems 3, 4.

THEOREM 3. *Let* $p, \gamma \in \mathbb{N}$, $l \in \mathbb{Z}_+$, $h > 0$, $x \in \mathbb{R}$.
(i) *If* $f \in C^{(l)}$, *then* $U_{h,p,\gamma}(f) \in C^{(l)}$, $U_{h,p,\gamma}^{(l)}(f,x) = U_{h,p,\gamma}(f^{(l)}, x)$.
(ii) *If* $f \in C_q$, *then* $q(U_{h,p,\gamma}(f)) \leqslant \mathscr{D}(p,\gamma) h^{-\gamma} \overline{\omega}_\gamma(f,h)$.

THEOREM 4. *Let* $p \in \mathbb{N}$, $\gamma, l \in \mathbb{Z}_+$, $h > 0$, $x \in \mathbb{R}$.
(i) *If* $f \in C^{(l)}$, *then* $V_{h,p,\gamma}(f) \in C^{(l)}$, $V_{h,p,\gamma}^{(l)}(f,x) = V_{h,p,\gamma}(f^{(l)}, x)$.
(ii) *If* $f \in C_q$, *then* $q(V_{h,p,\gamma}(f)) \leqslant \mathscr{E}(p,\gamma) h^{-\gamma} \underline{\omega}_\gamma(f,h)$.

§5. Mean functions of the type of the Stirling and Bessel numerical differentiation formulas

DEFINITION 1. Let $f : \mathbb{R} \to \mathbb{C}$ be integrable over every finite interval, $p, \gamma \in \mathbb{N}$, $\gamma \leqslant 2p$, $h > 0$, $x \in \mathbb{R}$. The function

$$S_{h,p,\gamma}(f, x) = \begin{cases} \displaystyle\sum_{k=\gamma/2}^{p} \frac{Q_k^{(\gamma)}(0)}{(2k)!} \mathcal{S}_{h,2k}^{(2k-\gamma)}(f, x) h^{2k-\gamma}, & \text{for } \frac{\gamma}{2} \in \mathbb{N}, \\ \displaystyle\sum_{k=(\gamma-1)/2}^{p-1} \frac{\mathcal{T}_k^{(\gamma)}(0)}{(2k+1)!} \frac{1}{2} \\ \qquad \times \left\{ S_{h,2k+1}^{(2k+1-\gamma)}(f, x + h/2) + S_{h,2k+1}^{(2k+1-\gamma)}(f, x - h/2) \right\} h^{2k+1-\gamma}, \\ & \text{for } \frac{\gamma+1}{2} \in \mathbb{N} \end{cases}$$

is called the mean function for f of order (p, γ) with step h of the type of the Stirling numerical differentiation formula.

DEFINITION 2. Let $f : \mathbb{R} \to \mathbb{C}$ be integrable over every finite interval, $p, \gamma \in \mathbb{N}$, $\gamma \leqslant 2p - 1$, $h > 0$, $x \in \mathbb{R}$. The function

$$\mathfrak{B}_{h,p,\gamma}(f, x) = \begin{cases} \displaystyle\sum_{k=\gamma/2}^{p-1} \frac{\Lambda_k^{(\gamma)}(0)}{(2k)!} \frac{\mathcal{S}_{h,2k}^{(2k-\gamma)}(f, x + h/2) + \mathcal{S}_{h,2k}^{(2k-\gamma)}(f, x - h/2)}{2} h^{2k-\gamma}, \\ & \text{for } \frac{\gamma}{2} \in \mathbb{N}, \\ \displaystyle\sum_{k=(\gamma-1)/2}^{p-1} \frac{M_k^{(\gamma)}(0)}{(2k+1)!} S_{h,2k+1}^{(2k+1-\gamma)}(f, x) h^{2k+1-\gamma}, & \text{for } \frac{\gamma+1}{2} \in \mathbb{N} \end{cases}$$

is called the mean function for f of order (p, γ) with step h of the type of the Bessel numerical differentiation formula.

We set

$$\lambda_{p,\gamma}(t) := \begin{cases} \displaystyle\frac{1}{t^\gamma} \sum_{k=\gamma/2}^{p} \frac{Q_k^{(\gamma)}(0)}{(2k)!} (-1)^{k-\gamma/2} 2^{2k-\gamma} \sin^{2k} t, & \text{for } \frac{\gamma}{2} \in \mathbb{N}, \\ \displaystyle\frac{\cos t}{t^\gamma} \sum_{k=(\gamma-1)/2}^{p-1} \frac{\mathcal{T}_k^{(\gamma)}(0)}{(2k+1)!} (-1)^{\frac{2k+1-\gamma}{2}} 2^{2k+1-\gamma} \sin^{2k+1} t, & \text{for } \frac{\gamma+1}{2} \in \mathbb{N}, \end{cases}$$

$$\mu_{p,\gamma}(t) := \begin{cases} \dfrac{\cos t}{t^\gamma} \displaystyle\sum_{k=\gamma/2}^{p-1} \dfrac{\Lambda_k^{(\gamma)}(0)}{(2k)!}(-1)^{k-\gamma/2} 2^{2k-\gamma} \sin^{2k} t, \\ \hfill \text{for } \dfrac{\gamma}{2} \in \mathbb{N}, \\ \dfrac{1}{t^\gamma} \displaystyle\sum_{k=(\gamma-1)/2}^{p-1} \dfrac{M_k^{(\gamma)}(0)}{(2k+1)!}(-1)^{\frac{2k+1-\gamma}{2}} 2^{2k+1-\gamma} \sin^{2k+1} t, \\ \hfill \text{for } \dfrac{\gamma+1}{2} \in \mathbb{N}. \end{cases}$$

Taking (4.7)–(4.10) and the definition of $S_{h,p,\gamma}(f)$ and $\mathfrak{B}_{h,p,\gamma}(f)$ into account, one can easily obtain

$$S_{h,p,\gamma}(f,x) = c_0(f) + U_{h,p,\gamma}(f^{(-\gamma)},x), \tag{1}$$

$$\mathfrak{B}_{h,p,\gamma}(f,x) = c_0(f) + V_{h,p,\gamma}(f^{(-\gamma)},x), \tag{2}$$

for $f \in C$, $x \in \mathbb{R}$.

THEOREM 1. *Let* $p, \gamma \in \mathbb{N}$, $l \in \mathbb{Z}_+$, $\gamma \leqslant 2p$, $k = 2[(\gamma+1)/2]$, $0 \leqslant \alpha \leqslant 2p+2-k$, $h > 0$, $x \in \mathbb{R}$.

(i) *If* $f \in L_1$, *then*

$$S_{h,p,\gamma}(f,x) = \sum_{m=-\infty}^{+\infty} \lambda_{p,\gamma}(mh/2) c_m(f) e^{imx}. \tag{3}$$

(ii) *If* $f \in C^{(l)}$, *then*

$$S_{h,p,\gamma}(f) \in C^{(l+\gamma)}, \quad S_{h,p,\gamma}^{(l)}(f,x) = S_{h,p,\gamma}(f^{(l)},x).$$

(iii) *If* $f \in C_q$, *then*

$$\begin{aligned} &q(f - S_{h,p,\gamma}(f)) \leqslant \mathscr{A}(p,\gamma,\alpha) h^\alpha \Omega_{2p+2-k,\alpha}(f,h), \\ &q(f - S_{h,p,\gamma}(f)) \leqslant \mathscr{A}(p,\gamma) \omega_{2p+2-k}(f,h), \\ &q(f - S_{h,p,\gamma}(f)) \leqslant \gamma \int_0^1 q\Big(\sum_{m=-p}^{p} \mathscr{K}(p,m,\gamma) f(\cdot + mhu)\Big)(1-u)^{\gamma-1}\, du, \\ &q(S_{h,p,\gamma}^{(\gamma)}(f)) \leqslant \mathscr{D}(p,\gamma) h^{-\gamma} \overline{\omega}_\gamma(f,h). \end{aligned} \tag{4}$$

THEOREM 2. *Let* $p, \gamma \in \mathbb{N}$, $l \in \mathbb{Z}_+$, $\gamma \leqslant 2p-1$, $k = 2[\gamma/2]$, $0 \leqslant \alpha \leqslant 2p-1-k$, $h > 0$, $x \in \mathbb{R}$.

(i) *If* $f \in L_1$, *then*

$$\mathfrak{B}_{h,p,\gamma}(f,x) = \sum_{m=-\infty}^{+\infty} \mu_{p,\gamma}(mh/2) c_m(f) e^{imx}. \tag{5}$$

(ii) *If* $f \in C^{(l)}$, *then*

$$\mathfrak{B}_{h,p,\gamma}(f) \in C^{(l+\gamma)}, \quad \mathfrak{B}_{h,p,\gamma}^{(l)}(f,x) = \mathfrak{B}_{h,p,\gamma}(f^{(l)},x).$$

(iii) *If* $f \in C_q^{(1)}$, *then*

$$q(f - \mathfrak{B}_{h,p,\gamma}(f)) \leqslant \mathscr{B}(p,\gamma,\alpha)h^{\alpha+1}\Omega_{2p-1-k,\alpha}(f',h),$$

$$q(f - \mathfrak{B}_{h,p,\gamma}(f)) \leqslant \mathscr{B}(p,\gamma)h\omega_{2p-1-k}(f',h),$$

(6)
$$q(f - \mathfrak{B}_{h,p,\gamma}(f)) \leqslant h\int_0^1 q\Bigg(\sum_{m=-p}^{p-1}\mathfrak{L}(p,m,\gamma)f'(\cdot + (m+1/2)hu)\Bigg)(1-u)^\gamma\,du,$$

$$q(\mathfrak{B}_{h,p,\gamma}^{(\gamma)}(f)) \leqslant \mathscr{E}(p,\gamma)h^{-\gamma}\underline{\omega}_\gamma(f,h).$$

The proofs of Theorems 1 and 2 (we omit them in view of their simplicity) are based (with the exception of (3)–(6)) on the material presented in §4 and on equalities (1), (2). The relations (3) and (5) follow from the formula

$$\mathcal{S}_{2h,r}(f,x) = \sum_{k=-\infty}^{\infty}\left(\frac{\sin kh}{kh}\right)^r c_k(f)e^{ikx}$$

(see, e.g., [**2**, p. 177]). In the proof of (4) and (6) we use (3.12), (3.13), Corollary 3.6, and Lemma 2.A.

COROLLARY 1. *Let* $f \in C_q$, $h > 0$.
(i) *If* $\gamma/2 \in \mathbb{N}$, *then*

$$q(f - S_{h,\gamma-1,\gamma}(f)) \leqslant \mathscr{A}(\gamma-1,\gamma)\omega_\gamma(f,h),$$

$$q(S_{h,\gamma-1,\gamma}^{(\gamma)}(f)) \leqslant \mathscr{D}(\gamma-1,\gamma)h^{-\gamma}\omega_\gamma(f,h).$$

(ii) *If* $(\gamma+1)/2 \in \mathbb{N}$, *then*

$$q(f - S_{h,\gamma,\gamma}(f)) \leqslant \mathscr{A}(\gamma,\gamma)\omega_{\gamma+1}(f,h),$$

$$q(S_{h,\gamma,\gamma}^{(\gamma)}(f)) \leqslant \mathscr{D}(\gamma,\gamma)h^{-\gamma}\omega_\gamma^*(f,h).$$

REMARK 1. It is not hard to show that $\mathscr{A}(\gamma,\gamma+1) = \mathscr{A}(\gamma,\gamma)$ for $(\gamma+1)/2 \in \mathbb{N}$.

Direct calculations give (we write $\mathscr{A}(\gamma,\gamma) = A_\gamma$, $\mathscr{D}(\gamma,\gamma+1) = d_\gamma$, $\mathscr{D}(\gamma,\gamma) = \mathscr{D}_\gamma$): $A_1 = \frac{1}{2}$; $A_3 = \frac{121}{360}$; $A_5 < 0.1605181$; $A_7 < 0.07002784$; $A_9 < 2.980903 \cdot 10^{-2}$; $A_{11} < 1.2695365 \cdot 10^{-2}$; $A_{13} < 5.4631472 \cdot 10^{-3}$; $A_{15} < 2.3822723 \cdot 10^{-3}$; $A_{17} < 1.0522661 \cdot 10^{-3}$; $A_{19} < 4.70012 \cdot 10^{-4}$; $A_{21} < 2.1187047 \cdot 10^{-4}$; $A_{23} < 9.6206605 \cdot 10^{-5}$; $A_{25} < 4.3938352 \cdot 10^{-5}$; $A_{27} < 2.0158662 \cdot 10^{-5}$; $A_{29} < 9.2823394 \cdot 10^{-6}$; $A_{31} < 4.2867506 \cdot 10^{-6}$; $A_{33} < 1.9844739 \cdot 10^{-6}$; $A_{35} < 9.2052484 \cdot 10^{-7}$; $A_{37} < 4.2772573 \cdot 10^{-7}$; $A_{39} < 1.9903572 \cdot 10^{-7}$; $A_{41} < 9.2736387 \cdot 10^{-8}$; $A_{43} < 4.3257019 \cdot 10^{-8}$; $A_{45} < 2.0197431 \cdot 10^{-8}$; $A_{47} < 9.4389511 \cdot 10^{-9}$; $A_{49} < 4.4147051 \cdot 10^{-9}$; $A_{51} < 2.0663212 \cdot 10^{-9}$;

$d_1 = 1$; $d_3 = \frac{5}{3}$; $d_5 < 2.86667$; $d_7 < 5.02752$; $d_9 < 8.93228$; $d_{11} < 16.0179$; $d_{13} < 28.9261$; $d_{15} < 52.5237$; $d_{17} < 95.7946$; $d_{19} < 175.3516$; $d_{21} < 321.964$; $d_{23} < 592.704$; $d_{25} < 1093.57$; $d_{27} < 2021.66$; $d_{29} < 3743.87$; $d_{31} < 6943.85$; $d_{33} < 12896.7$; $d_{35} < 23982.2$; $d_{37} < 44646.3$; $d_{39} < 83200.1$; $d_{41} < 155191$; $d_{43} < 289718$; $d_{45} < 541285$; $d_{47} < 1012024$; $d_{49} < 1893415$; $d_{51} < 3544632$;

$\mathscr{D}_1 = 1$; $\mathscr{D}_3 = 2$; $\mathscr{D}_5 < 3.77778$; $\mathscr{D}_7 < 7.03704$; $\mathscr{D}_9 < 13.0528$; $\mathscr{D}_{11} < 24.1856$; $\mathscr{D}_{13} < 44.8226$; $\mathscr{D}_{15} < 83.1281$; $\mathscr{D}_{17} < 154.312$; $\mathscr{D}_{19} < 286.734$; $\mathscr{D}_{21} < 533.32$; $\mathscr{D}_{23} < 992.914$; $\mathscr{D}_{25} < 1850.26$; $\mathscr{D}_{27} < 3450.82$; $\mathscr{D}_{29} < 6441.14$; $\mathscr{D}_{31} < 12031.8$;

$\mathscr{D}_{33} < 22490.8$; $\mathscr{D}_{35} < 42069.3$; $\mathscr{D}_{37} < 78739.6$; $\mathscr{D}_{39} < 147460$; $\mathscr{D}_{41} < 276306$; $\mathscr{D}_{43} < 517997$; $\mathscr{D}_{45} < 971569$; $\mathscr{D}_{47} < 1823127$; $\mathscr{D}_{49} < 3422516$; $\mathscr{D}_{51} < 6427608$.

§6. Estimates from above of the values of functionals by means of moduli of continuity (model cases)

We denote by $\mathfrak{M}_q$ the set of functionals $\Phi\colon C_q \to \mathbb{R}_+$ such that $\Phi(f+g) \leqslant \Phi(f) + \Phi(g)$ for all $f, g \in C_q$. If $\Phi \in \mathfrak{M}_q$, then we set $m_l(\Phi) = \sup_{f \in C_q^{(l)}} \{\Phi(f)/q(f^{(l)})\}$.

1. Estimates of $\Phi(f^{(\gamma)} - U_{h.p.\gamma}(f))$, $\gamma = \overline{1,4}$, and of $\Phi(f^{(\gamma)} - V_{h.p.\gamma}(f))$, $\gamma = \overline{0,3}$.

LEMMA 1. *If $n \in \mathbb{Z}_+$, then*

$$\mathscr{T}_n'(0) = (-1)^n (n!)^2, \tag{1}$$

$$\mathscr{T}_n'''(0) = 6(-1)^{n+1}(n!)^2 \sum_{k=1}^{n} k^{-2}. \tag{2}$$

Also, if $n \in \mathbb{N}$, then

$$\Lambda_n(0) = (-1)^n 2^{-2n}((2n-1)!!)^2, \tag{3}$$

$$\Lambda_n''(0) = (-1)^{n+1} 2^{3-2n}((2n-1)!!)^2 \sum_{k=1}^{n} (2k-1)^{-2}. \tag{4}$$

This Lemma is verified by direct calculations.

LEMMA 2. *Let $\Phi \in \mathfrak{M}_q$, $p \in \mathbb{N}$, $\gamma \in \{3,4\}$, $h > 0$,*

$$g(u) = g(\Phi, u, h) = \min\{4m_2(\Phi), m_0(\Phi)(uh)^2\}.$$

Then for every $f \in C_q^{(\gamma)}$ the following inequality holds:

$$\begin{aligned} \Phi(f^{(\gamma)} - U_{h.p.\gamma}(f)) &\leqslant \left(\gamma \Big/ \binom{2p}{p}\right) \int_0^1 \omega_{2p-2}(f^{(\gamma)}, uh) \\ &\times \left\{4m_0(\Phi)(1-u)^{\gamma-1} + (\gamma-1)(\gamma-2)\Big(\sum_{k=1}^{p} k^{-2}\Big) g(u) h^{-2}(1-u)^{\gamma-3}\right\} du. \end{aligned} \tag{5}$$

PROOF. By (4.1)

$$\begin{aligned} f^{(\gamma)}(x) - U_{h.p.\gamma}(f,x) &= ((2p)!)^{-1} \sum_{l=1}^{2} \binom{\gamma}{2l-1} h^{2-2l} \\ &\times \left(\int_0^1 \delta_{uh}^{2p}(f^{(\gamma+2-2l)}, x)(1-u)^{\gamma+1-2l}\, du\right) \mathscr{T}_p^{(2l-1)}(0). \end{aligned} \tag{6}$$

Set $\alpha_l = ((2p)!)^{-1} \binom{\gamma}{2l-1} h^{2-2l} \mathscr{T}_P^{(2l-1)}(0)$. Denote by a the largest $u \in [0,1]$ such that

$m_0(\Phi)(uh)^2 \leqslant 4m_2(\Phi)$. Using Lemma 2.A and Property 2.M.6 we obtain

$$\begin{aligned} A &= \Phi(\alpha_1 \int_0^1 \delta_{uh}^{2p}(f^{(\gamma)}, \cdot)(1-u)^{\gamma-1}\,du) \\ &\leqslant m_0(\Phi) q(\alpha_1 \int_0^1 \delta_{uh}^{2p}(f^{(\gamma)}, \cdot)(1-u)^{\gamma-1}\,du) \\ &\leqslant m_0(\Phi)|\alpha_1| \int_0^1 q(\delta_{uh}^{2p}(f^{(\gamma)}, \cdot))(1-u)^{\gamma-1}\,du \\ &\leqslant m_0(\Phi)|\alpha_1| \int_0^1 \omega_{2p}(f^{(\gamma)}, uh)(1-u)^{\gamma-1}\,du \\ &\leqslant 4m_0(\Phi)|\alpha_1| \int_0^1 \omega_{2p-2}(f^{(\gamma)}, uh)(1-u)^{\gamma-1}\,du. \end{aligned} \tag{7}$$

Next, apply Lemma 2.A and Properties 2.M.7, 2.M.6 of moduli of continuity to get

$$\begin{aligned} B &= \Phi(\alpha_2 \int_0^1 \delta_{uh}^{2p}(f^{(\gamma-2)}, \cdot)(1-u)^{\gamma-3}\,du) \\ &\leqslant |\alpha_2| \Big\{ m_0(\Phi) q\Big(\int_0^a \delta_{uh}^{2p}(f^{(\gamma-2)}, \cdot)(1-u)^{\gamma-3}\,du\Big) \\ &\qquad + m_2(\Phi) q\Big(\int_a^1 \delta_{uh}^{2p}(f^{(\gamma)}, \cdot)(1-u)^{\gamma-3}\,du\Big)\Big\} \\ &\leqslant |\alpha_2| \Big\{ m_0(\Phi) \int_0^a \omega_{2p-2}(f^{(\gamma)}, uh)(uh)^2(1-u)^{\gamma-3}\,du \\ &\qquad + 4m_2(\Phi) \int_a^1 \omega_{2p-2}(f^{(\gamma)}, uh)(1-u)^{\gamma-3}\,du \Big\} \\ &= |\alpha_2| \int_0^1 \omega_{2p-2}(f^{(\gamma)}, uh) g(u)(1-u)^{\gamma-3}\,du. \end{aligned} \tag{8}$$

It follows from (6) that $\Phi(f^{(\gamma)} - U_{h,p,\gamma}(f)) \leqslant A + B$. Now we apply (7), (8) and recall (1), (2), which gives (5). □

LEMMA 3. *Let* $\Phi \in \mathfrak{M}_q$, $p-1 \in \mathbb{N}$, $\gamma \in \{2,3\}$, $h > 0$,

$$g(u) = g(\Phi, u, h) = \min\{4m_2(\Phi), m_0(\Phi)(uh)^2\}.$$

Then for all $f \in C_q^{(\gamma+1)}$

(9)

$$\Phi(f^{(\gamma)} - V_{h,p,\gamma}(f)) \leqslant \frac{((2p-1)!!)^2 2^{2-2p}}{(2p-1)!} h \int_0^1 \omega_{2p-3}(f^{(\gamma+1)}, uh) \times \Big\{ m_0(\Phi)(1-u)^{\gamma} + \gamma(\gamma-1)\Big(\sum_{k=1}^{p}(2k-1)^{-2}\Big) g(u) h^{-2} (1-u)^{\gamma-2} \Big\}\,du.$$

PROOF. According to (4.4)

$$f^{(\gamma)}(x) - V_{h.p.\gamma}(f, x) = \frac{1}{(2p-1)!} \times \sum_{l=0}^{1} \binom{\gamma}{2l} \left(\int_0^1 \delta_{uh}^{2p-1}(f^{(\gamma+1-2l)}, x)(1-u)^{\gamma-2l}\, du \right) h^{1-2l} \Lambda_p^{(2l)}(0). \tag{10}$$

Set $\alpha_l = ((2p-1)!)^{-1}\binom{\gamma}{2l} h^{1-2l}\Lambda_p^{(2l)}(0)$. Denote by a the largest $u \in [0,1]$ such that $m_0(\Phi)(uh)^2 \leqslant 4m_2(\Phi)$. By Lemma 2.A and Property 2.M.6 we obtain

$$\begin{aligned} A &= \Phi\left(\alpha_0 \int_0^1 \delta_{uh}^{2p-1}(f^{(\gamma+1)}, \cdot)(1-u)^{\gamma}\, du\right) \\ &\leqslant m_0(\Phi) q\left(\alpha_0 \int_0^1 \delta_{uh}^{2p-1}(f^{(\gamma+1)}, \cdot)(1-u)^{\gamma}\, du\right) \\ &\leqslant m_0(\Phi)|\alpha_0| \int_0^1 q(\delta_{uh}^{2p-1}(f^{(\gamma+1)}, \cdot))(1-u)^{\gamma}\, du \\ &\leqslant m_0(\Phi)|\alpha_0| \int_0^1 \omega_{2p-1}(f^{(\gamma+1)}, uh)(1-u)^{\gamma}\, du \\ &\leqslant 4m_0(\Phi)|\alpha_0| \int_0^1 \omega_{2p-3}(f^{(\gamma+1)}, uh)(1-u)^{\gamma}\, du. \end{aligned} \tag{11}$$

Next, apply Lemma 2.A and Properties 2.M.7, 2.M.6 of moduli of continuity to get

$$\begin{aligned} B &= \Phi\left(\alpha_1 \int_0^1 \delta_{uh}^{2p-1}(f^{(\gamma-1)}, \cdot)(1-u)^{\gamma-2}\, du\right) \\ &\leqslant |\alpha_1| \left\{ m_0(\Phi) q\left(\int_0^a \delta_{uh}^{2p-1}(f^{(\gamma-1)}, \cdot)(1-u)^{\gamma-2}\, du\right) \right. \\ &\qquad \left. + m_2(\Phi) q\left(\int_a^1 \delta_{uh}^{2p-1}(f^{(\gamma+1)}, \cdot)(1-u)^{\gamma-2}\, du\right)\right\} \\ &\leqslant |\alpha_1| \left\{ m_0(\Phi) \int_0^a \omega_{2p-3}(f^{(\gamma+1)}, uh)(uh)^2(1-u)^{\gamma-2}\, du \right. \\ &\qquad \left. + 4m_2(\Phi) \int_a^1 \omega_{2p-3}(f^{(\gamma+1)}, uh)(1-u)^{\gamma-2}\, du \right\} \\ &= |\alpha_1| \int_0^1 \omega_{2p-3}(f^{(\gamma+1)}, uh) g(u)(1-u)^{\gamma-2}\, du. \end{aligned} \tag{12}$$

It follows from (10) that $\Phi(f^{(\gamma)} - V_{h.p.\gamma}(f)) \leqslant A + B$. Finally, applying (11), (12) with the help of (3), (4) we come to (9). □

COROLLARY 1. *Let* $\Phi \in \mathfrak{M}_q$, $p \in \mathbb{N}$, $\gamma \in \{3,4\}$, $h > 0$, $b \in [0,1]$. *Then for an arbitrary* $f \in C_q^{(\gamma)}$

$$\Phi(f^{(\gamma)} - U_{h.p.\gamma}(f))$$

$$\leqslant \frac{\omega_{2p-2}(f^{(\gamma)},h)}{\binom{2p}{p}}\Big\{m_0(\Phi)\Big(4+2\binom{\gamma}{3}b^3\Big(1-\frac{3(\gamma-3)b}{4}\Big)\sum_{k=1}^{p}\frac{1}{k^2}\Big) \tag{13}$$

$$+\binom{\gamma}{3}\frac{24m_2(\Phi)(1-b)^{\gamma-2}}{(\gamma-2)h^2}\sum_{k=1}^{p}\frac{1}{k^2}\Big\}.$$

PROOF. It follows from (5) that

$$\Phi(f^{(\gamma)} - U_{h.p.\gamma}(f)) \leqslant \omega_{2p-2}(f^{(\gamma)},h)\left(\gamma \Big/ \binom{2p}{p}\right)$$

$$\times\Big\{4m_0(\Phi)\int_0^1(1-u)^{\gamma-1}\,du + (\gamma-1)(\gamma-2)\Big(\sum_{k=1}^{p}k^{-2}\Big)h^{-2}\int_0^1 g(u)(1-u)^{\gamma-3}\,du\Big\}.$$

Now, since $\int_0^1(1-u)^{\gamma-1}\,du = 1/\gamma$, we have

$$\int_0^1 g(u)(1-u)^{\gamma-3}\,du$$

$$\leqslant m_0(\Phi)h^2\int_0^b u^2(1-u)^{\gamma-3}\,du + 4m_2(\Phi)\int_b^1(1-u)^{\gamma-3}\,du$$

$$= \frac{m_0(\Phi)h^2b^3}{3}\{1-\tfrac{3}{4}(\gamma-3)b\} + \frac{4m_2(\Phi)}{\gamma-2}(1-b)^{\gamma-2}.$$

It remains to compare these relations. □

COROLLARY 2. *Let* $\Phi \in \mathfrak{M}_q$, $p-1 \in \mathbb{N}$, $\gamma \in \{2,3\}$, $h > 0$, $b \in [0,1]$. *Then for an arbitrary* $f \in C_q^{(\gamma+1)}$

$$\Phi(f^{(\gamma)} - V_{h.p.\gamma}(f)) \leqslant h\omega_{2p-3}(f^{(\gamma+1)},h)\frac{((2p-1)!!)^2 2^{2-2p}}{(2p-1)!}$$

$$\times\Big\{m_0(\Phi)\Big\{\frac{1}{\gamma+1}+\frac{\gamma(\gamma-1)b^3}{3}\Big(1-\frac{3(\gamma-2)b}{4}\Big)\sum_{k=1}^{p}\frac{1}{(2k-1)^2}\Big\} \tag{14}$$

$$+4\gamma m_2(\Phi)h^{-2}(1-b)^{\gamma-1}\sum_{k=1}^{p}(2k-1)^{-2}\Big\}.$$

PROOF. It follows from (9) that

$$\Phi(f^{(\gamma)} - V_{h.p.\gamma}(f)) \leqslant h\omega_{2p-3}(f^{(\gamma+1)},h)\frac{((2p-1)!!)^2 2^{2-2p}}{(2p-1)!}$$

$$\times\Big\{m_0(\Phi)\int_0^1(1-u)^{\gamma}\,du$$

$$+\gamma(\gamma-1)\Big(\sum_{k=1}^{p}(2k-1)^{-2}\Big)h^{-2}\int_0^1 g(u)(1-u)^{\gamma-2}\,du\Big\}.$$

Now, since $\int_0^1 (1-u)^\gamma\,du = (\gamma+1)^{-1}$, we have

$$\begin{aligned}\int_0^1 & g(u)(1-u)^{\gamma-2}\,du \\ &\leqslant m_0(\Phi)h^2\int_0^b u^2(1-u)^{\gamma-2}\,du + 4m_2(\Phi)\int_b^1(1-u)^{\gamma-2}\,du \\ &= \frac{m_0(\Phi)h^2b^3}{3}\{1-\tfrac{3}{4}(\gamma-2)b\} + \frac{4m_2(\Phi)}{\gamma-1}(1-b)^{\gamma-1}.\end{aligned}$$

It remains to compare these relations. □

REMARK 1. Let $a = (4m_2(\Phi))^{1/2}(m_0(\Phi)h^2)^{-1/2}$. The best way to apply the estimates (13), (14) is the following: if $a \geqslant 1$, one must set $b = 1$ there, and if $a < 1$, then take $b = a$.

LEMMA 4. *Let* $\Phi \in \mathfrak{M}_q$, $p \in \mathbb{Z}_+$, $\gamma \in \{1,2\}$, $h > 0$, $f \in C_q^{(\gamma)}$. *Then*

$$\Phi(f^{(\gamma)} - U_{h,p,\gamma}(f)) \leqslant m_0(\Phi)\left(\gamma \Big/ \binom{2p}{p}\right)\int_0^1 \omega_{2p}(f^{(\gamma)}, uh)(1-u)^{\gamma-1}\,du.$$

LEMMA 5. *Let* $\Phi \in \mathfrak{M}_q$, $p \in \mathbb{N}$, $\gamma \in \{0,1\}$, $h > 0$, $f \in C_q^{(\gamma+1)}$. *Then*

$$\begin{aligned}\Phi(f^{(\gamma)} &- V_{h,p,\gamma}(f)) \\ &\leqslant m_0(\Phi)\frac{h((2p-1)!!)^2}{(2p-1)!2^{2p}}\int_0^1 \omega_{2p-1}(f^{(\gamma+1)}, uh)(1-u)^\gamma\,du.\end{aligned}$$

These two lemmas are proved in the same way as Lemmas 2, 3 (and even somewhat more easily).

2. Estimates of $\Phi(U_{h,3,\gamma}(f))$, $\gamma = \overline{3,4}$, and of $\Phi(V_{h,3,\gamma}(f))$, $\gamma = \overline{2,3}$.

LEMMA 6. *Let* $\Phi \in \mathfrak{M}_q$, $r \in \mathbb{Z}_+$, $h > 0$, $\Omega_r = \{0,1,2\} \cap \{0,1,\dots,r\}$, $f \in C_q^{(r)}$. *Then*

$$\begin{aligned}\Phi(U_{h,3,4}(f)) &\leqslant \omega_4(f^{(r)}, h)h^{-4} \\ &\quad\times \inf_{\gamma\in\Omega_r}\{m_r(\Phi) + (2^{2-\gamma}/6)m_{r-\gamma}(\Phi)h^\gamma\},\end{aligned} \tag{15}$$

$$\begin{aligned}\Phi(U_{h,3,3}(f)) &\leqslant \omega_3^*(f^{(r)}, h)h^{-3} \\ &\quad\times \inf_{\gamma\in\Omega_r}\{m_r(\Phi) + (2^{-\gamma})m_{r-\gamma}(\Phi)h^\gamma\}.\end{aligned} \tag{16}$$

PROOF. On the basis of (4.7) we have

$$U_{h,3,4}(f,x) = h^{-4}\{\Delta_h^4(f, x-2h) - (1/6)\Delta_h^6(f, x-3h)\}.$$

Together with Properties 2.M.7 and 2.M.6 of moduli of continuity this implies that for every $\gamma \in \Omega_r$

$$\begin{aligned}\Phi(U_{h,3,4}(f)) &\leqslant h^{-4}\{m_r(\Phi)q(\Delta_h^4(f^{(r)})) + (m_{r-\gamma}(\Phi)/6)q(\Delta_h^6(f^{(r-\gamma)}))\} \\ &\leqslant h^{-4}\{m_r(\Phi)\omega_4(f^{(r)}, h) + (m_{r-\gamma}(\Phi)h^\gamma/6)\omega_{6-\gamma}(f^{(r)}, h)\} \\ &\leqslant \omega_4(f^{(r)}, h)h^{-4}\{m_r(\Phi) + (2^{2-\gamma}m_{r-\gamma}(\Phi)h^\gamma/6)\}.\end{aligned}$$

Thus (15) is established. Now pass to (16). On the basis of (4.8) we have

$$U_{h.3.3}(f,x) = h^{-3}\{2^{-1}(\Delta_h^3(f,x-h)+\Delta_h^3(f,x-2h)) \\ -(1/8)(\Delta_h^5(f,x-2h)+\Delta_h^5(f,x-3h))\}.$$

Together with Properties 2.M*.7 and 2.M*.6 of modified moduli of continuity this gives that for every $\gamma \in \Omega_r$

$$\begin{aligned}\Phi(U_{h.3.3}(f)) &\leqslant h^{-3}\{m_r(\Phi)\omega_3^*(f^{(r)},h)+(m_{r-\gamma}(\Phi)/4)\omega_5^*(f^{(r-\gamma)},h)\} \\ &\leqslant h^{-3}\{m_r(\Phi)\omega_3^*(f^{(r)},h)+(m_{r-\gamma}(\Phi)h^\gamma/4)\omega_{5-\gamma}^*(f^{(r)},h)\} \\ &\leqslant \omega_3^*(f^{(r)},h)h^{-3}\{m_r(\Phi)+2^{-\gamma}m_{r-\gamma}(\Phi)h^\gamma\}.\end{aligned}$$ □

LEMMA 7. *Let* $\Phi \in \mathfrak{M}_q$, $r \in \mathbb{Z}_+$, $h > 0$, $\Omega_r = \{0,1,2\} \cap \{0,1,\dots,r\}$, $f \in C_q^{(r)}$. *Then*

$$(17)\quad \begin{aligned}\Phi(V_{h.3.3}(f)) &\leqslant \omega_3(f^{(r)},h)h^{-3}\inf_{\gamma\in\Omega_r}\{m_r(\Phi)+2^{-\gamma-1}m_{r-\gamma}(\Phi)h^\gamma\}, \\ \Phi(V_{h.3.2}(f)) &\leqslant \omega_2^*(f^{(r)},h)h^{-2}\inf_{\gamma\in\Omega_r}\{m_r(\Phi)+(5/6)2^{-\gamma}m_{r-\gamma}(\Phi)h^\gamma\}.\end{aligned}$$

The proof of this lemma is similar to that of Lemma 6.

3. Inequalities of Jackson type.

THEOREM 1. *Let* $\Phi \in \mathfrak{M}_q$, $h > 0$, $\Omega = \{0,1,2\}$,

$$g(u,h) = \min\{4m_2(\Phi), m_0(\Phi)(uh)^2\}.$$

Then for an arbitrary $f \in C_q$

$$(18)\quad \begin{aligned}\Phi(f) \leqslant \frac{1}{5}\int_0^1 &\omega_4(f,uh)\{4m_0(\Phi)(1-u)^3+(49/6)g(u,h)h^{-2}(1-u)\}\,du \\ &+\omega_4(f,h)h^{-4}\inf_{\gamma\in\Omega}\{m_4(\Phi)+(2^{2-\gamma}/6)m_{4-\gamma}(\Phi)h^\gamma\}.\end{aligned}$$

PROOF. If we suppose that $f \in C_q^{(4)}$, we can use the inequality $\Phi(f^{(4)}) \leqslant \Phi(f^{(4)} - U_{h.3.4}(f)) + \Phi(U_{h.3.4}(f))$ and apply (5) (for $p = 3$, $\gamma = 4$) and (15) (with $r = 4$) which gives us the inequality (we denote it by (18′)) obtained from (18) by replacing f by $f^{(4)}$.

Now, on the basis of (18′), we establish (18). If $m_\gamma(\Phi) = \infty$ for all $\gamma \in \{2,3,4\}$, then (18) is obvious. So we can suppose that $m_\gamma(\Phi)$ is finite for at least one $\gamma \in \{2,3,4\}$. In this case $\Phi(l) = 0$ for every constant function l. Together with the semiadditivity of Φ this implies that $\Phi(f) = \Phi(c_0(f) + (f^{(-4)})^{(4)}) = \Phi((f^{(-4)})^{(4)})$. Now obtain (18) by applying (18′) to the function $f^{(-4)}$. □

REMARK 2. Inequality (18′) is equivalent to (18) restricted to the set of functions $f \in C$ such that $\int_{-\pi}^{\pi} f(t)\,dt = 0$. The way of passing from (18′) to (18) given above in full is typical. Therefore later we shall not give corresponding reasoning in similar situations and restrict ourselves merely to the task of establishing inequalities of the type (18′).

THEOREM 2. *Let* $\Phi \in \mathfrak{M}_q$, $h > 0$, $\Omega = \{0, 1, 2\}$,

$$g(u,h) = \min\{4m_2(\Phi), m_0(\Phi)(uh)^2\}.$$

Then for every $f \in C_q^{(1)}$

$$\begin{aligned}\Phi(f) \leqslant h \int_0^1 \omega_3(f', uh)\{(15/128)m_0(\Phi)(1-u)^3 \\ + (259/320)g(u,h)h^{-2}(1-u)\}\, du \\ + \omega_3(f', h)h^{-3} \inf_{\gamma\in\Omega}\{m_4(\Phi) + 2^{-\gamma-1}m_{4-\gamma}(\Phi)h^\gamma\}.\end{aligned} \tag{19}$$

PROOF. Suppose first that $f \in C_q^{(4)}$. Taking into account that

$$\Phi(f^{(3)}) \leqslant \Phi(f^{(3)} - V_{h,3,3}(f)) + \Phi(V_{h,3,3}(f)),$$

and applying (9) (with $p = \gamma = 3$) and (17), we arrive at an inequality (which we denote by (19′)), obtained from (19) by replacing f by $f^{(4)}$. Now it remains to recall Remark 2. □

COROLLARY 3. *Let* $\Phi \in \mathfrak{M}_q$, $h > 0$, $b \in [0, 1]$. *Then for* $f \in C_q$

$$\Phi(f) \leqslant A(h,b)\omega_4(f,h), \tag{20}$$

where

$$\begin{aligned}A(h,b) = m_0(\Phi)((1/5) + (49/360)b^3(4-3b)) \\ + (49/15)m_2(\Phi)h^{-2}(1-b)^2 \\ + h^{-4} \inf_{\gamma\in\{0,1,2\}} (m_4(\Phi) + (2^{2-\gamma}/6)m_{4-\gamma}(\Phi)h^\gamma).\end{aligned}$$

PROOF. Inequality (20) follows from (18) if we apply to it the estimate $\omega_4(f, uh) \leqslant \omega_4(f, h)$ for $u \in [0, 1]$ and extract $\omega_4(f, h)$ from under the integral sign (see also Corollary 9 for $p = 3$, $\gamma = 4$). □

COROLLARY 4. *Let* $\Phi \in \mathfrak{M}_q$, $h > 0$, $b \in [0, 1]$. *Then for* $f \in C_q^{(1)}$

$$\begin{aligned}\Phi(f) \leqslant h\omega_3(f', h)\Big\{m_0(\Phi)(15/512 + (259/3840)b^3(4-3b)) \\ + (259/160)m_2(\Phi)h^{-2}(1-b)^2 \\ + h^{-4} \inf_{\gamma\in\{0,1,2\}} (m_4(\Phi) + 2^{-\gamma-1}m_{4-\gamma}(\Phi)h^\gamma)\Big\}.\end{aligned} \tag{21}$$

PROOF. Inequality (21) follows from (19) if we apply to it the estimate $\omega_3(f', uh) \leqslant \omega_3(f', h)$ for $u \in [0, 1]$ and extract $\omega_3(f', h)$ from under the integral sign (see also Corollary 2 for $\gamma = p = 3$). □

4. Inequalities of Jackson type for the best approximations.

THEOREM 3. *Let* $n \in \mathbb{Z}_+$, $f \in C_q$. *Then*
(i) *for* $0 < \gamma \leqslant \sqrt{2}$

$$E_n(f) \leqslant \Big(\frac{121}{360} + \frac{15 + 2\min\{5, 3\gamma^2\}}{72\gamma^4}\Big)\omega_4\Big(f, \frac{\pi\gamma}{2(n+1)}\Big); \tag{22}$$

(ii) *for* $\gamma \geqslant \sqrt{2}$

$$E_n(f) \leqslant \Big(\frac{1}{5} + \frac{588\gamma^2 - 784\sqrt{2}\gamma + 713}{360\gamma^4}\Big)\omega_4\Big(f, \frac{\pi\gamma}{2(n+1)}\Big). \tag{23}$$

In order to prove Theorem 3 we set $\Phi(f) = E_n(f)$, $h = \pi\gamma/(2n+2)$, $b = 1$ if $\gamma \leqslant \sqrt{2}$, $b = \sqrt{2}/\gamma$ if $\gamma \geqslant \sqrt{2}$ in Corollary 3 and recall that in this case

$$\begin{aligned} m_0(\Phi) &\leqslant 1, & m_2(\Phi) &\leqslant \frac{\pi^2}{8(n+1)^2}, \\ m_3(\Phi) &\leqslant \frac{\pi^3}{24(n+1)^3}, & m_4(\Phi) &\leqslant \frac{5\pi^4}{384(n+1)^4} \end{aligned} \tag{24}$$

(see §2).

COROLLARY 5. *Let* $n \in \mathbb{Z}_+$, $f \in C_q$. *Then*

$$E_n(f) \leqslant \frac{113}{80}\omega_4\Big(f, \frac{\pi}{2(n+1)}\Big), \tag{25}$$

$$E_n(f) \leqslant \frac{25}{72}\omega_4\Big(f, \frac{\pi}{n+1}\Big). \tag{26}$$

The estimate (25) is obtained from (22) for $\gamma = 1$; to establish (26) set $\gamma = 2$ in (23).

THEOREM 4. *Let* $n \in \mathbb{Z}_+$, $f \in C_q^{(1)}$. *Then*
(i) *for* $0 < \gamma \leqslant \sqrt{2}$

$$E_n(f) \leqslant \frac{\pi\gamma}{16(n+1)}\Big(\frac{743}{5!8} + \frac{10 + \min\{5, 3\gamma^2\}}{6\gamma^4}\Big)\omega_3\Big(f', \frac{\pi\gamma}{2(n+1)}\Big);$$

(ii) *for* $\gamma \geqslant \sqrt{2}$

$$E_n(f) \leqslant \frac{\pi}{16(n+1)}\Big(\frac{15\gamma}{64} + \frac{259}{40\gamma} - \frac{259\sqrt{2}}{30\gamma^2} + \frac{359}{40\gamma^3}\Big)\omega_3\Big(f', \frac{\pi\gamma}{2(n+1)}\Big).$$

To prove Theorem 4 set $\Phi(f) = E_n(f)$, $h = \pi\gamma/(2n+2)$; $b = 1$ if $\gamma \leqslant \sqrt{2}$ and $b = \sqrt{2}/\gamma$ if $\gamma \geqslant \sqrt{2}$ in Corollary 4 and use (24).

5. An addition to Property 2.M.5 of moduli of continuity. In view of Property 2.M.5 of moduli of continuity, for $f \in C_q$, $h, \lambda \in \mathbb{R}_+$ the inequality $\omega_4(f, \lambda h) \leqslant ([\lambda] + 1)^4\omega_4(f, h)$ holds. The following theorem supplements it.

THEOREM 5. *Let* $\lambda \geqslant 1$, $h \in \mathbb{R}_+$. *Then for* $f \in C_q$

$$\omega_4(f, \lambda h) \leqslant (\lambda^4 + (2/3)\lambda^2 + (242/45))\omega_4(f, h).$$

To prove Theorem 5 set $\Phi(f) = \omega_4(f, \lambda h)$, $b = 1$ in Corollary 3 and recall that $m_k(\Phi) \leqslant 2^{4-k}(\lambda h)^k$ by virtue of Properties 2.M.6 and 2.M.7 of moduli of continuity.

§7. Upper estimates of the values of functionals in terms of moduli of continuity (the general case)

1. Estimates of $\Phi(f^{(\gamma)} - U_{h.p.\gamma}(f))$ and $\Phi(f^{(\gamma)} - V_{h.p.\gamma}(f))$.

LEMMA 1. *Let $p, \gamma \in \mathbb{N}$, $\gamma \leqslant 2p+1$, $0 \leqslant \alpha \leqslant 2p$, $h > 0$, $\Phi \in \mathfrak{M}_q$,*

$$(1)\quad Z(p,\gamma,h,\alpha) := \frac{1}{(2p)!}\sum_{l=1}^{\gamma}\binom{\gamma}{l} B(\alpha+1, \gamma-l+1)|\mathscr{T}_p^{(l)}(0)|m_{l-1}(\Phi)h^{\alpha+1-l},$$

where, as usual, $B(a,b)$ is the Beta function. Then for all $f \in C_q^{(\gamma)}$

$$(2)\quad \begin{aligned}\Phi(f^{(\gamma)} - U_{h.p.\gamma}(f)) &\leqslant \frac{1}{(2p)!}\\ &\times \int_0^1 \omega_{2p}(f^{(\gamma)}, uh)\Big(\sum_{l=1}^{\gamma}\binom{\gamma}{l} m_{l-1}(\Phi)h^{1-l}|\mathscr{T}_p^{(l)}(0)|(1-u)^{\gamma-l}\Big)\,du,\end{aligned}$$

$$(3)\quad \Phi(f^{(\gamma)} - U_{h.p.\gamma}(f)) \leqslant Z(p,\gamma,h,\alpha)\Omega_{2p.\alpha}(f^{(\gamma)}, h).$$

PROOF. Set $\alpha_l = ((2p)!)^{-1}\binom{\gamma}{l}h^{1-l}\mathscr{T}_p^{(l)}(0)$. Formula (4.1) gives

$$\begin{aligned}\Phi(f^{(\gamma)} - U_{h.p.\gamma}(f)) &\leqslant \sum_{l=1}^{\gamma}\Phi\Big(\alpha_l\int_0^1 \delta_{uh}^{2p}(f^{(\gamma+1-l)}, \cdot)(1-u)^{\gamma-l}\,du\Big)\\ &\leqslant \sum_{l=1}^{\gamma}|\alpha_l| m_{l-1}(\Phi) q\Big(\int_0^1 \delta_{uh}^{2p}(f^{(\gamma)}, \cdot)(1-u)^{\gamma-l}\,du\Big).\end{aligned}$$

Applying Lemma 2.A we obtain

$$\begin{aligned}\Phi(f^{(\gamma)} - U_{h.p.\gamma}(f)) &\leqslant \sum_{l=1}^{\gamma}|\alpha_l| m_{l-1}(\Phi)\int_0^1 q(\Delta_{uh}^{2p}(f^{(\gamma)}))(1-u)^{\gamma-l}\,du\\ &\leqslant \sum_{l-1}^{\gamma}|\alpha_l| m_{l-1}(\Phi)\int_0^1 \omega_{2p}(f^{(\gamma)}, uh)(1-u)^{\gamma-l}\,du.\end{aligned}$$

Thus, we have proved (2). Now

$$\begin{aligned}\int_0^1 &\omega_{2p}(f^{(\gamma)}, uh)(1-u)^{\gamma-l}\,du\\ &\leqslant h^{\alpha}\Omega_{2p.\alpha}(f^{(\gamma)}, h)\int_0^1 u^{\alpha}(1-u)^{\gamma-l}\,du\\ &= B(\alpha+1, \gamma-l+1)h^{\alpha}\Omega_{2p.\alpha}(f^{(\gamma)}, h).\end{aligned}$$

Together with (2) this gives (3). □

LEMMA 2. *Let* $p \in \mathbb{N}$, $\gamma \in \mathbb{Z}_+$, $\gamma \leqslant 2p$, $0 \leqslant \alpha \leqslant 2p-1$, $h > 0$, $\Phi \in \mathfrak{M}_q$,

$$\Upsilon(p,\gamma,h,\alpha) := \frac{1}{(2p-1)!}\sum_{l=0}^{\gamma}\binom{\gamma}{l}B(\alpha+1,\gamma-l+1)|\Lambda_p^{(l)}(0)|m_l(\Phi)h^{\alpha+1-l}; \tag{4}$$

then for all $f \in C_q^{(\gamma+1)}$

$$\begin{aligned}\Phi(f^{(\gamma)} - V_{h.p.\gamma}(f)) &\leqslant \frac{1}{(2p-1)!}\\ &\times \int_0^1 \omega_{2p-1}(f^{(\gamma+1)},uh)\left(\sum_{l=0}^{\gamma}\binom{\gamma}{l}m_l(\Phi)h^{1-l}|\Lambda_p^{(l)}(0)|(1-u)^{\gamma-l}\right)du,\end{aligned} \tag{5}$$

$$\Phi(f^{(\gamma)} - V_{h.p.\gamma}(f)) \leqslant \Upsilon(p,\gamma,h,\alpha)\Omega_{2p-1.\alpha}(f^{(\gamma+1)},h). \tag{6}$$

PROOF. Set $\alpha_l = ((2p-1)!)^{-1}\binom{\gamma}{l}h^{1-l}\Lambda_p^{(l)}(0)$. Formula (4.4) gives

$$\begin{aligned}\Phi(f^{(\gamma)} - V_{h.p.\gamma}(f)) &\leqslant \sum_{l=0}^{\gamma}\Phi(\alpha_l\int_0^1 \delta_{uh}^{2p-1}(f^{(\gamma+1-l)},\cdot)(1-u)^{\gamma-l}\,du)\\ &\leqslant \sum_{l=0}^{\gamma}|\alpha_l|m_l(\Phi)q\left(\int_0^1 \delta_{uh}^{2p-1}(f^{(\gamma+1)},\cdot)(1-u)^{\gamma-l}\,du\right).\end{aligned}$$

Applying Lemma 2.A we obtain

$$\Phi(f^{(\gamma)} - V_{h.p.\gamma}(f)) \leqslant \sum_{l=0}^{\gamma}|\alpha_l|m_l(\Phi)\int_0^1 \omega_{2p-1}(f^{(\gamma+1)},uh)(1-u)^{\gamma-l}\,du.$$

Thus, we have proved (5). Now

$$\begin{aligned}&\int_0^1 \omega_{2p-1}(f^{(\gamma+1)},uh)(1-u)^{\gamma-l}\,du\\ &\qquad\leqslant h^{\alpha}\Omega_{2p-1.\alpha}(f^{(\gamma+1)},h)\int_0^1 u^{\alpha}(1-u)^{\gamma-l}\,du\\ &\qquad= B(\alpha+1,\gamma-l+1)h^{\alpha}\Omega_{2p-1.\alpha}(f^{(\gamma+1)},h).\end{aligned}$$

Together with (5) this gives (6). □

2. Estimates of $\Phi(U_{h.p.\gamma}(f))$ and $\Phi(V_{h.p.\gamma}(f))$.

LEMMA 3. *Let* $p, \gamma \in \mathbb{N}$, $\gamma \leqslant 2p$, $k = 2[(\gamma+1)/2]$, $s = 2[\gamma/2]$, $h > 0$, $\Phi \in \mathfrak{M}_q$,

$$a(\gamma, l, h) := \begin{cases} \min\{m_\gamma(\Phi)2^{2l-s}, m_{s+\gamma-2l}(\Phi)h^{2l-s}\}, \\ \qquad \text{for } l = \overline{s/2, \gamma-1}, \\ \min\{m_\gamma(\Phi)2^{2l-s}, m_0(\Phi)2^{2l-s-\gamma}h^\gamma\}, \\ \qquad \text{for } l \geqslant \gamma; \end{cases}$$

$$b(p, \gamma, h) := \begin{cases} \sum_{l=\gamma/2}^{p} \frac{|Q_l^{(\gamma)}(0)|}{(2l)!} a(\gamma, l, h), & \text{for } \frac{\gamma}{2} \in \mathbb{N}, \\ \sum_{l=(\gamma-1)/2}^{p-1} \frac{|\mathcal{T}_l^{(\gamma)}(0)|}{(2l+1)!} a(\gamma, l, h), & \text{for } \frac{\gamma+1}{2} \in \mathbb{N}. \end{cases}$$

Then for all $f \in C_q$

$$\Phi(U_{h.p.\gamma}(f)) \leqslant b(p, \gamma, h)h^{-\gamma}\overline{\omega}_\gamma(f^{(\gamma)}, h). \tag{7}$$

PROOF. First let $\gamma/2 \in \mathbb{N}$. Set $\alpha_l = h^{-\gamma}Q_l^{(\gamma)}(0)/((2l)!)$, $\chi_l(f, x) = \alpha_l \delta_h^{2l}(f, x)$. We are going to estimate $\Phi(\chi_l(f))$ for $l = \overline{\gamma/2, p}$. Here we shall use Properties 2.M.6 and 2.M.7 of moduli of continuity. First of all, note that

$$\begin{aligned} \Phi(\chi_l(f)) &\leqslant m_\gamma(\Phi)q(\chi_l(f^{(\gamma)})) \\ &= m_\gamma(\Phi)|\alpha_l|q(\delta_h^{2l}(f^{(\gamma)})) \\ &\leqslant m_\gamma(\Phi)|\alpha_l|2^{2l-\gamma}\omega_\gamma(f^{(\gamma)}, h). \end{aligned}$$

If $l = \overline{\gamma/2, \gamma-1}$, then one can estimate $\Phi(\chi_l(f))$ also in the following way:

$$\begin{aligned} \Phi(\chi_l(f)) &\leqslant m_{2\gamma-2l}(\Phi)q(\chi_l(f^{(2\gamma-2l)})) \\ &\leqslant m_{2\gamma-2l}(\Phi)|\alpha_l|\omega_{2l}(f^{(2\gamma-2l)}, h) \\ &\leqslant m_{2\gamma-2l}(\Phi)|\alpha_l|h^{2l-\gamma}\omega_\gamma(f^{(\gamma)}, h). \end{aligned}$$

For $l \geqslant \gamma$ we have

$$\begin{aligned} \Phi(\chi_l(f)) &\leqslant m_0(\Phi)|\alpha_l|q(\delta_h^{2l}(f)) \\ &\leqslant m_0(\Phi)|\alpha_l|h^\gamma\omega_{2l-\gamma}(f^{(\gamma)}, h) \\ &\leqslant m_0(\Phi)|\alpha_l|2^{2l-2\gamma}h^\gamma\omega_\gamma(f^{(\gamma)}, h). \end{aligned}$$

Thus, $\Phi(\chi_l(f)) \leqslant |\alpha_l|a(\gamma, l, h)\omega_\gamma(f^{(\gamma)}, h)$ and

$$\begin{aligned} \Phi(U_{h.p.\gamma}(f)) &\leqslant \sum_{l=\gamma/2}^{p} \Phi(\chi_l(f)) \\ &\leqslant \sum_{l=\gamma/2}^{p} |\alpha_l|a(\gamma, l, h)\omega_\gamma(f^{(\gamma)}, h) = b(p, \gamma, h)h^{-\gamma}\omega_\gamma(f^{(\gamma)}, h). \end{aligned}$$

So we have proved (7) for $\gamma/2 \in \mathbb{N}$. Now we go over to the case $(\gamma+1)/2 \in \mathbb{N}$. Assume $l = \overline{(\gamma-1)/2, p-1}$, set $\beta_l = h^{-\gamma}\mathcal{T}_l^{(\gamma)}(0)/((2l+1)!)$,

$$\eta_l(f,x) = \beta_l 2^{-1}\{\Delta_h^{2l+1}(f, x-lh) + \Delta_h^{2l+1}(f, x-(l+1)h)\}$$

and estimate $\Phi(\eta_l(f))$. Here we shall use Properties 2.M*.6 and 2.M*.7 of the modified moduli of continuity. Note, first of all, that

$$\begin{aligned}\Phi(\eta_l(f)) &\leqslant m_\gamma(\Phi) q(\eta_l(f^{(\gamma)}))\\ &\leqslant m_\gamma(\Phi)|\beta_l|\omega_{2l+1}^*(f^{(\gamma)}, h)\\ &\leqslant m_\gamma(\Phi)|\beta_l| 2^{2l+1-\gamma}\omega_\gamma^*(f^{(\gamma)}, h).\end{aligned}$$

If $l = \overline{(\gamma-1)/2, \gamma-1}$, one can estimate $\Phi(\eta_l(f))$ also in the following way:

$$\begin{aligned}\Phi(\eta_l(f)) &\leqslant m_{2\gamma-2l-1}(\Phi)|\beta_l|\omega_{2l+1}^*(f^{(2\gamma-2l-1)}, h)\\ &\leqslant m_{2\gamma-2l-1}(\Phi)|\beta_l| h^{2l+1-\gamma}\omega_\gamma^*(f^{(\gamma)}, h).\end{aligned}$$

For $l \geqslant \gamma$ we have

$$\begin{aligned}\Phi(\eta_l(f)) &\leqslant m_0(\Phi)|\beta_l|\omega_{2l+1}^*(f, h)\\ &\leqslant m_0(\Phi)|\beta_l| h^\gamma \omega_{2l+1-\gamma}^*(f^{(\gamma)}, h)\\ &\leqslant m_0(\Phi)|\beta_l| 2^{2l+1-2\gamma} h^\gamma \omega_\gamma^*(f^{(\gamma)}, h).\end{aligned}$$

Thus $\Phi(\eta_l(f)) \leqslant |\beta_l| a(\gamma, l, h)\omega_\gamma^*(f^{(\gamma)}, h)$ and

$$\begin{aligned}\Phi(U_{h.p.\gamma}(f)) &\leqslant \sum_{l=(\gamma-1)/2}^{p-1} \Phi(\eta_l(f))\\ &\leqslant \sum_{l=(\gamma-1)/2}^{p-1} |\beta_l| a(\gamma, l, h)\omega_\gamma^*(f^{(\gamma)}, h) = b(p,\gamma,h) h^{-\gamma}\omega_\gamma^*(f^{(\gamma)}, h). \qquad \square\end{aligned}$$

LEMMA 4. *Let* $p \in \mathbb{N}$, $\gamma \in \mathbb{Z}_+$, $\gamma \leqslant 2p-1$, $k = 2[\gamma/2]$, $s = 2[(\gamma+1)/2]$, $h > 0$, $\Phi \in \mathfrak{M}_q$,

$$c(\gamma, l, h) := \begin{cases} \min\{m_{\gamma+1}(\Phi)2^{2l-k}, m_{k+\gamma+1-2l}(\Phi)h^{2l-k}\}, \\ \qquad \textit{for } l = \overline{k/2, \gamma}, \\ \min\{m_{\gamma+1}(\Phi)2^{2l-k}, m_0(\Phi)2^{2l-1-k-\gamma}h^{\gamma+1}\}, \\ \qquad \textit{for } l \geqslant \gamma+1; \end{cases}$$

$$d(p,\gamma,h) := \begin{cases} \displaystyle\sum_{l=\gamma/2}^{p-1} \frac{|\Lambda_l^{(\gamma)}(0)|}{(2l)!} c(\gamma, l, h), & \textit{for } \dfrac{\gamma}{2} \in \mathbb{N}, \\ \displaystyle\sum_{l=(\gamma-1)/2}^{p-1} \frac{|M_l^{(\gamma)}(0)|}{(2l+1)!} c(\gamma, l, h), & \textit{for } \dfrac{\gamma+1}{2} \in \mathbb{N}. \end{cases}$$

Then for all $f \in C_q^{(1)}$

$$\Phi(V_{h.p.\gamma}(f)) \leqslant d(p,\gamma,h) h^{-\gamma}\underline{\omega}_\gamma(f^{(\gamma+1)}, h). \tag{8}$$

PROOF. Consider first the case $(\gamma+1)/2 \in \mathbb{N}$. Set

$$\alpha_l = h^{-\gamma} M_l^{(\gamma)}(0)/((2l+1)!),$$
$$\chi_l(f,x) = \alpha_l \Delta_h^{2l+1}(f, x-(l+\tfrac{1}{2})h)$$

and estimate $\Phi(\chi_l(f))$ for $l = \overline{(\gamma-1)/2, p-1}$. We shall use here Properties 2.M.6 and 2.M.7 of moduli of continuity. Note first of all that

$$\begin{aligned}\Phi(\chi_l(f)) &\leqslant m_{\gamma+1}(\Phi) q(\chi_l^{(\gamma+1)}(f)) \\ &= m_{\gamma+1}(\Phi)|\alpha_l| q(\Delta_h^{2l+1}(f^{(\gamma+1)})) \\ &\leqslant m_{\gamma+1}(\Phi)|\alpha_l| 2^{2l+1-\gamma}\omega_\gamma(f^{(\gamma+1)}, h).\end{aligned}$$

If $l = \overline{(\gamma-1)/2, \gamma}$, one can estimate $\Phi(\chi_l(f))$ also in the following way:

$$\begin{aligned}\Phi(\chi_l(f)) &\leqslant m_{2\gamma-2l}(\Phi) q(\chi_l(f^{(2\gamma-2l)})) \\ &\leqslant m_{2\gamma-2l}(\Phi)|\alpha_l|\omega_{2l+1}(f^{(2\gamma-2l)}, h) \\ &\leqslant m_{2\gamma-2l}(\Phi)|\alpha_l| h^{2l+1-\gamma}\omega_\gamma(f^{(\gamma+1)}, h).\end{aligned}$$

For $l \geqslant \gamma+1$ we have

$$\begin{aligned}\Phi(\chi_l(f)) &\leqslant m_0(\Phi)|\alpha_l| q(\Delta_h^{2l+1}(f)) \\ &\leqslant m_0(\Phi)|\alpha_l| h^{\gamma+1}\omega_{2l-\gamma}(f^{(\gamma+1)}, h) \\ &\leqslant m_0(\Phi)|\alpha_l| 2^{2l-2\gamma} h^{\gamma+1}\omega_\gamma(f^{(\gamma+1)}, h).\end{aligned}$$

Thus, $\Phi(\chi_l(f)) \leqslant |\alpha_l| c(\gamma, l, h)\omega_\gamma(f^{(\gamma+1)}, h)$ and

$$\Phi(V_{h,p,\gamma}(f)) \leqslant \sum_{l=(\gamma-1)/2}^{p-1} \Phi(\chi_l(f)) \leqslant d(p,\gamma,h) h^{-\gamma}\omega_\gamma(f^{(\gamma+1)}, h).$$

So we have proved (8) for $(\gamma+1)/2 \in \mathbb{N}$. Now we go over to the case $\gamma/2 \in \mathbb{N}$. Suppose $l = \overline{\gamma/2, p-1}$, and set

$$\beta_l = h^{-\gamma}\Lambda_l^{(\gamma)}(0)/((2l)!),$$
$$\eta_l(f,x) = \beta_l 2^{-1}\{\Delta_h^{2l}(f, x-(l+\tfrac{1}{2})h) + \Delta_h^{2l}(f, x-(l-\tfrac{1}{2})h)\}$$

and estimate $\Phi(\eta_l(f))$. We shall use here Properties 2.M*.6 and 2.M*.7 of the modified moduli of continuity. Note, first of all, that

$$\begin{aligned}\Phi(\eta_l(f)) &\leqslant m_{\gamma+1}(\Phi) q(\eta_l(f^{(\gamma+1)})) \\ &\leqslant m_{\gamma+1}(\Phi)|\beta_l|\omega_{2l}^*(f^{(\gamma+1)}, h) \\ &\leqslant m_{\gamma+1}(\Phi)|\beta_l| 2^{2l-\gamma}\omega_\gamma^*(f^{(\gamma+1)}, h).\end{aligned}$$

If $l = \overline{\gamma/2, \gamma}$, then $\Phi(\eta_l(f))$ can be estimated also in the following way:

$$\begin{aligned}\Phi(\eta_l(f)) &\leqslant m_{2\gamma+1-2l}(\Phi)|\beta_l|\omega_{2l}^*(f^{(2\gamma+1-2l)}, h) \\ &\leqslant m_{2\gamma+1-2l}(\Phi)|\beta_l| h^{2l-\gamma}\omega_\gamma^*(f^{(\gamma+1)}, h).\end{aligned}$$

For $l \geqslant \gamma + 1$ we have

$$\begin{aligned}\Phi(\eta_l(f)) &\leqslant m_0(\Phi)|\beta_l|\omega_{2l}^*(f,h)\\ &\leqslant m_0(\Phi)|\beta_l|h^{\gamma+1}\omega_{2l-\gamma-1}^*(f^{(\gamma+1)},h)\\ &\leqslant m_0(\Phi)|\beta_l|2^{2l-1-2\gamma}h^{\gamma+1}\omega_\gamma^*(f^{(\gamma+1)},h).\end{aligned}$$

Thus $\Phi(\eta_l(f)) \leqslant |\beta_l|c(\gamma,l,h)\omega_\gamma^*(f^{(\gamma+1)},h)$ and

$$\begin{aligned}\Phi(V_{h.p.\gamma}(f)) &\leqslant \sum_{l=\gamma/2}^{p-1} \Phi(\eta_l(f))\\ &\leqslant \sum_{l=\gamma/2}^{p-1} |\beta_l|c(\gamma,l,h)\omega_\gamma^*(f^{(\gamma+1)},h)\\ &= d(p,\gamma,h)h^{-\gamma}\omega_\gamma^*(f^{(\gamma+1)},h).\end{aligned}$$ □

3. Inequalities of Jackson type.

THEOREM 1. *Let the conditions of Lemma* 3 *be fulfilled*, $0 \leqslant \alpha \leqslant 2p+2-k$. *Then for* $f \in C_q$

(9) $$\Phi(f) \leqslant m_0(\Phi)\mathscr{A}(p,\gamma,\alpha)h^\alpha\Omega_{2p+2-k.\alpha}(f,h) + b(p,\gamma,h)h^{-\gamma}\overline{\omega}_\gamma(f,h).$$

PROOF. Denote by (9′) the inequality obtained from (9) by replacing f by $f^{(\gamma)}$. To prove (9) it is sufficient (cf. the proof of Theorem 6.1 and Remark 6.2) to establish (9′) for $f \in C_q^{(\gamma)}$. Apply Theorem 4.1 and Lemma 3 to such f, which gives

$$\begin{aligned}\Phi(f^{(\gamma)}) &\leqslant \Phi(f^{(\gamma)} - U_{h.p.\gamma}(f)) + \Phi(U_{h.p.\gamma}(f))\\ &\leqslant m_0(\Phi)q(f^{(\gamma)} - U_{h.p.\gamma}(f)) + \Phi(U_{h.p.\gamma}(f))\\ &\leqslant m_0(\Phi)\mathscr{A}(p,\gamma,\alpha)h^\alpha\Omega_{2p+2-k.\alpha}(f^{(\gamma)},h) + b(p,\gamma,h)h^{-\gamma}\overline{\omega}_\gamma(f^{(\gamma)},h).\end{aligned}$$ □

COROLLARY 1. *Let* $\Phi \in \mathfrak{M}_q$, $\gamma/2 \in \mathbb{N}$, $h > 0$, $f \in C_q$. *Then*

(10) $$\Phi(f) \leqslant (m_0(\Phi)\mathscr{A}(\gamma-1,\gamma) + b(\gamma-1,\gamma,h)h^{-\gamma})\omega_\gamma(f,h).$$

THEOREM 2. *Let the conditions of Lemma* 3 *be fulfilled*, $0 \leqslant \alpha \leqslant 2p$, *and let* $Z(p,\gamma,h,\alpha)$ *be defined by* (1). *Then for* $f \in C_q$

(11) $$\Phi(f) \leqslant Z(p,\gamma,h,\alpha)\Omega_{2p.\alpha}(f,h) + b(p,\gamma,h)h^{-\gamma}\overline{\omega}_\gamma(f,h).$$

PROOF. It is sufficient to establish (11′) obtained from (11) by replacing f by $f^{(\gamma)}$ under the condition $f \in C_q^{(\gamma)}$. For such f we apply Lemmas 1 and 3 to get

$$\begin{aligned}\Phi(f^{(\gamma)}) &\leqslant \Phi(f^{(\gamma)} - U_{h.p.\gamma}(f)) + \Phi(U_{h.p.\gamma}(f))\\ &\leqslant Z(p,\gamma,h,\alpha)\Omega_{2p.\alpha}(f^{(\gamma)},h) + b(p,\gamma,h)h^{-\gamma}\overline{\omega}_\gamma(f^{(\gamma)},h).\end{aligned}$$ □

COROLLARY 2. *Let* $p \in \mathbb{N}$, $0 \leqslant \alpha \leqslant 2p$, $h > 0$, $\Phi \in \mathfrak{M}_q$,

$$H(p,h,\alpha) = \sum_{l=1}^{2p+1} \frac{|\mathscr{T}_p^{(l)}(0)|}{l!}\left(\prod_{j=1}^{2p+1-l}(\alpha+j)\right)^{-1} m_{l-1}(\Phi)h^{\alpha+1-l}.$$

Then for $f \in C_q$

$$\Phi(f) \leqslant H(p,h,\alpha)\Omega_{2p.\alpha}(f,h), \tag{12}$$

$$\Phi(f) \leqslant \int_0^1 \omega_{2p}(f,uh) \times \frac{1}{(2p)!}\sum_{l=1}^{2p+1}\binom{2p+1}{l} h^{1-l}|\mathcal{T}_p^{(l)}(0)|m_{l-1}(\Phi)(1-u)^{2p+1-l}\,du. \tag{13}$$

PROOF. Set $\gamma = 2p$ in (11); it is clear that

$$\Phi(f) \leqslant (Z(p,2p,h,\alpha) + m_{2p}(\Phi)h^{\alpha-2p})\Omega_{2p.\alpha}(f,h)$$

for $f \in C_q$. Next, note that

$$\frac{1}{(2p)!}\binom{2p}{l}B(\alpha+1,2p+1-l) = \frac{1}{l!}\left(\prod_{j=1}^{2p+1-l}(\alpha+j)\right)^{-1},$$

which gives

$$Z(p,2p,h,\alpha) + m_{2p}(\Phi)h^{\alpha-2p} = H(p,h,\alpha).$$

Therefore $\Phi(f) \leqslant H(p,h,\alpha)\Omega_{2p.\alpha}(f,h)$. Now we go over to proving (13). Set $\gamma = 2p+1$ in (2) and note that $U_{h.p.2p+1}(f)$ vanishes identically. This gives us the inequality obtained from (13) by replacing f by $f^{(\gamma)}$, which holds for all $f \in C_q^{(\gamma)}$ and is equivalent to (13). □

COROLLARY 3. *Let* $\Phi \in \mathfrak{M}_q$, $\gamma/2 \in \mathbb{N}$, $h > 0$, $f \in C_q$. *Then*

$$\Phi(f) \leqslant \left(\frac{1}{(\gamma+1)!}\sum_{l=1}^{\gamma+1}\binom{\gamma+1}{l}|\mathcal{T}_{\gamma/2}^{(l)}(0)|m_{l-1}(\Phi)h^{1-l}\right)\omega_\gamma(f,h). \tag{14}$$

Inequality (14) is obtained from (12) if one sets $\alpha = 0$, $p = \gamma/2$ in (12).

THEOREM 3. *Let the conditions of Lemma* 4 *hold*, $0 \leqslant \alpha \leqslant 2p-1-k$. *Then for* $f \in C_q^{(1)}$

$$\Phi(f) \leqslant m_0(\Phi)\mathcal{B}(p,\gamma,\alpha)h^{\alpha+1}\Omega_{2p-1-k.\alpha}(f',h) + d(p,\gamma,h)h^{-\gamma}\underline{\omega}_\gamma(f',h). \tag{15}$$

PROOF. Denote by (15′) the inequality obtained from (15) by replacing f by $f^{(\gamma)}$. To prove (15) it is sufficient to establish (15′) for all $f \in C_q^{(\gamma+1)}$. For such f apply Theorem 4.2 and Lemma 4 to get

$$\begin{aligned}\Phi(f^{(\gamma)}) &\leqslant \Phi(f^{(\gamma)} - V_{h.p.\gamma}(f)) + \Phi(V_{h.p.\gamma}(f))\\ &\leqslant m_0(\Phi)q(f^{(\gamma)} - V_{h.p.\gamma}(f)) + \Phi(V_{h.p.\gamma}(f))\\ &\leqslant m_0(\Phi)\mathcal{B}(p,\gamma,\alpha)h^{\alpha+1}\Omega_{2p-1-k.\alpha}(f^{(\gamma+1)},h)\\ &\quad + d(p,\gamma,h)h^{-\gamma}\underline{\omega}_\gamma(f^{(\gamma+1)},h).\end{aligned}$$ □

COROLLARY 4. *Let* $\Phi \in \mathfrak{M}_q$, $(\gamma+1)/2 \in \mathbb{N}$, $h > 0$, $f \in C_q^{(1)}$. *Then*

$$\Phi(f) \leqslant (m_0(\Phi)\mathcal{B}(\gamma,\gamma)h + d(\gamma,\gamma,h)h^{-\gamma})\omega_\gamma(f',h).$$

THEOREM 4. *Let the conditions of Lemma* 4 *hold*, $0 \leqslant \alpha \leqslant 2p-1$, *and let* $\Upsilon(p,\gamma,h,\alpha)$ *be defined by* (4). *Then for* $f \in C_q^{(1)}$

$$\Phi(f) \leqslant \Upsilon(p,\gamma,h,\alpha)\Omega_{2p-1.\alpha}(f',h) + d(p,\gamma,h)h^{-\gamma}\underline{\omega}_\gamma(f',h). \tag{16}$$

PROOF. Again, it is sufficient to establish the inequality (16′) obtained from (16) by replacing f by $f^{(\gamma)}$ under the condition $f \in C_q^{(\gamma+1)}$. For such f we apply Lemmas 2 and 4 to get

$$\begin{aligned}\Phi(f^{(\gamma)}) &\leqslant \Phi(f^{(\gamma)} - V_{h.p.\gamma}(f)) + \Phi(V_{h.p.\gamma}(f))\\ &\leqslant \Upsilon(p,\gamma,h,\alpha)\Omega_{2p-1.\alpha}(f^{(\gamma+1)},h) + d(p,\gamma,h)h^{-\gamma}\underline{\omega}_\gamma(f^{(\gamma+1)},h). \qquad \square\end{aligned}$$

COROLLARY 5. *Let* $p \in \mathbb{N}$, $0 \leqslant \alpha \leqslant 2p-1$, $h > 0$, $\Phi \in \mathfrak{M}_q$,

$$K(p,h,\alpha) = \sum_{l=0}^{2p} \frac{|\Lambda_p^{(l)}0)|}{l!}\left(\prod_{j=1}^{2p-l}(\alpha+j)\right)^{-1} m_l(\Phi)h^{\alpha+1-l}.$$

Then for $f \in C_q^{(1)}$

$$\Phi(f) \leqslant K(p,h,\alpha)\Omega_{2p-1.\alpha}(f',h), \tag{17}$$

$$\begin{aligned}\Phi(f) &\leqslant \int_0^1 \omega_{2p-1}(f',uh)\\ &\quad\times\left\{\frac{1}{(2p-1)!}\sum_{l=0}^{2p}\binom{2p}{l}h^{1-l}|\Lambda_p^{(l)}(0)|m_l(\Phi)(1-u)^{2p-l}\right\}du.\end{aligned} \tag{18}$$

PROOF. Set $\gamma = 2p-1$ in (16). It is clear that then

$$\Phi(f) \leqslant (\Upsilon(p,2p-1,h,\alpha) + m_{2p}(\Phi)h^{\alpha+1-2p})\Omega_{2p-1.\alpha}(f',h)$$

for $f \in C_q^{(1)}$.

Further, use the identity

$$\frac{1}{(2p-1)!}\binom{2p-1}{l}B(\alpha+1,2p-l) = \frac{1}{l!}\left(\prod_{j=1}^{2p-l}(\alpha+j)\right)^{-1},$$

which gives

$$\Upsilon(p,2p-1,h,\alpha) + m_{2p}(\Phi)h^{\alpha+1-2p} = K(p,h,\alpha).$$

Thus, $\Phi(f) \leqslant K(p,h,\alpha)\Omega_{2p-1.\alpha}(f',h)$. Now go over to (18). Set $\gamma = 2p$ in (5) and note that $V_{h.p.2p}(f)$ vanishes identically. This leads us to the inequality obtained from (18) by replacing f by $f^{(\gamma)}$ which holds for all $f \in C_q^{(\gamma+1)}$ and which is equivalent to (18). $\square$

COROLLARY 6. *Let* $\Phi \in \mathfrak{M}_q$, $\gamma/2 \in \mathbb{N}$, $h > 0$, $f \in C_q^{(1)}$. *Then*

$$\Phi(f) \leqslant \left(\frac{1}{\gamma!}\sum_{l=0}^{\gamma}\binom{\gamma}{l}|\Lambda_{\gamma/2}^{(l)}(0)|m_l(\Phi)h^{1-l}\right)\omega_{\gamma-1}(f',h). \tag{19}$$

This inequality is obtained from (17) if one sets $\alpha = 0$ in it.

COROLLARY 7. *Let* $k \in \mathbb{N}$, $h > 0$, $\Phi \in \mathfrak{M}_q$,

$$\alpha_k(h) = \frac{1}{(2k)!} \sum_{l=1}^{2k+1} \binom{2k+1}{l} |\mathcal{T}_k^{(l)}(0)| \frac{1}{2k+2-l} m_{l-1}(\Phi) h^{-l},$$

$$\beta_k(h) = \frac{1}{(2k-1)!} \sum_{l=0}^{2k} \binom{2k}{l} |\Lambda_k^{(l)}(0)| \frac{1}{2k+1-l} m_l(\Phi) h^{-l};$$

then

(i) *for* $f \in C_q$

$$\Phi(f) \leqslant \alpha_k(h) \int_0^h \omega_{2k}(f, t)\, dt; \tag{20}$$

(ii) *for* $f \in C_q^{(1)}$

$$\Phi(f) \leqslant \beta_k(h) \int_0^h \omega_{2k-1}(f', t)\, dt. \tag{21}$$

PROOF. We verify (20). Apply the known Chebyshev inequality

$$\int_a^b l(t) g(t)\, dt \leqslant \frac{1}{b-a} \int_a^b l(t)\, dt \int_a^b g(t)\, dt,$$

(for increasing l and decreasing g on $[a, b]$) to the right-hand side of (13):

$$\Phi(f) \leqslant \int_0^1 \omega_{2k}(f, uh)\, du \frac{1}{(2k)!} \sum_{l=1}^{2k+1} \binom{2k+1}{l} h^{1-l} |\mathcal{T}_k^{(l)}(0)| m_{l-1}(\Phi)$$
$$\times \int_0^1 (1-u)^{2k+1-l}\, du = \alpha_k(h) \int_0^h omega_{2k}(f, t)\, dt.$$

Inequality (21) is established in a similar way, but instead of (13) we use (18). □

4. Inequalities of Jackson type for the best approximations. Set

$$r_k(\gamma) := \left(\frac{2}{\pi\gamma}\right)^{2k} \sum_{l=k}^{2k-1} \frac{|Q_l^{(2k)}(0)|}{(2l)!} \min\left\{K_{2k} 2^{2l-2k}, K_{4k-2l} \left(\frac{\pi\gamma}{2}\right)^{2l-2k}\right\},$$

$$r(k, \gamma) := \frac{1}{(2k+1)!} \sum_{l=0}^{k} \binom{2k+1}{2l+1} |\mathcal{T}_k^{(2l+1)}(0)| K_{2l} \left(\frac{2}{\pi\gamma}\right)^{2l},$$

$$\alpha(k, \gamma) := \frac{1}{(2k)!} \sum_{l=0}^{k} \binom{2k+1}{2l+1} |\mathcal{T}_k^{(2l+1)}(0)| \frac{K_{2l}}{2k-2l+1} \left(\frac{2}{\pi\gamma}\right)^{2l},$$

$$\beta(k, \gamma) := \frac{1}{(2k-1)!} \sum_{l=0}^{k} \binom{2k}{2l} |\Lambda_k^{(2l)}(0)| \frac{K_{2l}}{2k+1-2l} \left(\frac{2}{\pi\gamma}\right)^{2l}.$$

Recall that (see §2)

$$Q_l^{(2k)}(0) = (2k)!\, t(2l, 2k), \qquad \mathcal{T}_k^{(2l+1)}(0) = (2l+1)!\, t(2k+2, 2l+2),$$

$$\Lambda_k^{(2l)}(0) = (2l)!\, t(2k+1, 2l+1), \qquad K_{2l} = \frac{\pi^{2l}}{2^{2l}((2l)!)} |\mathcal{E}_{2l}|,$$

which gives us

$$r_k(\gamma) = \frac{1}{\gamma^{2k}} \sum_{l=k}^{2k-1} |t(2l, 2k)| \min\left\{ \frac{2^{2l-2k}|\mathscr{E}_{2k}|}{(2l)!}, \frac{\binom{4k}{2l}|\mathscr{E}_{4k-2l}|}{\binom{4k}{2k}((2k)!)\gamma^{2k-2l}} \right\},$$

$$r(k,\gamma) = \frac{1}{(2k)!} \sum_{l=0}^{k} \binom{2k}{2l} |\mathscr{E}_{2l} t(2k+2, 2l+2)| \gamma^{-2l},$$

$$\alpha(k,\gamma) = \frac{2k+1}{(2k)!} \sum_{l=0}^{k} \binom{2k}{2l} \frac{|\mathscr{E}_{2l} t(2k+2, 2l+2)|}{2k+1-2l} \gamma^{-2l},$$

$$\beta(k,\gamma) = \frac{1}{(2k-1)!} \sum_{l=0}^{k} \binom{2k}{2l} \frac{|\mathscr{E}_{2l} t(2k+1, 2l+1)|}{2k+1-2l} \gamma^{-2l}.$$

THEOREM 5. *Let $n \in \mathbb{Z}_+$, $k \in \mathbb{N}$, $\gamma > 0$. Then*
(i) *for $f \in C_q$*

$$E_n(f) \leqslant (\mathscr{A}(2k-1, 2k) + r_k(\gamma))\omega_{2k}\left(f, \frac{\pi\gamma}{2(n+1)}\right),$$

$$E_n(f) \leqslant r(k,\gamma)\omega_{2k}\left(f, \frac{\pi\gamma}{2(n+1)}\right),$$

$$E_n(f) \leqslant \frac{1}{(2k)!} \int_0^1 \omega_{2k}\left(f, \frac{\pi\gamma u}{2(n+1)}\right) \times \left\{ \sum_{l=0}^{k} \binom{2k+1}{2l+1} |\mathscr{T}_k^{(2l+1)}(0)| K_{2l} \left(\frac{2}{\pi\gamma}\right)^{2l} (1-u)^{2k-2l} \right\} du, \tag{22}$$

$$E_n(f) \leqslant \alpha(k,\gamma) \int_0^1 \omega_{2k}\left(f, \frac{\pi\gamma u}{2(n+1)}\right) du;$$

(ii) *for $f \in C_q^{(1)}$*

$$E_n(f) \leqslant \frac{\pi\gamma}{2(n+1)} \frac{1}{(2k-1)!} \int_0^1 \omega_{2k-1}\left(f', \frac{\pi\gamma u}{2(n+1)}\right) \times \left\{ \sum_{l=0}^{k} \binom{2k}{2l} |\Lambda_k^{(2l)}(0)| K_{2l} \left(\frac{2}{\pi\gamma}\right)^{2l} (1-u)^{2k-2l} \right\} du,$$

$$E_n(f) \leqslant \frac{\pi\gamma}{2(n+1)} \beta(k,\gamma) \int_0^1 \omega_{2k-1}\left(f', \frac{\pi\gamma u}{2(n+1)}\right) du.$$

In order to prove Theorem 5 one must apply inequalities (10), (14), (13), (20), (18), (21) to the functional $\Phi(f) = E_n(f)$. Here also we use Corollary 2.B.

COROLLARY 8. *Let $n \in \mathbb{Z}_+$, $k \in \mathbb{N}$. Then*

$$E_n(f) \leqslant C(k)\omega_{2k}\left(f, \frac{\pi}{2(n+1)}\right),$$

where

$$C(k)=\frac{1}{(4k-2)!}\sum_{l=0}^{k-1}2^{2k-2-2l}((2l)!)|t(4k,2l+2)|$$
$$+\frac{1}{\binom{4k}{2k}((2k)!)}\sum_{l=1}^{k}\binom{4k}{2l}|\mathscr{E}_{2l}||t(4k-2l,2k)|.$$

In particular, $C(1)=\frac{1}{2}+\frac{1}{2}=1$, $C(2)=121/360+105/360=113/180$, $C(3)=72811/453600+59/360<0.325$, $C(4)=\frac{2^4 381556653}{14!}+\frac{13^3 5}{8!3}<0.161$.

The result is obtained from (22) for $\gamma=1$.

5. Linear methods of approximation. Let $n\in\mathbb{Z}_+$, $k\in\mathbb{N}$, $\gamma>0$, $h=\pi\gamma/(2n+2)$, $x\in\mathbb{R}$, $f\in L_1$. Then we set

$$\mathscr{X}_{n,k,\gamma}(f,x)=\sum_{l=k}^{2k-1}\frac{Q_l^{(2k)}(0)}{(2l)!}X_{n,g(k,\gamma,l)}(\mathscr{S}_{h,2l}^{(2l-2k)}(f),x)h^{2l-2k},$$

where $\mathscr{S}_{h,2l}(\varphi)$ and $X_{n,p}(\varphi)$ are Steklov functions and Akhieser–Kreĭn–Favard sums as before (see §2), and

$$g(k,\gamma,l)=\begin{cases}2k, & \text{if } K_{2k}2^{2l-2k}\leqslant K_{4k-2l}(\pi\gamma/2)^{2l-2k},\\ 4k-2l, & \text{if } K_{4k-2l}(\pi\gamma/2)^{2l-2k}<K_{2k}2^{2l-2k}.\end{cases}$$

THEOREM 6. *Let* $n\in\mathbb{Z}_+$, $k\in\mathbb{N}$, $\gamma>0$, $h=\pi\gamma/(2n+2)$. *Then for all* $f\in C_q$

$$q(f-\mathscr{X}_{n,k,\gamma}(f))\leqslant(\mathscr{A}(2k-1,2k)+r_k(\gamma))\omega_{2k}\left(f,\frac{\pi\gamma}{2(n+1)}\right).$$

PROOF. We start with estimating

$$\alpha_l=q\big(\mathscr{S}_{h,2l}^{(2l-2k)}(f)-X_{n,g(k,\gamma,l)}(\mathscr{S}_{h,2l}^{(2l-2k)}(f))\big)h^{2l-2k}.$$

We shall use here Theorem 2.A, properties of Steklov functions, and Properties 2.M.6 and 2.M.7 of moduli of continuity. Consider first the case $g(k,\gamma,l)=2k$. Then

$$\begin{aligned}\alpha_l&\leqslant h^{2l-2k}K_{2k}(n+1)^{-2k}q(\mathscr{S}_{h,2l}^{(2l)}(f))\\&\leqslant h^{-2k}(n+1)^{-2k}K_{2k}q(\delta_h^{2l}(f))\\&\leqslant(\gamma\pi/2)^{-2k}K_{2k}2^{2l-2k}\omega_{2k}(f,h).\end{aligned}$$

Now suppose that $g(k,\gamma,l)=4k-2l$. In this case

$$\begin{aligned}\alpha_l&\leqslant h^{2l-2k}K_{4k-2l}(n+1)^{2l-4k}q(\mathscr{S}_{h,2l}^{(2k)}(f))\\&=h^{2l-2k}K_{4k-2l}(n+1)^{2l-4k}q(\mathscr{S}_{h,2l}^{(2l)}(f^{(2k-2l)}))\\&\leqslant h^{-2k}K_{4k-2l}(n+1)^{2l-4k}h^{2l-2k}\omega_{2k}(f,h)\\&\leqslant(\gamma\pi/2)^{-2k}K_{4k-2l}(\gamma\pi/2)^{2l-2k}\omega_{2k}(f,h).\end{aligned}$$

So we have

$$\alpha_l\leqslant(\gamma\pi/2)^{-2k}\min\Big\{K_{2k}2^{2l-2k},K_{4k-2l}(\gamma\pi/2)^{2l-2k}\Big\}\omega_{2k}(f,h).\tag{23}$$

Introduce $A = q(S_{h,2k-1,2k}(f) - \mathscr{X}_{n,k,\gamma}(f))$. With the help of (23) we obtain

$$
\begin{aligned}
(24) \qquad A &\leqslant (\gamma\pi/2)^{-2k} \sum_{l=k}^{2k-1} \frac{|Q_l^{(2k)}(0)|}{(2l)!} \min\Big\{K_{2k}2^{2l-2k}, K_{4k-2l}(\gamma\pi/2)^{2l-2k}\Big\}\omega_{2k}(f,h) \\
&= r_k(\gamma)\omega_{2k}(f,h).
\end{aligned}
$$

Now apply (24) and the inequality

$$q(f - S_{h,2k-1,2k}(f)) \leqslant \mathscr{A}(2k-1, 2k)\omega_{2k}(f,h)$$

(see Theorem 5.1 (3)) to get

$$
\begin{aligned}
&q(f - \mathscr{X}_{n,k,\gamma}(f)) \\
&\qquad \leqslant q(f - S_{h,2k-1,2k}(f)) + q(S_{h,2k-1,2k}(f) - \mathscr{X}_{n,k,\gamma}(f)) \\
&\qquad \leqslant (\mathscr{A}(2k-1,2k) + r_k(\gamma))\omega_{2k}(f, \pi\gamma/(2n+2)). \qquad \square
\end{aligned}
$$

To be specific, let $C_q = C$. The operator of best approximation associating to a function $f \in C$ its best approximation polynomial $T_n(f)$ is nonlinear, i.e., the condition $l = f + g$ does not imply $T_n(l) = T_n(f) + T_n(g)$. By (22)

$$(25) \qquad \|f - T_n(f)\| \leqslant (\mathscr{A}(2k-1,2k) + r_k(\gamma))\omega_{2k}(f, \pi\gamma/(2n+2)).$$

So the question arises whether it is possible to replace in (31) the operator of the best approximation by some linear operator mapping C into H_n. Since for every $f \in C$ there is the inclusion $\mathscr{X}_{n,k,\gamma}(f) \in H_n$ and the operator $\mathscr{X}_{n,k,\gamma}$ is linear in C, Theorem 6 gives an affirmative answer to the above question. Similar questions may be put with respect to other estimates contained in Theorem 5.

References

1. N. I. Akhiezer, *Lectures in the theory of approximation*, 2nd rev. ed., "Nauka", Moscow, 1965. (Russian)
2. A. F. Timan, *Theory of approximation of functions of a real variable*, Fizmatgiz, Moscow, 1960; English transl., Pergamon Press, Oxford, and Macmillan, New York, 1963.
3. N. P. Korneichuk, *Exact constants in approximation theory*, "Nauka", Moscow, 1987; English transl., Cambridge Univ. Press, Cambridge, 1991.
4. A. I. Stepanets, *Uniform approximations by trigonometric polynomials*, "Naukova Dumka", Kiev, 1981. (Russian)
5. ———, *Classification and approximation of periodic functions*, "Naukova Dumka", Kiev, 1987. (Russian)
6. V. V. Shalaev, *On approximation of continuous periodic functions by trigonometric polynomials*, Studies in Modern Problems of Summation and Approximation of Functions and Applications, Dnepropetrovsk, 1977, pp. 39–43. (Russian)
7. V. V. Zhuk, *Approximation of periodic functions*, Izdat. Leningrad. Gos. Univ., Leningrad, 1982. (Russian)
8. G. I. Natanson, *Second modulus of continuity*, Studies in the Theory of Functions of Several Real Variables, Yaroslavl. Gos. Univ., Yaroslavl′, 1984, pp. 76–82. (Russian)
9. S. B. Stechkin, *The approximation of periodic functions by Fejér sums*, Trudy Mat. Inst. Steklov. **62** (1961), 48–60; English transl. in Amer. Math. Soc. Transl. Ser. 2 **28** (1963).
10. Yu. A. Brudnyi, *On a theorem of best local approximations*, Kazan. Gos. Univ. Uchebn. Zap. **124** (1964), no. 2, 43–49. (Russian)
11. V. V. Zhuk, *Structural properties of functions and precision of approximation*, Leningrad, 1984. (Russian)
12. B. Sendov, *On a theorem of Ju. Brudnyi*, Math. Balkanica **1** (1987), 106–111.
13. N. S. Bakhvalov, N. P. Zhidkov, and G. M. Kobel′kov, *Numerical methods*, "Nauka", Moscow, 1987. (Russian)

14. I. S. Berezin and N. P. Zhidkov, *Computing methods. Vol.* I, Fizmatgiz, Moscow, 1962; English transl., Addison-Wesley, Reading, MA, and Pergamon Press, Oxford, 1965.
15. L. V. Kantorovich and V. I. Krylov, *Approximate methods of higher analysis*, Fizmatgiz, Moscow, 1962; 5th ed.; English transl. of 3rd ed., Interscience, New York, and Noordhoff, Groningen, 1958.
16. V. V. Zhuk and G. I. Natanson, *Trigonometric Fourier series and elements of approximation theory*, Izdat. Leningrad. Gos. Univ., Leningrad, 1983. (Russian)
17. I. S. Gradshtein and I. M. Ryzhik, *Tables of integrals, series and products*, 5th ed., "Nauka", Moscow, 1971; English transl. of 4th ed., Academic Press, New York, 1965; rev. aug. ed., 1980.
18. M. Abramowitz and I. A. Stegun (eds.), *Handbook of mathematical functions, with formulas, graphs and mathematical tables*, U. S. Govt. Printing Office, Washington, DC, 1964; Reprint, Dover, New York, 1966.
19. I. P. Mysovskikh, *Lectures on computing methods*, Fizmatgiz, Moscow, 1962; English transl., Noordhoff, Groningen.
20. L. Moser and M. Wyman, *Asymptotic development of the Stirling numbers of the first kind*, J. London Math. Soc. **33** (1958), 133–146.
21. K. A. Rybnikov (ed.), *Combinatorial analysis. Problems and exercises*, "Nauka", Moscow, 1982. (Russian)
22. V. N. Sachkov, *Introduction to combinatorial methods of discrete mathematics*, "Nauka", Moscow, 1982; English transl. in J. Soviet Math. (to appear).
23. A. Kaufmann, *Introduction to applied combinators*, "Nauka", Moscow, 1975. (Russian)
24. D. Riordan, *Combinatorial identities*, Wiley, New York, 1968.
25. V. V. Zhuk and G. I. Natanson, *On the theory of cubic periodic splines with equidistant nodes*, Vestnik Leningrad. Univ. Mat. Mekh. Astronom. **1984**, no. 1, 5–11; English transl. in Vestnik Leningrad Univ. Math. **16** (1984).
26. I. P. Natanson, *Constructive function theory*, GITTL, Moscow, 1949; English transl., Vol. I, Ungar, New York, 1964; Vols. II, III, Ungar, New York, 1965.

Department of Applied Mathematics, St. Petersburg State University, Stary Petergof, 2, Bibliotechnaya pl., St. Petersburg, 198904, Russia

Translated by G. ROZENBLUM

Amer. Math. Soc. Transl.
(2) Vol. **159**, 1994

Hyperstonean Preimage of Alexandrov Spaces

V. K. Zakharov

§0. Introduction

In various branches of mathematics a number of extensions of the ring C of all bounded continuous functions on a space T has emerged; some of them have already become classical: the rings BM and BM^0 of Borel and Baire measurable functions,[1] respectively [**1**, § 31; **2**, IV.1; **3**, 18.1.2], the rings B and B^0 of classes of functions with the Baire property [**3**, 15.6] and with the Baire property with respect to cozero sets [**1**, § 32], the ring R_μ of classes of Riemann μ-integrable functions [**4**, IV, § 5, ex. 16, 17], the ring L_μ of Lebesgue μ-measurable functions [**4**, IV, § 6.3], the ring UM of universally measurable functions [**4**, V, § 3.4], the Arens second dual ring C'' [**5**, 3, 27.2], etc. It turned out that all these rings are isomorphic to rings of continuous functions on corresponding compact spaces, i.e., they are c-rings. Since all these extensions have appeared outside of algebra, from the algebraic point of view the nature of the connection between C and the above rings remains completely unknown. This is the reason why the problem of algebraic description of classical c-extensions of the ring C arises.

To every classical c-extension $C \rightarrowtail A$ of the ring C there corresponds a certain preimage $T \twoheadleftarrow H$ of the space T; it is the preimage of the maximal ideals of the c-extension realizing this extension by means of continuous functions. This means that there exists a functor from the category of c-extensions $C \rightarrowtail A$ to the category of completely regular preimages $T \twoheadleftarrow H$ or, more precisely, to the category of a-preimages of a-spaces (A. D. Alexandrov spaces [**6**]) which is better suited for the compact case than the coarser category of totally regular preimages. With respect to this functor the problem of algebraic description of the classical c-extensions generates the parallel problem of topological description of the classical realizing a-preimages of the space T.

In [**7**] (see also [**8**]) it was established that the realizing preimage of the Baire extension $C \rightarrowtail B$ is the Gleason-Ponomarev absolute $T \twoheadleftarrow pT$ of the space T [**9**]. In [**10**] and [**11**] the study of the realizing preimage $T \twoheadleftarrow sT$ for the small Baire extension $C \rightarrowtail B^0$ was begun. A topological description of this a-preimage was given in [**12**].

Later investigations have shown that the approach based on the classical theory of enclosable preimages, which was successfully used in the cases of the absolute $T \twoheadleftarrow pT$

1991 *Mathematics Subject Classification*. Primary 16S60, 16W80; Secondary 16W99.

[1] All the functions are supposed to be bounded.

and the sequential absolute $T \twoheadleftarrow sT$, failed in the cases of the remaining realizing a-preimages. The reason is the fact that these a-preimages have three characteristic peculiarities: a) almost all of them possess the staircase property of enclosability which consists of a sequence of elementary properties. Each of these properties is not characteristic, if taken separately, and is defined consecutively, which means that each subsequent property is defined only after all previous properties have been defined; b) the enclosability properties mentioned above are not only globally spatial, but also possess some local subspatial effects without which global versions of these properties are not characteristic; c) all the elements of Alexandrov bases of these preimages are divided into two categories, namely, generators having enclosability properties and the rest of the elements, representable in terms of generators; for the latter the enclosability character vanishes entirely.

For this reason, in 1976 Flachsmeyer **[13]** singled out the *problem of the existence of a purely topological description of the hyperstonean preimage* $T \twoheadleftarrow hT$ realizing Arens extension $C \rightarrowtail C''$. Later, in 1986 Fremlin **[14]** posed *the problem of the existence of a direct topological construction of the preimage* $T \twoheadleftarrow hT$ *only in terms of the space* T *itself.*

Another approach, taking all these nonclassical peculiarities into account, was developed by the author **[15–17]**.[2] In solving Flachsmeyer's problem the author has distinguished a new topological structure which was called a "screen". The a-spaces with a screen were called *as-spaces*. The presence of a screen makes it possible to control the tightness of the enclosure not only in the a-space itself, but also in every subspace of the screen, which enables one to define enclosability properties in as-spaces in a more subtle way. Taking the three aforementioned peculiarities into account, the author has introduced a very general notion of an *enclosable as-cover of the type* $Z^{\pi_1} \cdots Z^{\pi_k}$ *for the as-space* T as such an as-preimage $T \twoheadleftarrow H$ for which a sequence of elementary enclosability properties Z^{π_i} for the as-space H and the process of consequent transition from the a-foundation $\mathscr{T}$ in T to the a-foundation $\mathscr{H}$ in H via properties of Z^{π_i}-enclosability are described. By means of this notion we succeeded in characterizing the hyperstonean preimage $T \twoheadleftarrow hT$ as an enclosable as-cover of the type ZZZ^c of the space T (Theorem 1). Note also that this approach to enclosable covers has revealed a complete parallelism in the description of the preimages (with different screens) $T \twoheadleftarrow s_\mu T$ and $T \twoheadleftarrow sT$ corresponding to the Riemann extension $C \rightarrowtail R_\mu$ and the small Baire extension $C \rightarrowtail B^0$ connected with the classical ring of quotients of C. So the preimage of the Riemann extension proved to be parallel to the classical ring of quotients $C \rightarrowtail C^{\mathrm{cl}}$.

The main technical difficulty in characterizing the classical realizing as-preimages results from the third before-mentioned peculiarity of these preimages. To be specific, intermediate a-foundations $Z^{\pi_i}(\mathscr{K}_{i-1})$ between $r^{-1}\mathscr{T}$ and $\mathscr{H}$ are defined in terms of generators. It is possible to transfer these generators into other as-preimages in a sufficiently unique way. However, when we begin transferring arbitrary elements in $Z^{\pi_i}(\mathscr{K}_{i-1})$ some serious difficulties arise, connected with nonuniqueness of representation of the elements in terms of generators. We can prove that the transfer of arbitrary elements is independent of their representation with the help of the extensions connected with as-preimages. The families of all bounded functions with small variations on as-spaces are considered as c-rings furnished with a new ring

[2] See also V. K. Zakharov, *Topological preimages corresponding to nonclassical extensions of rings of continuous functions*, Vestnik Moscow Univ., Ser. 1, 1990, N 1, 44–47.

structure parallel to the screen and we call this structure a "refinement" [18]. A c-ring with refinement is called a cr-ring [18]. Thus to the category of as-preimages of the as-space T we associate the dual category of cr-extensions of a cr-ring C. Therefore, the main technical tools here are: 1) the passage from as-preimages to cr-extensions associated to as-preimages and 2) utilization of the topological connection modulo Boolean ideals between the elements of cr-extensions associated to as-preimages and the elements of the classical functional factor cr-extensions.

The aim of this paper is to give a detailed description of this approach as applied to one of the most complicated classical realizing preimages, namely, to the hyperstonean preimage $T \twoheadleftarrow hT$ of the second dual Arens extension $C \rightarrowtail C''$. In [3, 19.9, 27.3.3] and [4, V, § 5, ex. 11, 12] several important properties of the hyperstonean covering were already singled out; however, most of them are of a nontopological nature and therefore they do not provide its topological description. This paper continues the author's work [17], in which we gave only a phenomenological description of the preimage $T \twoheadleftarrow hT$ without clarifying its internal structure. In order to demonstrate the possibilities of our method, at the end of the paper we give (without proof) a characterization of the Gordon preimage $T \twoheadleftarrow gT$ realizing the universal measurable Caplan extension $C \rightarrowtail UM$ (Theorem 2).

§1. cr-extensions and as-preimages

1.1. cr-extensions.

1.1.1. *c-rings and c-extensions.* Henceforth all rings are assumed to be commutative and unital and all ring homomorphisms to be unital. We call a ring A a *c-ring* if A possesses the following properties: a) for all a and b there exists c such that $a^2 + b^2 = c^2$; b) for each a there exist b and c such that $a = b^2 - c^2$ and $bc = 0$; c) if for given a there exists a sequence $\{b_n\}$ such that $n(a^2 + b_n^2) = 1$, then $a = 0$; d) for each a there exists $(1 + a^2)^{-1}$; e) for each a there exist b and $n \in \mathbb{N}$ such that $a^2 + b^2 = n1$; f) if $\{a_n\}$ is a sequence for which there exists a sequence $\{m_k\} \subset \mathbb{N}$ such that $k((a_m - a_n)^2 + b^2) = 1$ for all $m, n \geqslant m_k$ and for corresponding $b = b(k, m, n)$, then there exists a for which there exists a sequence $\{n_k\} \subset \mathbb{N}$ such that $k((a - a_n)^2 + c^2) = 1$ for all $n \geqslant n_k$ and for corresponding $c = c(k, n)$. The system of properties a)—e) was stated by Delfosse [19]. An ideal E of a c-ring A will be called *closed* if for each sequence $\{a_n\}$ and every a for which there exists a sequence $\{n_k\} \subset \mathbb{N}$ such that $k((a - a_n)^2 + c^2) = 1$ for all $n \geqslant n_k$ and for corresponding $c = c(n, k)$, the condition $\{a_n\} \subset E$ implies $a \in E$. The importance of the class of c-rings is demonstrated by the following theorem due to Delfosse [19].

THEOREM. *A ring is a c-ring if and only if it is isomorphic to the ring of all bounded continuous functions on some compact space. An ideal of a c-ring is closed if and only if its image under the Delfosse isomorphism is closed with respect to uniform convergence.*

Let C be some fixed c-ring. An injective homomorphism $u \colon C \rightarrowtail A$, where A is a c-ring, will be called a c-extension of the c-ring C.

1.1.2. *Kelley ideals.* Let $\mathscr{P}$ denote the Boolean algebra of all subsets of a completely regular space T. Denote by $\mathscr{C}_\sigma$ the family of all subsets $E \subset T$ such that $E = \bigcup F_k$ for some sequence of compact subsets F_k.

A σ-ideal N in $\mathscr{P}$ is called *regular* if: a) for each $P \in N$ there exists a sequence of open sets G_k such that $P \subset \bigcap G_k \in N$; b) for each open $G \subset T$ there exists $E \in \mathscr{C}_\sigma$

such that $E \subset G$ and $G \setminus E \in N$; c) for each $E \in \mathscr{C}_\sigma$ there exists a set $E' \in \mathscr{C}_\sigma$, $E' \cap E = \varnothing$, such that $T \setminus (E \cup E') \in N$.

Let $\mathscr{E}$ be some subset in $\mathscr{P}$. $\mathscr{E}$ is said to have *nonzero intersection number modulo* N if there exists $r \in \mathbb{N}$ such that $i_N\{E_p\} \equiv \max\{l/m \mid \exists\, 1 \leqslant p_1 < \cdots < p_l \leqslant m(E_{p_1} \cap \cdots \cap E_{p_l} \notin N)\} \geqslant 1/r$ for every finite sequence $\{E_p \in \mathscr{E} \mid 1 \leqslant p \leqslant m\}$ **[21]**.

A regular σ-ideal N is a *Kelley ideal* if $\mathscr{C}_\sigma \setminus N$ is the union of countable subfamilies $\{\mathscr{C}_k\}$ such that a) if $E \in \mathscr{C}_k$, $E' \in C_\sigma$, and $E \Delta E' \in N$, then $E' \in \mathscr{C}_k$; b) each $\mathscr{C}_k$ has nonzero intersection number modulo N; c) if $\{E_p \mid p \in \mathbb{N}\}$ is an increasing sequence in $\mathscr{C}_\sigma \setminus N$ and $\bigcup E_p \in \mathscr{C}_k$, then $E_{p_0} \in \mathscr{C}_k$ for some p_0. The set of all Kelley ideals is denoted by $\mathscr{A}$. We equip it with the ordering opposite to the ordering by inclusions.

Recall that a Radon measure on T is the σ-additive compact-regular real-valued function ν on the σ-field $\mathscr{B}$ of all Borel sets in T. Let n denote the class of all bounded Radon measures on T jointly absolutely continuous with the measure ν, i.e., having the same family of negligible sets **[4,** V. § 5, 6]. The set of all such classes for all bounded Radon measures on T is denoted by $\mathscr{N}$. The importance of the notion of Kelley ideals is demonstrated by the following lemma proved in **[17,** 1.1.3].

LEMMA. *For every Kelley ideal N there exists a Radon measure ν such that $N = \{P \in \mathscr{P} \mid \exists B \in \mathscr{B}(P \subset B \& \nu B = 0)\}$. The mapping $\zeta\colon N \mapsto n$ is a bijection between $\mathscr{A}$ and $\mathscr{N}$.*

In **[17,** 2.1, Lemma 1] the following result was also proved.

LEMMA. *If N_1 and N_2 are Kelley ideals, then $N_1 \cap N_2$ is also a Kelley ideal. Therefore, the ordered set $\mathscr{A}$ is a semilattice with the operation $N_1 \vee N_2 = N_1 \cap N_2$.*

LEMMA 1. *The semilattice $\mathscr{A}$ is a complete lattice with the smallest element $\mathscr{P}$ and with relative complements.*

PROOF. It follows from **[4,** IX, § 3.2; **2,** VIII, 1.4.3; and **4,** IX, § 1.7] that the family $\mathscr{M}$ of all bounded Radon measures on T is a Dedekind complete vector lattice. By 19.1.5 in **[3]** two measures ν_1 and ν_2 belong to n if and only if the principal bands ν_1^{dd} and ν_2^{dd} on $\mathscr{M}$ coincide. Therefore the family $\mathscr{N}$ coincides with the family of all principal bands on $\mathscr{M}$. Introduce the ordering by inclusion in $\mathscr{N}$. Thus $\mathscr{N}$ becomes a complete lattice with zero element and relative complements. According to 19.1.4 in **[3]** the mappings ζ and ζ^{-1} preserve the order. By the bijectivity of ζ this means that ζ is a lattice isomorphism. The lemma is proved. □

Consider on T the lattice $\mathscr{U}\mathscr{M}$ of all universally measurable sets, i.e., sets measurable with respect to all bounded Radon measures **[4,** V, § 3.4]. For $X \in \mathscr{U}\mathscr{M}$ and $N \in \mathscr{A}$ we consider the ideal $N(X) \equiv \{P \in \mathscr{P} \mid P \cap X \in N\}$. In **[17,** 2.1, Lemma 6] the following statement was proved.

LEMMA. *The ideal $N(X)$ is a Kelley ideal.*

1.1.3. *cr-rings and cr-extensions.* Let T be a fixed completely regular space and let C be the c-ring of all bounded continuous functions on T.

For N and $\{N_\xi\}$ in $\mathscr{A}$ we shall write $N = \operatorname{top}\{N_\xi\}$ if $N_\xi \supset N$ and for each proper Kelley ideal $M \supset N$ there exist ξ_0 and a proper Kelley ideal L such that $L \supset M \cup N_{\xi_0}$.

Denote by $\mathscr{C}(A)$ the family of all closed ideals of a c-ring A. The mapping $\mathfrak{A}\colon \mathscr{A} \to \mathscr{C}(A)$, where $\mathscr{A}$ is defined in 1.1.2, will be called a *refinement of the c-ring A* if a) $\mathfrak{A}(N) \doteq A$ if and only if $N = \mathscr{P}$; b) $\bigcap \mathfrak{A}(N) = \{0\}$; c) $N_1 \supset N_2$ implies

$\mathfrak{A}(N_1) \supset \mathfrak{A}(N_2)$; and d) $N = \operatorname{top} N_\xi$ implies $\mathfrak{A}(N) = \bigcap \mathfrak{A}(_\xi)$. A c-ring A equipped with a refinement $\mathfrak{A}$ will be called a *cr-ring* and denoted by $(A, \mathfrak{A})$. Denote $\mathfrak{A}(N)$ by A_N. Then $\mathfrak{A} = \{A_N \mid N \in \mathscr{A}\}$.

On a c-ring C consider a fixed refinement $\mathfrak{L}\colon \mathscr{A} \to \mathscr{C}(C)$ such that $\mathfrak{L}(N) \equiv \{c \in C \mid \operatorname{coz} c \in N\}$. A c-extension $u\colon C \rightarrowtail A$, where $(A, \mathfrak{A})$ is a cr-ring, is called a *cr-extension of the cr-ring* $(C, \mathfrak{L})$ if $C_N = u^{-1}A_N$. This extension will be denoted by $u\colon (C, \mathfrak{L}) \rightarrowtail (A, \mathfrak{A})$. A *morphism from* $u\colon (C, \mathfrak{L}) \rightarrowtail (A, \mathfrak{A})$ *to* $\hat{u}\colon (C, \mathfrak{L}) \rightarrowtail (\hat{A}, \hat{\mathfrak{A}})$ is a ring homomorphism $v\colon A \to \hat{A}$ such that $v \circ u = \hat{u}$ and $vA_N \subset \hat{A}_N$. If, in addition, v is injective and $A_N = v^{-1}\hat{A}_N$, we say that the second cr-extension is *greater* than the first one, or that the first cr-extension is *embedded* in the second one.

A mapping $\mathfrak{A}\colon \mathscr{A} \to \mathscr{C}(A)$ will be called *saturated* if for every proper ideal A_N and for every proper annihilator ideal E such that $E^* \not\subset A_N$ there exists a proper ideal A_M such that $A_N \cup E \subset A_M$ and $M \supset N$. If the mapping $\mathfrak{A}$ possesses properties a)—c) in the definition of refinement and is saturated, then $\mathfrak{A}$ is a refinement. A cr-ring $(A, \mathfrak{A})$ and its cr-extension $u\colon (C, \mathfrak{L}) \rightarrowtail (A, \mathfrak{A})$ will be called *saturated* if the refinement $\mathfrak{A}$ is saturated.

1.2. ***as*-preimages and enclosable *as*-coverings.**

1.2.1. *a-spaces and a-preimages.* Let H be some set. A family $\mathscr{K}$ of subsets in H will be called a *foundation* if it contains $\varnothing$ and H and is closed with respect to finite intersections. A foundation $\mathscr{K}$ is called *separating* if for arbitrary points s_1 and s_2 in H there exists an element $G \in \mathscr{K}$ such that $s_1 \in G$ and $s_2 \notin G$. Let co-$\mathscr{K} \equiv \{H \setminus G \mid G \in \mathscr{K}\}$. A foundation $\mathscr{H}$ is called an *Alexandrov foundation* or an *a-foundation* if a) $\mathscr{H}$ is closed with respect to finite unions; b) disjoint elements in co-$\mathscr{H}$ are contained in disjoint elements in $\mathscr{H}$; c) each element in $\mathscr{H}$ is a finite union of elements in co-$\mathscr{H}$. A pair $(H, \mathscr{H})$ with a separating a-foundation $\mathscr{H}$ is called an *Alexandrov space* or an *a-space*. These spaces were introduced by A. D. Alexandrov in **[6]**.

Let $\mathscr{K}$ be a subfoundation of an a-foundation $\mathscr{H}$. Consider the families $\mathscr{G}(\mathscr{K})$ and $\mathscr{G}^0(\mathscr{K})$, consisting of the unions of the elements of all, respectively, all countable, subfamilies in $\mathscr{K}$, and also $\mathscr{F}(\mathscr{K}) \equiv$ co-$\mathscr{G}(\mathscr{K})$. For every function $f\colon H \to \mathbb{R}$ the set $\operatorname{coz} f \equiv \{s \in H \mid f(s) \neq 0\}$ is called the cozero set of f. In the sequel we assume $\mathbb{R}$ to be equipped with the canonical a-foundation consisting of all open sets. If $(H, \mathscr{H})$ and $(\hat{H}, \hat{\mathscr{H}})$ are a-spaces, the mapping $\gamma\colon H \to \hat{H}$ is called an *a-mapping* if $\gamma^{-1}\hat{\mathscr{H}} \equiv \{\gamma^{-1}G \mid \tilde{G} \in \hat{\mathscr{H}}\} \subset \mathscr{H}$. In **[6]** the following theorem was proved.

THEOREM. *Let $(H, \mathscr{H})$ be an a-space. Then $(H, \mathscr{G}(\mathscr{H}))$ is a completely regular space and for every $G \in \mathscr{H}$ there exists a bounded a-function $f\colon H \to \mathbb{R}$ such that $G = \operatorname{coz} f$.*

Let $(T, \mathscr{T})$ be some fixed a-space. An a-space $(H, \mathscr{H})$ with a surjective perfect **[9,** VI, § 2] a-mapping $\tau\colon H \twoheadrightarrow T$ will be called an *a-preimage of the a-space* $(T, \mathscr{T})$ and denoted by $\tau\colon (T, \mathscr{T}) \twoheadleftarrow (H, \mathscr{H})$.

1.2.2. *c-rings and a-subfoundations on a-spaces.* Let $(H, \mathscr{H})$ be an a-space and $\mathscr{K}$ a subfoundation in $\mathscr{H}$. The function $f\colon H \to \mathbb{R}$ will be called a *function with small oscillation with respect to the subfoundation* $\mathscr{K}$ if there exists a sequence of finite coverings $\varkappa_n$ of H such that the oscillation $\omega(f, Q) \equiv \sup\{|f(s_1) - f(s_2)| \,\big|\, \{s_1, s_2\} \subset Q\}$ of f on each element in $Q \in \varkappa_n \subset \mathscr{K}$ is less than $1/n$. The c-ring of all such bounded functions is denoted by $O(H, \mathscr{K})$. The unit element in $O(H, \mathscr{K})$ is the function $\mathbf{1}$ which assumes the value 1 for every $s \in H$. If $\mathscr{K}$ is closed with respect to countable unions, then $O(H, \mathscr{K})$ coincides with the c-ring of all bounded functions $f: H \to \mathbb{R}$ such that $f^{-1}(]a,b[) \in \mathscr{K}$ for every $]a,b[$. The c-ring $O(H, \mathscr{K})$ will be called a *c-ring of the a-space $(H, \mathscr{H})$*.

Let $\mathscr{C}$ be some family of finite coverings $\varkappa \subset \mathscr{H}$ of the set H closed with respect to the operation $\Lambda\{\varkappa_k \mid k = 1, \dots, m\} \equiv \{Q^1 \cap \dots \cap Q^m \mid Q^k \in \varkappa_k, k = 1, \dots, m\}$ and including the covering $\{H\}$. The function $f: H \to \mathbb{R}$ will be called a *function with small oscillations with respect to $\mathscr{C}$* if there exists a sequence of coverings $\{\varkappa_n\} \subset \mathscr{C}$ such that $\omega(f, Q) < 1/n$ for every $Q \in \varkappa_n$. The c-ring of all such bounded functions is denoted by $O(H, \mathscr{C})$. Consider the a-subfoundation $Z(\mathscr{C}) \subset \mathscr{H}$ consisting of the cozero sets of all functions $f \in O(H, \mathscr{C})$. It will be called an *a-subfoundation generated by the family $\mathscr{C}$*. It is possible to give an intrinsic description of it.

1.2.3. *as-spaces and as-preimages.* Let T and S be the same as in Section 1.1.3. Consider the canonical a-base $\mathscr{T}$ on T consisting of the cozero sets of all functions $f \in C$. Then $C = O(T, \mathscr{T})$ and therefore C is the c-ring of the a-space $(T, \mathscr{T})$.

The mapping $\mathfrak{H}: \mathscr{A} \to \mathscr{F}(\mathscr{H})$ will be called a *screen of the a-space $(H, \mathscr{H})$* if a) $\mathfrak{H}(N) = \varnothing$ iff $N = \mathscr{P}$; b) $\bigcup \mathfrak{H}(N)$ is dense in H; c) $N_1 \supset N_2$ implies $\mathfrak{H}(N_1) \subset \mathfrak{H}(N_2)$; d) $N = \operatorname{top} N_\xi$ implies $\mathfrak{H}(N) = \operatorname{cl} \cup \mathfrak{H}(N_\xi)$.

An a-space $(H, \mathscr{H})$ equipped with a screen $\mathfrak{H}$ will be called an *as-space* and denoted by $(H, \mathscr{H}, \mathfrak{H})$. From now on when dealing with as-spaces the symbols $\mathscr{H}$ and $\mathfrak{H}$ will sometimes be omitted.

On an a-space T consider a fixed screen $\mathfrak{T} \equiv \{T_N \mid N \in \mathscr{A}\}$ such that $T_N \equiv T \setminus \bigcup \{G \in \mathscr{T} \mid G \in N\}$. The a-preimage $\tau: T \twoheadleftarrow H$, where $(H, \mathfrak{H})$ is an as-space, will be called an *as-preimage of the as-space $(T, \mathfrak{T})$* if $\tau H_N = T_N$. Such a preimage will be denoted by $\tau: (T, \mathfrak{T}) \twoheadleftarrow (H, \mathfrak{H})$. The *morphism from $\tau: (T, \mathfrak{T}) \twoheadleftarrow (H, \mathfrak{H})$ to $\hat{\tau}: (T, \mathfrak{T}) \twoheadleftarrow (\hat{H}, \mathfrak{H})$* is a perfect a-mapping $\gamma: H \to \hat{H}$ such that $\hat{\tau} \circ \gamma = \tau$ and $\gamma H_N \subset \hat{H}_N$. If, in addition, γ is surjective and $\gamma H_N = \hat{H}_N$ we say that the first as-preimage is *greater* then the second one.

A mapping $\mathfrak{H}: \mathscr{A} \to \mathscr{F}(\mathscr{H})$ is called *saturated* if for each $H_N \neq \varnothing$ and each $G \in \mathscr{H}$ intersecting H_N there exists $H_M \neq \varnothing$ such that $H_M \subset H_N \cap G$ and $M \supset N$. If the mapping $\mathfrak{H}$ possesses the properties a)—c) in the definition of screens and is saturated then it is a screen. An as-space $(H, \mathfrak{H})$ and its as-preimage $\tau: (T, \mathfrak{T}) \twoheadleftarrow (H, \mathfrak{H})$ will be called *saturated* if the screen $\mathfrak{H}$ is saturated.

1.2.4. *Elementary types of enclosability.* For P and Q in $\mathscr{G}(\mathscr{H})$ we shall write $Q = sP$ if $P \subset Q$ and from $N \in \mathscr{A}$, $G \in \mathscr{H}$ and $G \cap P \cap H_N = \varnothing$ it follows that $G \cap Q \cap H_N = \varnothing$. For a family $\varkappa \subset \mathscr{G}(\mathscr{H})$ consider the sets $\operatorname{bod} \varkappa \equiv \bigcup \{Q \mid Q \in \varkappa\}$ and $\operatorname{cobod} \varkappa \equiv H \setminus \operatorname{bod} \varkappa$. A family $\varkappa' \subset \mathscr{G}(\mathscr{H})$ is an *s-dense enclosure of the family* $\varkappa \equiv \{Q_i\} \subset \mathscr{G}(\mathscr{H})$ if $Q_i' = sQ_i$ and $\operatorname{bod} \varkappa' \in \mathscr{H} \cap \text{co-}\mathscr{H}$. In this case we shall write $\varkappa' = s\varkappa$. A family $\varkappa \subset \mathscr{G}(\mathscr{H})$ is said to be *complemented by a family $\rho \subset \mathscr{G}(\mathscr{H})$* if $(\operatorname{bod} \varkappa) \cap (\operatorname{bod} \rho) = \varnothing$ and $H = s(\operatorname{bod} \varkappa \cup \operatorname{bod} \rho)$.

Let a subfoundation $\mathscr{K}$ be fixed in $\mathscr{H}$. The set of all finite families $\varkappa \equiv \{Q_i\} \subset \mathscr{G}(\mathscr{K})$ such that from $G \in \mathscr{K}$ and $G \subset Q_i$ it follows that $G \cap Q_j \in \mathscr{K}$ for all j will be denoted by $K(\mathscr{K})$. The set of all families $\varkappa \in K(\mathscr{K})$ complemented by some families $\rho \in K(\mathscr{K})$ is denoted by $K^c(\mathscr{K})$. Henceforth π will denote one of the symbols $\varnothing$ or c, and the symbol $\varnothing$ will be omitted. So we can write $K^\pi(\mathscr{K})$ instead of $K(\mathscr{K})$ and $K^c(\mathscr{K})$.

An as-space H will be called *Z^π-enclosable with respect to the subfoundation $\mathscr{K}$* if every family $\varkappa \in K^\pi(\mathscr{K})$ possesses an s-dense enclosure. For $\mathscr{K} = \mathscr{H}$ we obtain the definition of a *${}^aZ^\pi$-enclosable as-space H*. The as-preimage $T \twoheadleftarrow H$ is *${}^aZ^\pi$-enclosable* if the as-space H is ${}^aZ^\pi$-enclosable.

1.2.5. *Composite types of enclosability.* We define the many-place operation $\wedge$ on the set of all families $\varkappa \subset \mathcal{G}(\mathcal{H})$ by setting $\wedge\{\varkappa_k \mid k = 1, \dots, m\} \equiv \{Q^1 \cap \dots \cap Q^m \mid Q^k \in \varkappa_k, k = 1, \dots, m\}$. Consider the set $S^\pi(\mathcal{K})$ of all finite families $\varkappa' \subset \mathcal{H}$ that are s-dense enclosures of some families $\varkappa \in K^\pi(\mathcal{K})$. Each enclosure $\varkappa' \in S^\pi(\mathcal{K})$ generates a covering $\varkappa' \cup \{\operatorname{cobod} \varkappa'\}$ of H. Consider the set $C^\pi(\mathcal{K})$ of coverings of H generated by the set $\{\varkappa' \cup \{\operatorname{cobod} \varkappa'\} \mid \varkappa' \in S^\pi(\mathcal{K})\}$ and by the operation $\wedge$. Denote by $Z^\pi(\mathcal{K})$ the a-subfoundations of $Z(C^\pi(\mathcal{K}))$ generated by $C^\pi(\mathcal{K})$. Then $\mathcal{K} \subset Z^\pi(\mathcal{K})$.

Let $\tau \colon T \twoheadleftarrow H$ be an as-preimage. For a word $\pi_1 \cdots \pi_k$ we define by induction the sequence of subfoundations $\mathcal{K}_0 \equiv \tau^{-1}\mathcal{T}$, $\mathcal{K}_1 \equiv Z^{\pi_1}(\mathcal{K}_0)$, and $\mathcal{K}_i \equiv Z^{\pi_i}(\mathcal{K}_{i-1})$. The as-preimage $\tau \colon T \twoheadleftarrow H$ will be called *$Z^{\pi_1} \cdots Z^{\pi_k}$-enclosable* if the as-space H is Z^{π_i}-enclosable with respect to the subbase $\mathcal{K}_{i-1}$ for all $i = 1, \dots, k$. The as-preimage $\tau \colon T \twoheadleftarrow H$ is called the *as-space of enclosures of the type $Z^{\pi_1} \cdots Z^{\pi_k}$ of the as-space T* if $\mathcal{H} = \mathcal{K}_k$. An *enclosable as-cover of the type $Z^{\pi_1} \cdots Z^{\pi_k} |^a Z^\pi$ of the as-space T* is an as-preimage which a) is the greatest of all as-spaces of enclosures of the type $Z^{\pi_1} \cdots Z^{\pi_k}$ of the as-space T; b) is the smallest of all $Z^{\pi_1} \cdots Z^{\pi_k}$-enclosable as-preimages of the as-space T; c) is ${}^aZ^\pi$-enclosable. Any as-space T can have only one (up to isomorphism) enclosable as-cover. It is the unique $Z^{\pi_1} \cdots Z^{\pi_k}$-enclosable as-space of enclosures of the type $Z^{\pi_1} \cdots Z^{\pi_k}$ of the as-space T; this means that it is totally characterized only by the relative properties entering into the left-hand side of $Z^{\pi_1} \cdots Z^{\pi_k} |^a Z^\pi$. In this paper only the types $ZZZ^c |^a Z$ and $ZZ^c |^a Z^c$ will be used.

1.3. cr-extensions associated with as-preimages.

1.3.1. *cr-extensions corresponding to as-preimages.* Consider on an as-space $(H, \mathcal{H}, \mathfrak{H})$ the c-ring $O \equiv O(H, \mathcal{H})$ of the a-space $(H, \mathcal{H})$ and a refinement $\mathfrak{O} \equiv \{O_N \mid N \in \mathcal{A}\}$ such that $O_N \equiv \{f \in O \mid H_N \cap \operatorname{coz} f = \varnothing\}$. For an as-preimage $\tau \colon T \twoheadleftarrow H$ consider a cr-extension $u_\tau \colon (C, \mathfrak{L}) \rightarrowtail (O, \mathfrak{O})$ such that $u_\tau c \equiv c \circ \tau$. We shall call it a *$cr$-extension corresponding to (or associated with) the as-preimage* $\tau \colon T \twoheadleftarrow H$. It is clear that $\hat{\tau} \colon T \rightarrowtail \hat{H} \leqslant \tilde{\tau} \colon T \twoheadleftarrow \tilde{H}$ implies $\hat{u} \colon C \rightarrowtail \hat{O} \leqslant \tilde{u} \colon C \rightarrowtail \tilde{O}$. Now we prove the converse.

Proposition 1. *If $\hat{u} \colon C \rightarrowtail \hat{O} \leqslant \tilde{u} \colon C \rightarrowtail \tilde{O}$, then $\hat{\tau} \colon T \twoheadleftarrow \hat{H} \leqslant \tilde{\tau} \colon T \twoheadleftarrow \tilde{H}$.*

Proof. Let $v \colon \hat{O} \rightarrowtail \tilde{O}$ be the morphism imbedding the first cr-extension into the second one. For a fixed point $s \in \tilde{H}$ consider the sets $\Gamma \equiv \{f \in \hat{O} \mid s \in \operatorname{coz} v f\}$ and $P \equiv \hat{\tau}^{-1}\tilde{\tau}s$. Since $P \cap \operatorname{cl} \operatorname{coz} f \neq \varnothing$ for all $f \in \Gamma$, we have $P_s \equiv \bigcap\{P \cap \operatorname{cl} \operatorname{coz} f \mid f \in \Gamma\} \neq \varnothing$ by compactness of P. Since P_s consists of a single point, we can define the mapping $\gamma \colon \tilde{H} \to \hat{H}$ correctly by setting $\gamma s \equiv P_s$. It is continuous. It follows from $\hat{\tau} \circ \gamma = \tilde{\tau}$ that γ is perfect [**9**, VI, § 2.56].

The relation $vf = f \circ \gamma$ implies that $\gamma^{-1}G \in \tilde{\mathcal{H}}$ for all $G \in \hat{\mathcal{H}}$ and that the mapping γ is surjective. Besides this, it implies $\gamma \tilde{H}_N = \hat{H}_N$. Proposition 1 is proved. □

1.3.2. *Realization of cr-extensions.* We shall say that an as-preimage $\tau \colon T \twoheadleftarrow H$ *realizes the cr-extension* $u \colon C \rightarrowtail A$ if there exists a morphism r embedding this cr-extension into a cr-extension $u_\tau \colon C \rightarrowtail O$ such that $\mathcal{H} = \{\operatorname{coz} ra \mid a \in A\}$ and $H_N = \{s \in H \mid \forall a \in A_N (ra(s) = 0)\}$. This morphism r is called a *realizing morphism*. If $rA = O$ and $r^{-1}O_N = A_N$, then r is called a *realizing isomorphism*. Any cr-extension can have only one (up to isomorphism) realizing as-preimage.

1.4. Functional polyfactor cr-extensions.

1.4.1. *Definition of functional polyfactor cr-extensions.* Let $\mathscr{I}$ be some ideal in the Boolean algebra $\mathscr{P}$ **[1,** § 1.VII]. Real functions f and g on T are called equivalent modulo $\mathscr{I}$ $(f \sim g \bmod \mathscr{I})$ if $\{t \in T \mid |f(t) - g(t)| \geqslant 1/n\} \in \mathscr{I}$ for every $n \in \mathbb{N}$ **[3,** 15.1.3]. If $\mathscr{G}$ and $\mathscr{K}$ are some families of subsets in T closed with respect to union we denote by $<\mathscr{J}, \mathscr{K}>$ the ideal $\{P \in \mathscr{P} \mid \exists J \in \mathscr{J} \exists K \in \mathscr{K}(P \subset J \cup K)\}$.

A family $\mathfrak{J} \equiv \{\mathscr{I}_N \mid N \in \mathscr{A}\}$ of ideals on the as-space T will be called a *polyideal on* T if $\mathscr{I}_N \subset \mathscr{I}_M$ for arbitrary $M \supset N$. A polyideal $\mathfrak{J}$ is called *s-meager* if $G \in \mathscr{T}$ and $G \cap T_N \neq \varnothing$ imply $G \cap T_N \notin \mathscr{I}_N$. The family $\mathscr{A}$ is an s-meager polyideal on T. We call it the *Kelley polyideal.*

Let F be a c-subring of the c-ring of all bounded real functions on an as-space T and $C \subset F$. A mapping $\varphi\colon \mathscr{A} \to F$ such that $\varphi(N) \sim \varphi(M) \bmod M$ for all $M \supset N$ is called an *F-polyfunction on* T **[3,** 27.2.1]. Denote $\varphi(N)$ by f_N. Then $\varphi = \{f_N \mid N \in \mathscr{A}\}$. An F-polyfunction φ is called *bounded* if there exists a number $r \in \mathbb{N}$ such that $|f_N| \leqslant r\mathbf{1}$ for all N. F-polyfunctions $\varphi' \equiv \{f'_N\}$ and $\varphi'' \equiv \{f''_N\}$ are called *equivalent modulo the polyideal* $\mathscr{A}$ $(\varphi' \sim \varphi'' \bmod \mathscr{A})$ if $f'_N \sim f''_N \bmod N$ for every N. The set of equivalence classes $\bar{\varphi}$ of all bounded F-polyfunctions φ is denoted by $F/\mathscr{A}$. It is not hard to verify that $F/\mathscr{A}$ is a c-ring. Let $f \in F$. Consider the stable polyfunction $\varphi_f \equiv \{f_N\}$ such that $f_N \equiv f$ for every N. Introduce the homomorphism $u\colon C \rightarrowtail F/\mathscr{A}$ such that $uc \equiv \bar{\varphi}_c$. The c-extension $u\colon C \rightarrowtail F/\mathscr{A}$ thus obtained is called a *functional polyfactor c-extension.*

For $f \in F$ and $n \in \mathbb{N}$ consider the sets $\operatorname{coz}_n f \equiv \overline{|f|^{-1}(]1/n, +\infty[)}$. Consider also the mapping $\mathfrak{A}\colon \mathscr{A} \to \mathscr{C}(F/\mathscr{A})$ such that $\mathfrak{A}(N) \equiv \{\overline{\{f_N\}} \in F/\mathscr{A} \mid \forall n(T_N \cap \operatorname{coz}_n f_N \in N)\}$.

The triple $(\mathfrak{T}, \mathscr{A}, F)$ will be called *consistent* if a) for all $n \in \mathbb{N}$ and $f \in F$ subject to $\operatorname{coz}_n f \notin N$ we have $T_N \cap \operatorname{coz}_{2n} f \notin N$; b) for all n, N, and f such that $T_N \cap \operatorname{coz}_n f \notin N$, there exists $M \in \mathscr{A}$ such that $N \subset M \neq \mathscr{P}$ and $T_M \setminus \operatorname{coz}_{2n} f \in M$. If the triple $(\mathfrak{T}, \mathscr{A}, F)$ is consistent, the mapping $\mathfrak{A}$ is a saturated refinement of the c-ring $F/\mathscr{A}$. The cr-extension $u\colon (C, \mathfrak{L}) \rightarrowtail (F/\mathscr{A}, \mathfrak{A})$ corresponding to the consistent triple $(\mathfrak{T}, \mathscr{A}, F)$ is called *a functional-polyfactor cr-extension.*

1.4.2. *Topological connection of elements of functional-polyfactor cr-extensions and of cr-extensions associated to as-preimages.* Let $u_\tau\colon C \rightarrowtail O$ be the cr-extension corresponding to an as-preimage $\tau\colon T \twoheadleftarrow H$; let $u\colon C \rightarrowtail F/\mathscr{A}$ be a functional-polyfactor cr-extension and let $\mathfrak{J} \equiv \{\mathscr{I}_N \mid N \in \mathscr{A}\}$ be a polyideal on H. A function $a \in O$ and an element $p = \overline{\{f_N\}} \in F/\mathscr{A}$ will be called *topologically connected modulo* $\mathfrak{J}$ $(a \mapsfrom p \bmod \mathfrak{J})$ if $a \sim f_N \circ \tau \bmod <\tau^{-1}N, \mathscr{I}_N>$. The relation $a \mapsfrom p \bmod \mathfrak{J}$ is multivalued in general. The multivalued mappings $a \mapsto p$ and $p \mapsto a$ will be called *topologically connecting modulo* $\mathfrak{J}$. The former mapping is called *lowering*, the latter is called *lifting*. They preserve the ring operations. Consider the ideals $<\tau^{-1}N, \mathscr{I}_N>$ and the polyideal $\{<\tau^{-1}N, \mathscr{I}_N> \mid N \in \mathscr{A}\}$. If this polyideal is s-meager, then the relation $a \mapsfrom p \bmod \mathfrak{J}$ is one-to-one and $a \in O_N$ if and only if $p \in (F/\mathscr{A})_N$.

1.4.3. *Concrete s-meager ideals in as-preimages.* Let $\tau\colon T \rightarrowtail H$ be an as-preimage; let P and Q be subsets in H and $P \subset Q$. The set P is called *N-dense* in Q if the conditions $M \supset N$, $G \in \mathscr{H}$, and $G \cap P \cap H_M = \varnothing$ imply $G \cap Q \cap H_M = \varnothing$. The family of all open sets U that are N-dense in H will be denoted by $\mathscr{U}_N(H)$.

The ideal $\{R \subset H \mid \exists U \in \mathscr{U}_N(H)(R \subset H \setminus U)\}$ is denoted by $\mathscr{R}_N(H)$. The family $\mathfrak{R}(H) \equiv \{\mathscr{R}_N(H) \mid N \in \mathscr{A}\}$ is an s-meager polyideal on H.

Denote by $\mathcal{S}_N(H)$ the family of all sets of the form $\bigcap\{U_i \in \mathcal{U}_N(H) \mid i \in \mathbb{N}\}$ and by $\mathcal{F}_N(H)$ the ideal $\{F \subset H \mid \exists S \in \mathcal{S}_N(H)(F \subset H \setminus S)\}$. The family $\mathfrak{F}(H) \equiv \{\mathcal{F}_N(H) \mid N \in \mathcal{A}\}$ is a polyideal on H.

LEMMA 2. *Let $\tau\colon T \twoheadleftarrow H$ be an as-preimage, $S \in \mathcal{S}_N(H)$, and $P \in N$. Then the set $S \setminus \tau^{-1}P$ is N-dense in H.*

PROOF. Let $S = \bigcap U_i$ and $M \supset N$. For $T_M \setminus P$ there exists a sequence $\{K_j\}$ of compact sets such that $\bigcup K_j \subset T_M \setminus P$ and $T_M \setminus P \setminus \bigcup K_j \in M$. Consider the ideals $M_j \equiv M(K_j)$. Then $M = \operatorname{top} M_j$ implies $H_M = \operatorname{cl} \cup H_{M_j}$. The sets $V_{ji} \equiv U_i \cap H_{M_j}$ are dense in H_{M_j} by definition. Since the space H_{M_j} is compact, the set $S_j \equiv \bigcap\{V_{ji} \mid i\}$ is dense in H_{M_j}. Let $G \cap (S \setminus \tau^{-1}P) \cap H_M = \varnothing$. Since $H_{M_j} \cap \tau^{-1}P = \varnothing$, it follows from $(G \cap H_{M_j}) \cap S_j = \varnothing$ that $G \cap H_M = \varnothing$. The lemma is proved. □

COROLLARY. *The polyideals $\{<\tau^{-1}N, \mathcal{F}_N(H)> \mid N \in \mathcal{A}\}$ and $\{<\tau^{-1}N, \mathcal{R}_N(H)> \mid N \in \mathcal{A}\}$ are s-meager.*

§2. Description of hyperstonean and basically hyperstonean preimages

2.1. Definition of the Arens extension and of the hyperstonean preimage. A continuous functional φ on a Banach space C is called tight if there exists a sequence $\{K_n\}$ of compact sets in T such that $c \in C$ and $0 < c \leqslant \chi(T \setminus K_n)$ imply $|\varphi c| < 1/n$ [**2**, p. 1.10.4]. Denote by C^t the normed space of all continuous tight functionals on C. This space is remarkable by the fact that according to the Riesz-Prokhorov theorem [**4**, IX, § 5.2] it coincides with the space of all integrals i_μ for all bounded Radon measures μ on T. Further denote by C^{tt} the Banach space of all continuous linear functionals on C^t. In the case of a compact space T it coincides with the usual second dual C''.

We equip C^{tt} with the *Arens product* in the following way. For $b, c \in C$ and $\xi \in C^t$ define $\xi \cdot b \in C^t$ by setting $(\xi \cdot b)c \equiv \xi(bc)$. For $\psi \in C^{tt}$ define $\psi \cdot \xi \in C^t$ by setting $(\psi \cdot \xi) \equiv \psi(\xi \cdot b)$. Thus C^{tt} becomes a c-ring. Let v be the canonical imbedding of C into C^{tt} such that $(vc)\xi \equiv \xi(c)$ for all $\xi \in C^t$. The c-extension $v\colon C \rightarrowtail C^{tt}$ is called the (*second dual*) *Arens extension of the ring* C.

Consider on an as-space $(T, \mathcal{T}, \mathfrak{T})$ the lattice $\mathcal{UM}$ of all universally measurable sets (described in 1.1.2). A function $f\colon T \to \mathbb{R}$ is called (Lebesgue) universally measurable if $f^{-1}(]x, y[) \in \mathcal{UM}$ for every interval $]x, y[$. The c-ring of such bounded functions will be denoted by UM. It is clear that $UM = O(T, \mathcal{UM})$. The functional c-extension $u\colon C \rightarrowtail UM$ is called *the universally measurable Kaplan extension*. It was studied by Kaplan in [**21**]. According to 1.4.1 the set $\mathbb{UM} \equiv UM \setminus \mathcal{A}$ is a functional-polyfactor c-ring. The c-extension $u\colon C \rightarrowtail \mathbb{UM}$ will be called the *functional-polyfactor Arens extension*. It was introduced by Arens in [**5**] where it was proved that the c-extensions $v\colon C \rightarrowtail C^{tt}$ and $u\colon C \rightarrowtail \mathbb{UM}$ are isomorphic. Therefore, we shall study only the extension $u\colon C \rightarrowtail \mathbb{UM}$ here.

The triple $(\mathfrak{T}, \mathcal{A}, \mathcal{UM})$ possesses the following properties: a) $T_N \cap X \notin N$ for every $X \in \mathcal{UM} \setminus N$; b) for arbitrary N and X satisfying $T_N \cap X \notin N$ there exists $M \in \mathcal{A}$ such that $N \subset M \neq \mathcal{P}$ and $T_M \setminus X \in M$. It follows from these properties that the triple $(\mathfrak{T}, \mathcal{A}, UM)$ is consistent. According to 1.4.1 the mapping $\mathfrak{A}\colon \mathcal{A} \to \mathcal{C}(\mathbb{UM})$ defined by $\mathfrak{A}(N) \equiv \{\overline{\{f_N\}} \in \mathbb{UM} \mid \forall n(T_N \cap \operatorname{coz}_n f_N \in N\}$ is a saturated refinement of the c-ring $\mathbb{UM}$ and $u\colon (C, \mathfrak{C}) \rightarrowtail (\mathbb{UM}, \mathfrak{A})$ is a cr-extension. A description of the extensions $C \rightarrowtail UM$ and $C \rightarrowtail \mathbb{UM}$ was announced by the author in [**22**].

The realizing as-preimage $\tau : T \twoheadleftarrow hT$ of the Arens cr-extension $u \colon C \rightarrowtail \mathbb{UM}$ is called *the hyperstonean preimage of the space* T **[13]**. Here we prove its existence.

Denote by $\hat{\mathscr{P}}$ the class of all as-spaces of enclosures $\hat{\tau} \colon T \twoheadleftarrow \hat{H}$ of the type ZZZ^c and by $\tilde{\mathscr{P}}$ the class of all ZZZ^c-enclosable as-preimages $\tilde{\tau} \colon T \twoheadleftarrow H$. Let $\hat{u} \colon C \rightarrowtail \hat{O}$ and $\tilde{u} \colon C \rightarrowtail \tilde{O}$ denote cr-extensions corresponding to these preimages.

2.2. Construction of the preimage in $\hat{\mathscr{P}} \cap \tilde{\mathscr{P}}$. A mapping $\eta \colon \mathscr{A} \to \mathscr{U}\mathscr{M}$ such that $\eta(N) \sim \eta(M) \bmod M$ for all $M \supset N$ will be called a *universally measurable polyset on* T. Denote $\eta(N)$ by X_N. Then $\eta = \{X_N \mid N \in \mathscr{A}\}$. Polysets $\eta \equiv \{X_N\}$ and $\theta \equiv \{Y_N\}$ will be called *equivalent modulo* $\mathscr{A}$ ($\eta \sim \theta \bmod \mathscr{A}$) if $X_N \sim Y_N \bmod N$ for every N. The set of equivalence classes $p \equiv \bar{\eta}$ of all polysets η will be denoted by $\mathscr{L}$. It is not hard to verify that $\mathscr{L}$ is a Boolean algebra. For $X \in \mathscr{U}\mathscr{M}$ consider the stable polyset $\eta_X \equiv \{X_N\}$ such that $X_N \equiv X$ for each N.

In the lattice $\mathscr{L}$ we consider the set H_0 of all maximal ideals θ that do not contain the unit. To each element $\overline{\{X_N\}}$ we associate the set $\sigma_0\overline{\{X_N\}}$ of all maximal ideals θ such that $\overline{\{X_N\}} \notin \theta$. Since $\bigcup \sigma_0\overline{\{X_N^k\}} = \sigma_0\overline{\{\bigcup X_N^k\}}$ and $\bigcap \sigma_0\overline{\{X_N^k\}} = \sigma_0\overline{\{\bigcap X_N^k\}}$ for every finite family $\overline{\{X_N^k\}} \subset \mathscr{L}$ we can introduce a topology on H_0 by taking $\{\sigma_0\overline{\{X_N\}} \mid \overline{\{X_N\}} \in \mathscr{L}\}$ as a base of open sets.

LEMMA 3. *The space H_0 is compact.*

PROOF. Let θ_1 and θ_2 be different maximal ideals. Then there exist $\overline{\{X_n^i\}} \in \theta_i \setminus \theta_j$. Consider $Y_N^i \equiv X_N^i \setminus (X_N^1 \cap X_N^2)$. So $\theta_i \in \sigma_0\overline{\{Y_N^j\}}$ and $\sigma_0\overline{\{Y_n^1\}} \cap \sigma_0\overline{\{Y_N^2\}} = \varnothing$. Thus H_0 is Hausdorff.

Let $\{\sigma_0 p_\gamma \mid \gamma \in \Gamma\}$ be a covering of H_0. Suppose that $\sup p_{\gamma_i} \neq \bar{\eta}_T$ for every finite subset $\{\gamma_i\} \subset \Gamma$. Consider the ideal $\theta' \equiv \{q \in \mathscr{L} \mid \exists\{\gamma_i\}(q \leqslant \sup p_{\gamma_i})\}$. It can be imbedded into a maximal ideal θ. Since $\theta \in \sigma_0 p_\gamma$ for some γ, then $\theta \not\ni p_\gamma \in \theta' \subset \theta$. This contradiction means that $\bigcup_i \sigma_0 p_{\gamma_i} = H_0$ for some $\{\gamma_i\}$. The lemma is proved. □

LEMMA 4. *If $p \neq \bar{\eta}_\varnothing$ then $\sigma_0 p \neq \varnothing$.*

PROOF. For $\{X_N\} \in p$ take $Z_N \equiv T \setminus X_N$. Consider the ideal $\theta' \equiv \{\overline{\{R_N\}} \mid R_N \cap X_N \in N\}$ without unit. Imbed it into some maximal ideal θ corresponding to some point $s \in H_0$. Since $q \equiv \overline{\{Z_N\}} \in \theta$ and $p \vee q \equiv \bar{\eta}_T$ it follows that $p \notin \theta$. Therefore $s \in \sigma_0 p$. The lemma is proved. □

For given T and $\mathscr{T}$ there exists a compact space T_0 such that T is dense in T_0 and $\mathscr{T}$ coincides with the trace on T of the family $\mathscr{T}_0$ of cozero sets of all continuous functions on T_0.

For the point $s \in H_0$ corresponding to the maximal ideal θ_s we consider the sets $P_{0s} \equiv \bigcap\{\mathrm{cl}_{T_0} \cup \{X_N \cap T_N \mid N \in \mathscr{A}\} \mid \overline{\{X_N\}} \notin \theta_s\}$ and $P_s \equiv \bigcap\{\mathrm{cl} \cup \{X_N \cap T_N \mid N \in \mathscr{A}\} \mid \overline{\{X_N\}} \notin \theta_s\}$.

LEMMA 5. *The set P_{0s} is a singleton. The set P_s cannot contain more than one point.*

PROOF. Suppose that there exist two different points t_1 and t_2 in P_s. Take G_1, G_2 in $\mathscr{T}$ such that $t_i \notin \mathrm{cl}\, G_j$ and $G_1 \cup G_2 = T$. Then $\bar{\eta}_{G_j} \notin \theta_s$ for some j. But this would imply $t_i \in \mathrm{cl} \cup \{G_j \cap T_N \mid N\} = \mathrm{cl}\, G_j$ which is wrong. So $|P_s| \leqslant 1$. In a similar way it turns out that $|P_{0s}| \leqslant 1$. Let a finite family $\{p_k\}$ be disjoint with θ_s and let $\{X_N^k\} \in p_k$. Then $s \in \sigma_0 \wedge p_k$ means that $Y_N \equiv \bigcap X_N^k \notin N$ for some N. By properties of the corresponding triple discussed in §2.1 we have $Y_N \cap T_N \neq \varnothing$.

Thus, the relation $\bigcap\{\mathrm{cl}_{T_0}\bigcup\{X_N^k \cap T_N \mid N\}\} \supset \mathrm{cl}_{T_0}\bigcup\{Y_N \cap T_N \mid N\} \neq \varnothing$ means that the family of closed sets $\{\mathrm{cl}_{T_0}\bigcup\{X_N \cap T_N \mid \overline{\{X_N\}} \notin \theta_s\}$ has a finite intersection property. Therefore $P_{0s} \neq \varnothing$ and so $|P_{0s}| = 1$. The lemma is proved. □

Consider the subspace $H \equiv \{s \in H_0 \mid P_s \neq \varnothing\}$. The sets $\sigma p \equiv H \cap \sigma_0 p$ form a base of open sets in H. Consider the a-foundation $\mathscr{H}_0$ on H consisting of the cozero sets of all continuous functions on H_0 and the a-foundation $\mathscr{H} \equiv \{G \cap H \mid G \in \mathscr{H}_0\}$ on H. Then $(H_0, \mathscr{H}_0)$ and $(H, \mathscr{H})$ are a-spaces.

Introduce the mappings $\tau_0\colon H_0 \to T_0$ and $\tau\colon H \to T$ by $\tau_0 s \equiv P_{0s}$ and $\tau s \equiv P_s$.

LEMMA 6. *The mappings τ_0 and τ are surjective and perfect, $\tau = \tau_0|H$, $H = \tau_0^{-1}T$.*

PROOF. Let $t \in T$. Associate to t the ideal $\theta' \equiv \{p \mid \exists\{X_N\} \in p(t \notin \mathrm{cl}\bigcup\{X_N \cap T_N \mid N\})\}$. Suppose that it contains the unit. Then $t \notin \mathrm{cl}\bigcup\{Y_N \cap T_N\} \equiv F$ for some $\{Y_N\} \in \eta_T$. Consider the open set $G \equiv T \setminus F$. Since the polyideal $\mathscr{A}$ is s-meager, $G \cap T_N = \varnothing$ for all N, which is impossible. Thus θ' does not contain the unit. Imbed it into some maximal ideal θ corresponding to some point $s \in H_0$. Then $t \in P_s = \tau s$, which means that τ is surjective. In a similar way we establish the surjectivity of τ_0.

Let Q be an open neighborhood in T_0 of the point $t \equiv \tau_0 s \in T_0$. Choose G_0 and D_0 in $\mathscr{T}_0$ so that $t \in G_0 \subset \mathrm{cl}\, G_0 \subset Q$, $t \notin \mathrm{cl}\, D_0$ and $G_0 \cup D_0 = T_0$. Consider $G \equiv G_0 \cap T$ and $D \equiv D_0 \cap T$. If $r \in \sigma_0 \eta_G$, then $\tau_0 r \in \mathrm{cl}\bigcup\{G \cap T_N \mid N\} = \mathrm{cl}\, G \subset Q$. If $\bar\eta_D \notin \theta_s$ we would come to $t \in \mathrm{cl}\, D_0$, which is wrong. So $s \in \sigma_0 \bar\eta_G$, which means that τ_0 is continuous.

It is clear that $\tau = \tau_0|H$ and $H \subset \tau_0^{-1}T$. Let $\tau_0 s \in T$. Then $P_s = T \cap \tau_0 s$ implies $s \in H$. Thus $H = \tau_0^{-1}T$. It follows from the last relation that τ is perfect. The lemma is proved. □

It follows from this lemma that $\tau_0\colon (T_0, \mathscr{T}_0) \twoheadleftarrow (H_0, \mathscr{H}_0)$ is an a-preimage . The equality $\tau = \tau_0|H$ implies $\tau^{-1}\mathscr{T} \subset \mathscr{H}$. Therefore $\tau\colon (T, \mathscr{T}) \twoheadleftarrow (H, \mathscr{H})$ is an a-preimage.

Define the mapping $\mathfrak{H}\colon \mathscr{A} \to \mathscr{F}(\mathscr{H})$ by setting $\mathfrak{H}(N) \equiv H \setminus \bigcup\{\sigma\overline{\{X_N\}} \mid T_N \cap X_N \in N\}$. The mapping $\mathfrak{H}_0$ is defined in a similar way.

LEMMA 7. *The mappings $\mathfrak{H}_0$ and $\mathfrak{H}$ are saturated.*

PROOF. Let $G_0 \in \mathscr{H}_0$ and $G_0 \cap H_0 \neq \varnothing$. Take an open set $U_0 \equiv \sigma_0\overline{\{X_N\}}$ such that $V_0 \equiv \mathrm{cl}\, U_0 \subset G_0$ and $U_0 \cap H_{0N} \neq \varnothing$. Then $T_N \cap X_N \notin N$ for some N. This means that $T_L \cap X_L \notin L$ for all $L \subset N$. Therefore one may assume that $N \neq \mathscr{P}$. So there exists $M \supset N$ such that $M \neq \mathscr{P}$ and $T_M \setminus X_N \in M$. Let $s \notin V_0$. Then $s \in \sigma_0 q \subset H_0 \setminus V_0$ for some $q \equiv \overline{\{Y_N\}}$. According to Lemma 4, $Y_M \cap X_M \in M$ implies $T_M \cap Y_M \in M$ and, hence, $H_{0M} \cap \sigma_0 q = \varnothing$. Consequently, $s \notin H_{0M}$ and therefore $H_{0M} \subset V_0 \subset G_0$, which proves the lemma. □

LEMMA 8. *The mapping $\mathfrak{H}_0$ is a screen.*

PROOF. If $N = \mathscr{P}$, then $H_{0N} = \varnothing$ and there is nothing to prove. Let $N \neq \mathscr{P}$. Then the set $\theta' \equiv \{\overline{\{X_N\}} \mid X_N \cap T_N \in N\}$ is an ideal not containing the unit. Imbed it in some maximal ideal θ corresponding to some point $s \in H_0$. Then $s \in H_{0N} \neq \varnothing$. Consider the nonempty set $U \equiv \sigma_0\overline{\{X_N\}}$. Then $X_N \notin N$ for some $N \neq \mathscr{P}$. So it follows from $T_N \cap X_N \notin N$ that $T_M \setminus X_N \in M$ for some $\mathscr{P} \neq M \supset N$. Consider the element q corresponding to the polyset $\{T \setminus X_N\}$. Since $\sigma_0 q \cap H_{0M} = \varnothing$, we have $H_{0M} \subset U$. Thus $\bigcup H_{0N}$ is dense in H_0. Now we have found that $\mathfrak{H}_0$ satisfies the conditions a)—c) and is saturated; so it is a screen. The lemma is proved. □

Denote $\mathrm{cl}_{T_0} T_N$ by T_{0N}.

LEMMA 9. $\tau_0 H_{0N} = T_{0N}$.

PROOF. Take some $s \in H_{0N}$ and suppose that $t \equiv \tau_0 s \notin T_{0N}$. Choose the sets G_0 and D_0 in $\mathscr{T}_0$ such that $t \in G_0 \subset T_0 \setminus T_{0N}$, $t \notin \mathrm{cl}\, D_0$ and $G_0 \cup D_0 = T_0$. Consider $G \equiv G_0 \cap T$ and $D \equiv D_0 \cap T$. Since $H_{0N} \cap \sigma_0 \bar{\eta}_G = \varnothing$, we have $s \in \sigma_0 \bar{\eta}_D$. This means that $t \in \mathrm{cl}\, D_0$, which is wrong. Thus $t \in T_{0N}$.

Now suppose that $Q \equiv T_{0N} \setminus \tau_0 H_{0N} \neq \varnothing$. In this case $Q \equiv U_0 \cap T_{0N}$ for some open U_0. Take $V_0 \in \mathscr{T}_0$ such that $V_0 \cap T_{0N} \neq \varnothing$ and $\mathrm{cl}\, V_0 \subset U_0$. Consider $V \equiv V_0 \cap T$. Then $T_M \setminus V \in M$ for some $\mathscr{P} \neq M \subset N$. Therefore $T_M \subset \mathrm{cl}\, V$ implies $T_{0M} \subset Q$. All this means that $\tau_0 H_{0M} \cap \tau_0 H_{0N} = \varnothing$, which contradicts the inclusion $M \supset N$. The lemma is proved. □

COROLLARY 1. $\tau H_N = T_N$.

COROLLARY 2. $H_N = \varnothing$ *if and only if* $N = \mathscr{P}$.

LEMMA 10. H_N *is dense in* H_{0N}, H *is dense in* H_0, *and* $\mathfrak{H}$ *is a screen.*

PROOF. Suppose that there exists an open set U in H_0 such that $U \cap H_{0N} \neq \varnothing$ and $U \cap H_N = \varnothing$. Consider an open set V such that $\mathrm{cl}\, V \subset U$ and $V \cap H_{0N} \neq \varnothing$, and also the set $F \equiv \tau_0(H_{0N} \cap \mathrm{cl}\, V)$. Then $F \subset T_0 \setminus T$. There exists a sequence $\{K_i\}$ of compact sets such that $\bigcup K_i \subset T_N$ and $T_N \setminus \bigcup K_i \in N$. Consider the ideals $N_i \equiv N(K_i)$; then $N = \mathrm{top}\, N_i$. Since $T_{0N_i} \subset K_i$, we have $V \cap H_{0N_i} = \varnothing$. Now, according to Lemma 8, $V \cap H_{0N} = \varnothing$, which is wrong.

Next let Q be an open set in H_0. Then $Q \cap H_{0N} \neq \varnothing$ for some N. Therefore $Q \cap H_N \neq \varnothing$; so H is dense in H_0. In a similar way we establish that $\bigcup H_N$ is dense in H. Since $\mathfrak{H}$ is saturated it is a screen. The lemma is proved. □

COROLLARY. *If* $p \neq \bar{\eta}_0$ *in* $\mathscr{L}$, *then* $\sigma p \neq \varnothing$.

Thus $\tau \colon (T, \mathscr{T}, \mathfrak{T}) \twoheadleftarrow (H, \mathscr{H}, \mathfrak{H})$ is an as-preimage. Denote $\tau^{-1}\mathscr{T}$ by $\mathscr{K}_0$, $Z(\mathscr{K}_0)$ by $\mathscr{K}_1$, and $Z(\mathscr{K}_1)$ by $\mathscr{K}_2$.

LEMMA 11. *Let* $G \in \mathscr{G}(\mathscr{T})$. *Then* $\sigma \bar{\eta}_G \in \mathscr{H} \cap \text{co-}\mathscr{H}$ *and* $\tau^{-1} G$ *is* s*-dense in* $\sigma \bar{\eta}_G$.

PROOF. Consider the set $F \equiv T \setminus G$. Then $\sigma \bar{\eta}_G \cup \sigma \bar{\eta}_F = H$ and $\sigma \bar{\eta}_G \cap \sigma \bar{\eta}_F = \varnothing$. Similar properties hold for $\sigma_0 \bar{\eta}_G$ and $\sigma_0 \bar{\eta}_F$. Therefore they belong to $\mathscr{H}_0$. So all the previous sets belong to $\mathscr{H}$. Suppose that there exists a point $s \in \tau^{-1} G \cap \sigma \bar{\eta}_F$. Then $\tau s \in G$ and $\tau s \in \mathrm{cl} \cup \{F \cap T_N \mid N\} \subset \mathrm{cl}(F \cap \mathrm{cl} \cup T_N) = F$, which is impossible. This means that $U \equiv \tau^{-1} G \subset \sigma \bar{\eta}_G \equiv V$. Let $Q \in \mathscr{H}$ and $Q \cap U \cap H_N = \varnothing$. By definition, $Q = \bigcup \sigma p_k$ for some sequence $p_k \equiv \{X_N^k\} \in \mathscr{L}$. Suppose that $\sigma p_k \cap V \cap H_N \neq \varnothing$. Then $R \equiv T_N \cap X_N^k \cap G \notin N$. Therefore one can find sequences $\{K_i\}$ and $\{L_j\}$ of compact sets such that $\varnothing \neq K_i \subset R$, $L_j \subset T \setminus R$, and $T \setminus (\bigcup K_i \cup \bigcup L_j) \in N$. Consider the ideals $N_i \equiv N(K_i)$ and $N_j \equiv N(L_j)$. Then $N = \mathrm{top}\{N_i, N_j \mid i, j\}$. Since $T_{N_j} \cap X_N^k \cap G = \varnothing$, we have $H_{N_j} \cap \sigma p_k \cap V = \varnothing$. Therefore $\sigma p_k \cap V \cap H_{N_i} \neq \varnothing$ for some i. Now the inclusion $H_{N_i} \subset \tau^{-1} K_i \subset U \subset V$ implies $\sigma p_k \cap U \cap H_{N_i} \neq \varnothing$ and, consequently, $\sigma p_k \cap U \cap H_N \neq \varnothing$, which is wrong. Thus $Q \cap V \cap H_N = \varnothing$ and the lemma is proved. □

COROLLARY. *Let* $F \in \mathscr{F}(\mathscr{T})$. *Then* $\sigma \bar{\eta}_F \in \mathscr{K}_1$, $\sigma \bar{\eta}_F = \mathrm{int}\, \tau^{-1} F$, *and* $H_N \subset \tau^{-1} F$ *implies that* $H_N \subset \sigma \bar{\eta}_F$.

PROOF. Consider the sets $G \equiv T \setminus F$, $U \equiv \sigma\bar{\eta}_G$, and $V \equiv \sigma\bar{\eta}_F$. It follows from the lemma that the family $\varkappa' \equiv \{U\}$ is an s-dense enclosure of the family $\{\tau^{-1}G\}$. Besides this, $V = \operatorname{cobod}\varkappa'$. Therefore $\{U, V\} \in C(\mathscr{K}_0)$. Since $\chi(V) \in O(H, C(\mathscr{K}_0))$ we have $V \in \mathscr{K}_1$. □

LEMMA 12. *Let $\sigma p \cap H_N = \varnothing$ and $\{X_N\} \in p$. Then $X_N \in N$.*

PROOF. Suppose that $X_N \notin N$. Then there exists a compact set $F \subset X_N$ such that $F \notin N$. Consider the ideal $M \equiv N(F) \neq \mathscr{P}$. Since $T_M \subset F$, the inclusion $H_M \subset \tau^{-1}F$ implies $H_M \subset \sigma\bar{\eta}_F$. Further, $(F \setminus X_N) \cap T_N \in N$ implies $(\sigma\bar{\eta}_F \setminus \sigma p) \cap H_N = \varnothing$. Hence, $\varnothing \neq H_M \subset \sigma\bar{\eta}_F \cap H_N = \varnothing$, which proves the lemma. □

LEMMA 13. *H_N and $H \setminus H_N$ belong to $\mathscr{K}_2 \cap \text{co-}\mathscr{H}$.*

PROOF. For an arbitrary ideal M consider the ideals M_N and M_N^* such that $M_N = M \wedge N$, $M_N \wedge M_N^* = \mathscr{P}$, and $M_N \cap M_N^* = M$ (see Lemma 1). Since arbitrary representatives of the classes $n \equiv \zeta N$ and $m_n^* \equiv \zeta M_N^*$ are mutually singular, there exists a universally measurable set $S_{NM}^* \in N$ such that $S_{NM} \equiv T \setminus S_{NM}^* \in M_N^*$.

Let $L \supset M$; consider corresponding ideals for L. Then $L_N^* \supset M_N^*$. Therefore $S_{NL}^* \setminus S_{NM}^* \in L_N^* \cap N \subset L_N^* \cap L_N = L$. Further we have $S_{NM}^* \setminus S_{NL}^* \in L_N^* \cap N \subset L$. Thus $S_{NM}^* \sim S_{NL}^* \bmod L$. Consider the polysets $\eta_N \equiv \{S_{NM} \mid M\}$ and $\eta_N^* \equiv \{S_{NM}^* \mid M\}$ and the elements $e_N \equiv \bar{\eta}_N$ and $e_N^* \equiv \overline{\eta_N^*}$. It is clear that $Q_N^* \equiv \sigma e_N^* \subset H \setminus H_N$. We verify the inclusion $P_N \equiv \sigma e_N \subset H_N$. Let $\{X_N\} \in p$ and $T_N \cap X_N \in N$. Since $N \cup M \subset M_N$, we have $X_N \sim X_M \bmod M_N$. Therefore $T_N \cap X_M \subset M_N$ and $T \setminus T_N \in N \subset M_N$ imply $X_M \in M_N$. Besides this, $X_M \cap S_{NM} \in M_N^*$. Consequently, $X_M \cap S_{NM} \in M$ means that $\sigma p \cap P_N = \varnothing$, which gives us the required inclusion. Thus $P_N = H_N$ and $Q_N^* = H \setminus H_N$. This means that both sets belong to $\mathscr{H} \cap \text{co-}\mathscr{H}$.

Denote $T_{M_N^*}$ by T_*. Since $T_* \setminus (T_* \cap S_{NM}^*) \in M_N^*$ there exists a sequence $\{F_{NM}^i \mid i\}$ of closed sets such that $S \equiv \bigcap F_{NM}^i \subset T \cap S_{NM}^*$ and $T_* \setminus S \in M_N^*$. Consider the sets $Q_{NM}^i \equiv \operatorname{int}\tau^{-1}F_{NM}^i$ and $Q_{NM} \equiv \bigcup Q_{NM}^i \in \mathscr{K}_1$. Choose a sequence of closed sets $\{K_j\}$ in such a way that $\bigcup K_j \subset T \setminus S$ and $T \setminus \bigcup K_j \in N$. Consider the ideals $N_j \equiv N(K_j)$. It follows from $N = \operatorname{top} N_j$ and $Q_{NM}^i \cap H_{N_j} = \varnothing$ that $Q_{NM} \cap H_N = \varnothing$.

Now consider the set $Q_N \equiv \overline{\bigcup\{Q_{NM} \mid M\}} \in \mathscr{G}(\mathscr{K}_1)$. We are going to verify that $Q_N^* = sQ_N$. Let $p = \overline{\{X_N\}}$ and $\sigma p \cap Q_N \cap H_M = \varnothing$. According to Lemma 12, the relation $X_M \cap F_{NM}^i \in M \subset M_N^*$, which holds for all i, implies $X_M \cap T_* \in M_N^*$. Besides this, $X_M \cap S_{NM}^* \in N \subset M_N$. Therefore from $X_M \cap S_{NM}^* \in M$ it follows that $\sigma p \cap Q_N^* \cap H_M = \varnothing$. Thus the single-element family $\{Q_N^*\}$ is an s-dense enclosure of the single-element family $\{Q_N\}$. Since the functions $\chi(H_N)$ and $\chi(H \setminus H_N)$ belong to $O(H, C(\mathscr{K}_1))$, the initial sets belong to $\mathscr{K}_2$. The lemma is proved. □

PROPOSITION 2. *The as-preimage $\tau\colon T \twoheadleftarrow H$ is the as-space of enclosures of the type ZZZ^c of the as-space T.*

PROOF. We are using the notation of the previous lemma. Let $\{X_N\} \in p$. For given N there exist closed sets Y_N^k and Z_N^l such that $\bigcup Y_N^k \subset X_N \subset T \setminus \bigcup Z_N^l$ and $T \setminus \bigcup Y_N^k \cup \bigcup Z_N^l \in N$. Consider the elements $e_N^k \equiv \bar{\eta}_{Y_N^k} \wedge e_N$ and $f_N^l \equiv \bar{\eta}_{Z_N^l} \wedge e_N$ and the sets $Q_N^k \equiv \sigma e_N^k$ and $R_N^l \equiv \sigma f_N^l$ in $\mathscr{K}_2$.

Take arbitrary ideals N_1 and N_2 and consider the ideal $N \equiv N_1 \wedge N_2$. Then $M_N^* = M_{N_1}^* \cap M_{N_2}^*$ implies $S_{N_1M} \cap S_{N_2M} \in M_N^*$. Since $M_N \supset N \supset N_1 \cup N_2$, then $X_{N_1} \sim X_{N_2} \bmod M_N$. Therefore $Y_{N_1}^k \cap Z_{N_2}^l \in M_N$. Thus $Y_{N_1}^k \cap S_{N_1M} \cap Z_{N_2}^l \cap S_{N_2M} \in M_N \cap M_N^* = M$. This means that $Q_{N_1}^k \cap R_{N_2}^k \cap H_M = \varnothing$ for every M. Consequently,

$Q^k_{N_1} \cap R^l_{N_2} = \varnothing$. Consider the sets $Q \equiv \bigcup\{Q^k_N \mid k, N\}$ and $R \equiv \bigcup\{R^l_N \mid l, N\}$ in $\mathscr{G}(\mathscr{K}_2)$. According to what has already been proved, $Q \cap R = \varnothing$. Let $\{Y_N\} \in q$ and $\sigma q \cap (Q \cup R) \cap H_N = \varnothing$. Due to Lemma 12, $Y_N \cap ((Y^k_N \cap S_{NN}) \cap (Z^l_N \cap S_{NN})) \in N$. Together with $S^*_{NN} \in N$ it implies that $Y_N \cap (Y^k_N \cap Z^l_N) \in N$. So, the inclusion $Y_N \in N$ implies $\sigma p \cap H_N = \varnothing$. Therefore $H = s(Q \cup R)$, i.e., the single-element family $\varkappa \equiv \{Q\}$ is complemented.

Since $N \cup M \subset M_N$, $X_N \sim X_M \bmod M_N$. Therefore $Y^k_N \cap S_{NM} \setminus X_N \cap Y^k_N \cap S_{NM} = \varnothing \in M_N$ implies $Y^k_N \cap S_{NM} \setminus X_M \cap Y^k_N \cap S_{NM} \in M_N$. At the same time $S_{NM} \in M^*_N$, so that the latter difference belongs to M. Therefore the equality $e^k_N = p \wedge e^k_N$ implies $Q \subset \sigma p$. In a similar way $X_N \cap Z^l_N = \varnothing$ implies $X_M \cap Z^l_N \cap S_{NM} \in M$ for all M, i.e., $p \wedge f^l_N = 0$. This means that $\sigma p \cap R = \varnothing$. Thus $\sigma p = sQ$. Besides this, we have $\sigma p \in \mathscr{H} \cap \text{co-}\mathscr{H}$. So the single-element family $\varkappa' \equiv \{\sigma p\}$ is an s-dense enclosure of the family $\varkappa$. Since $\chi(\sigma p) \in O(H, C^c(\mathscr{K}_2))$, $\sigma p \in Z^c(\mathscr{K}_2)$.

Now let $G \in \mathscr{H}$. Then, by definition, $G = \bigcup \sigma p_i$ for some sequence p_i. It follows from what has already been proved that $G \in Z^c(\mathscr{K}_2)$. The proposition is proved. □

LEMMA 14. *The as-preimage $\tau\colon T \twoheadleftarrow H$ is ${}^a Z$-enclosable.*

PROOF. Let $Q \in \mathscr{G}(\mathscr{H})$. Then $Q = \bigcup\{\sigma e \mid e \in E\}$ for some set $E \subset \mathscr{L}$. Let $\{X^e_N\} \in e$. Fix an ideal N and consider the complete Boolean algebra $\mathscr{L}_N \equiv \mathscr{U}\mathscr{M}/N$. Then there exists a countable set $\{e_i \mid i \in I_N\}$ such that $\bar{X}_N = \sup\{\bar{X}^e_N \mid e \in E\}$ in $\mathscr{L}_N$, where $X_N \equiv \bigcup\{X^e_N \mid i \in I_N\}$. We shall write i instead of e_i in the subscript. Let $N \subset M$. Since $X_N \sim X_M \bmod M$ we can consider the element $p \equiv \overline{\{X_N\}}$ and the set $Q' \equiv \sigma p \in \mathscr{H} \cap \text{co-}\mathscr{H}$. It follows from $X_N \supset X^e_N \bmod N$ that $\sigma e \subset Q'$ for every $e \in E$. Consequently $Q \subset Q'$. Let $\{Y_N\} \in q$ and $\sigma q \cap Q \cap H_N = \varnothing$. According to Lemma 12, $Y_N \cap X^i_N \in N$ implies $\sigma q \cap Q' \cap H_N = \varnothing$. Hence $Q' = sQ$. From the property just proved it follows now that every family in $K(\mathscr{H})$ possesses an s-dense enclosure. The lemma is proved. □

2.3. Preimages in $\hat{\mathscr{P}}$ and corresponding extensions.

PROPOSITION 3. *Let $\tau\colon T \twoheadleftarrow H \in \hat{\mathscr{P}}$. Then for every function a such that $a \in O(H, C^c(Z(Z(\tau^{-1}\mathscr{T}))))$ there exists an element $p \in \mathbb{UM}$ such that $a \mapsto p \bmod \mathfrak{F}(H)$.*

PROOF. Denote $\tau^{-1}\mathscr{T}$ by $\mathscr{K}_0$, $Z(\mathscr{K}_0)$ by $\mathscr{K}_1$, $Z(\mathscr{K}_1)$ by $\mathscr{K}_2$, $O(H, C(\mathscr{K}_0))$ by O_0, $O(H, C(\mathscr{K}_1))$ by O_1, and $O(H, C^c(\mathscr{K}_2))$ by O_2.

Let $a \in O_0$. Then there exists a sequence of coverings $\{\rho'_m\} \subset C(\mathscr{K}_0)$ such that $\omega(a, Q) < 1/m$ for every $Q \in \rho'_m$. By definition, for ρ'_m there exists a finite set of coverings $\{\rho'_{mk} \mid k \in K_m\}$ such that $\rho'_{mk} = \varkappa'_{mk} \cup \{R'_{mk}\}$, $R'_{mk} = \operatorname{cobod} \varkappa'_{mk}$, and $\rho'_m = \bigwedge \rho'_{mk}$. Further, we have $\varkappa'_{mk} = s\varkappa_{mk}$ for some families $\varkappa_{mk} \in K(\mathscr{K}_0)$. Denote the set of all subsets p in K_m by P_m. Let $\varkappa'_{mk} = \{Q'_{mk\gamma} \mid \gamma \in \Gamma_{mk}\}$, $\varkappa_{mk} = \{Q_{mk\gamma} \mid \gamma \in \Gamma_{mk}\}$, $Q_{mk\gamma} = \tau^{-1}G_{mk\gamma}$ for some $G_{mk\gamma} \in \mathscr{G}$, and $F_{mk} \equiv T \setminus \bigcup\{G_{mk\gamma} \mid \gamma\}$. Denote by I_{mp} the set of all mappings $i\colon p \to \bigcup\{\Gamma_{mk} \mid k \in p\}$ subject to $i(k) \in \Gamma_{mk}$. Consider the sets $Q'_{mpi} \equiv \bigcap\{Q_{mki(k)} \mid k \in p\}$, $Q_{mpi} \equiv \{Q_{mki(k)} \mid k \in p\}$, $G_{mpi} \equiv \bigcap\{G_{mki(k)} \mid k \in p\}$, $R'_{mp} \equiv \bigcap\{R'_{mk} \mid k \in K_m \setminus p\}$, and $F_{mp} \equiv \bigcap\{F_{mk} \mid k \in K_m \setminus p\}$ for all $p \in P_m$. Introduce $S_{mpi} \equiv G_{mpi} \cap F_{mp} \in \mathscr{U}\mathscr{M}$ and $H_{mpi} \equiv Q'_{mpi} \cap R'_{mp} \in \mathscr{H}$. Then $Q'_{mpi} = sQ_{mpi}$, $Q_{mpi} = \tau^{-1}G_{mpi}$, $H \setminus R'_{mp} = s\tau^{-1}(T \setminus F_{mp})$, and $\rho'_m = \{H_{mpi} \mid p \in P_m \& i \in I_{mp}\}$.

Consider the numbers $\varkappa_{mpi} \equiv \inf\{a(s) \mid s \in H_{mpi}\}$ and $y_{mpi} \equiv \sup\{a(s) \mid s \in H_{mpi}\}$. Introduce the open sets $U_{mpi} \equiv Q_{mpi} \cap R'_{mp} \subset \tau^{-1}S_{mpi} \cap H_{mpi}$ and $U_m \equiv \bigcup\{U_{mpi} \mid i, p\}$. Then $H_{mpi} = sU_{mpi}$ and, therefore, U_m is s-dense in H. Let

a take its values in $]-z, z[$. Consider the universally measurable functions g_{mpi} and h_{mpi} such that $g_{mpi}(t) \equiv -z$ and $h_{mpi}(t) \equiv z$ for all $t \notin S_{mpi}$ and $g_{mpi}(t) \equiv x_{mpi}$, $h_{mpi}(t) \equiv y_{mpi}$ for all $t \in S_{mpi}$. Define the universally measurable functions g_m and h_m by $g_m(t) \equiv \max\{g_{mpi}(t)\}$ and $h_m(t) \equiv \min\{h_{mpi}(t)\}$.

Suppose that there exists some point $t \in T \setminus \bigcup S_{mpi}$ and consider the ideal $N \equiv \{P \in \mathscr{P} \mid t \notin P\}$. Then $H_{mpi} \cap H_N \neq \varnothing$ for some subscripts. Therefore $U_{mpi} \cap H_N \neq \varnothing$ implies $S_{mpi} \cap T_N \neq \varnothing$, which is wrong. Thus $T = \bigcup S_{mpi}$. Further we have $\omega(g, S_{mpi}) < 1/m$, $\omega(h, S_{mpi}) < 1/m$, and $h_m - g_m \leqslant (1/m)\mathbf{1}$. Let $s \in U_m$. Then $s \in U_{mpi}$ for some values of the subscripts. Therefore $a(s) - g_m(\tau s) < 1/m$ and $h_m(\tau s) - a(s) < 1/m$. Besides that, $a(s) - g_m(\tau s) \geqslant a(s) - h_m(\tau s) \geqslant a(s) - h_{mpi}(\tau s) > -1/m$ and $h_m(\tau s) - a(s) > -1/m$. Let $t \in T$. Then $t \in S_{mpi} \cap S_{nqj}$ for some values of the subscripts. Consider the ideal N. Then $H_N \subset Q_{mpi} \cap \tau^{-1} F_{mp} \cap Q_{nqj} \cap \tau^{-1} F_{nq}$. It follows from the s-density property that $H_N \subset R'_{mp} \cap R'_{nq}$. Therefore $H_N \subset U_{mpi} \cap U_{nqj}$. Fix a point $s \in H_N$. For this point we have $|g_m(t) - g_n(t)| \leqslant |g_m(t) - g_m(\tau s)| + |g_m(\tau s) - a(s)| + |a(s) - g_n(\tau s)| + |g_n(\tau s) - g_n(t)| < 4/m$ for every $n \geqslant m$. In view of the completeness of the ring UM with respect to uniform convergence, there exists $f \in UM$ such that $|f(t) - g_m(t)| < 5/m$ for all $t \in T$. Since $|a(s) - f(\tau s)| < 6/m$ for $s \in U_m$, we get $a(s) = f(\tau s)$ for every $s \in S \equiv \bigcap U_m \in \mathcal{S}_s(H) \equiv \{\bigcap\{U_m \mid m \in \mathbb{N}\} \mid H = sU_m\}$.

Now let $a \in O_1$. Then there exists a sequence of coverings $\{\rho'_m\} \subset C(\mathscr{K}_1)$ such that $\omega(a, Q) < 1/m$ for all $Q \in \rho'_m$. By the definition of ρ'_m there exists a finite set of coverings $\{\rho'_{mk} \mid k \in K_m\}$ such that $\rho'_{mk} = \varkappa'_{mk} \cup \{R_{mk}\}$, $R_{mk} = \operatorname{cobod} \varkappa'_{mk}$, and $\rho'_m = \bigwedge \rho'_{mk}$. We have $\varkappa'_{mk} = s\varkappa_{mk}$ for some $\varkappa_{mk} \in K(\mathscr{K}_1)$. Let p and P_m be the same as before, $\varkappa'_{mk} = \{Q'_{mk\gamma} \mid \gamma \in \Gamma_{mk}\}$, and $\varkappa_{mk} = \{Q_{mk\gamma} \mid \gamma \in \Gamma_{mk}\} \subset \mathscr{G}(\mathscr{K}_1)$. Our conditions mean that $Q_{mk\gamma} = \bigcup\{Q_{mk\gamma\alpha} \in \mathscr{K}_1 \mid \alpha\}$ and $Q_{mk\gamma\alpha} = \operatorname{coz} a_{mk\gamma\alpha}$ for some nonnegative functions $a_{mk\gamma\alpha} \in O_0$. According to what has already been proved, there exist nonnegative functions $f_{mk\gamma\alpha} \in UM$ and sets $S_{mk\gamma\alpha} \in \mathcal{S}_s(H)$ such that $a_{mk\gamma\alpha}(s) = f_{mk\gamma\alpha}(\tau s)$ for all $s \in S_{mk\gamma\alpha}$. Consider the sets $F_{mk\gamma\alpha} \equiv \operatorname{coz} f_{mk\gamma\alpha}$. It may be assumed that no function $a_{mk\gamma\alpha}$ and $f_{mk\gamma\alpha}$ exceeds one.

Fix the ideal N and consider the complete Boolean algebra $\mathscr{U}_N \equiv \mathscr{U}\mathscr{M}/N$. Then there exists a countable set of subscripts $\{\alpha_r \mid r \in \mathbb{N}\}$ such that $\bar{F}^N_{mk\gamma} = \sup\{\bar{F}_{mk\gamma\alpha} \mid \alpha\}$ in $\mathscr{U}_N$, where $F^N_{mk\gamma} \equiv \bigcup\{F_{mk\gamma\alpha_r} \mid r\}$. In what follows we write r instead of α_r in the subscript. Consider the sets $Q^N_{mk\gamma} \equiv \bigcup\{Q_{mk\gamma r} \mid r\}$ and $S^N_{mk\gamma} \equiv \bigcap S_{mk\gamma r}$. We verify that $Q^N_{mk\gamma}$ is N-dense in $Q_{mk\gamma}$. Let $G \in \mathscr{H}$, $N \subset M$, and $G \cap Q^N_{mk\gamma} \cap H_M = \varnothing$. Take a sequence $\{K_i \mid i\}$ such that $K_i \subset F_{mk\gamma\alpha} \cap F^N_{mk\gamma} \equiv P$ and $P \setminus \bigcup K_i \in M$. Choose also a sequence of closed sets $\{K_j \mid j\}$ such that $K_j \subset T \setminus F_{mk\gamma\alpha} \equiv Q$ and $q \setminus \bigcup K_j \in M$. Consider the ideals $M_i \equiv M(K_i)$ and $M_j \equiv M(K_j)$. Since $F_{mk\gamma\alpha} \setminus F^N_{mk\gamma} \in N$, we have $M = \operatorname{top}\{M_i, M_j \mid i, j\}$. If $s \in Q_{mk\gamma\alpha} \cap S^N_{mk\gamma} \cap H_{M_i}$, then the inclusion $\tau s \in F^N_{mk\gamma}$ implies $s \in Q^N_{mk\gamma} \cap H_M$. Therefore $G \cap Q_{mk\gamma\alpha} \cap S^N_{mk\gamma} \cap H_{M_i} = \varnothing$. By Lemma 2, $G \cap Q_{mk\gamma\alpha} \cap H_{M_i} = \varnothing$. Next, it follows from $Q_{mk\gamma\alpha} \cap S_{mk\gamma\alpha} \cap H_{M_j} = \varnothing$ that $Q_{mk\gamma\alpha} \cap H_{M_j} = \varnothing$. All these relations taken together imply now $G \cap Q_{mk\gamma\alpha} \cap H_M = \varnothing$, which proves the required property.

Consider the functions $a^N_{mk\gamma} \equiv \sum a_{mk\gamma r}/2^r$ and $f^N_{mk\gamma} \equiv \sum f_{mk\gamma r}/2^r$. Then $Q^N_{mk\gamma} = \operatorname{coz} a^N_{mk\gamma}$, $F^N_{mk\gamma} = \operatorname{coz} f^N_{mk\gamma}$, and $a^N_{mk\gamma}(s) = f^N_{mk\gamma}(\tau s)$ for every $s \in S^N_{mk\gamma} \equiv \bigcap S_{mk\gamma r}$. Consider the sets $Q^N_{mk} \equiv \bigcup\{Q^N_{mk\gamma} \mid \gamma\}$, $F^N_{mk} \equiv \bigcup\{F^N_{mk\gamma} \mid \gamma\}$, and $G^N_{mk} \equiv T \setminus F^N_{mk}$ and the functions $a^N_{mk} \equiv \sum\{a^N_{mk\gamma} \mid \gamma\}$ and $f^N_{mk} \equiv \sum f^N_{mk\gamma}$. Then $Q^N_{mk} = \operatorname{coz} a^N_{mk}$, $F^N_{mk} = \operatorname{coz} f^N_{mk}$, and Q^N_{mk} are N-dense in $H \setminus R_{mk}$. Besides this, we have

$a_{mk}^N(s) = f_{mk}^N(\tau s)$ for every $s \in S_{mk}^N \equiv \bigcap S_{mk\gamma}^N \in \mathcal{S}_s(H)$. Consider now $b_{mk} \equiv \chi(R_{mk}) \in O_1$, which is the function of zero oscillation on one covering ρ'_{mk}, and also $g_{mk}^N \equiv \chi(G_{mk}^N) \in UM$. For every point $s \in (Q_{mk}^N \cup R_{mk}) \cap S_{mk}^N \in \mathcal{S}_N(H)$ the equality $b_{mk}(s) = g_{mk}^N(\tau s)$ holds.

Next, we introduce the families $\varphi_{mk}^N \equiv \{F_{mk\gamma}^N \mid \gamma\}$ and the coverings $\psi_{mk}^N \equiv \varphi_{mk}^N \cup \{G_{mk}\}$ and $\psi_m^N \equiv \bigwedge \psi_{mk}^N$. By I_{mp} we denote the set of all mappings $i: p \to \bigcup\{\Gamma_{mk} \mid k \in p\}$ such that $i(k) \in \Gamma_{mk}$. Consider the sets $Q'_{mpi} \equiv \bigcap\{Q'_{mki(k)} \mid k \in p\}$, $Q_{mpi} \equiv \bigcap\{Q_{mki(k)} \mid k \in p\}$, $Q_{mpi}^N \equiv \bigcap\{Q_{mki(k)}^N \mid k \in p\}$, $F_{mpi}^N \equiv \bigcap\{F_{mki(k)}^N \mid k \in p\}$, $R_{mp} \equiv \bigcap\{R_{mk} \mid k \in K_m \setminus p\}$ and $G_{mp}^N \equiv \bigcap\{G_{mk}^N \mid k \in K_m \setminus p\}$ for all $p \in P_m$. Also introduce $X_{mpi}^N \equiv F_{mpi}^N \cap G_{mp}^N$ and $H'_{mpi} \equiv Q'_{mpi} \cap R_{mp} \in \mathcal{H}$. Then $Q'_{mpi} = sQ_{mpi}$, Q_{mpi}^N is N-dense in Q_{mpi}, $\rho'_m = \{H_{mpi} \mid p \in P_m \& i \in I_{mp}\}$, and $\psi_m^N = \{X_{mpi}^N \mid p \in P_m \& i \in I_{mp}\}$.

Now we consider the functions $f_{mpi}^N \equiv \inf\{f_{mki(k)}^N \mid k \in p\}$, $a_{mpi}^N \equiv \inf\{a_{mki(k)}^N \mid k \in p\}$, and also $b_{mp} \equiv \min\{b_{mk} \mid k \in K_m \setminus p\}$ and $g_{mp}^N \equiv \min\{g_{mk}^N \mid k \in K_m \setminus p\}$. Finally, consider the functions $c_{mpi}^N \equiv a_{mpi}^N \wedge b_{mp}^N$ and $h_{mpi}^N \equiv f_{mpi}^N \wedge g_{mp}^N$. Then $H_{mpi}^N \equiv Q_{mpi}^N \cap R_{mp} = \operatorname{coz} c_{mpi}^N$, $X_{mpi}^N = \operatorname{coz} h_{mpi}^N$, and H_{mpi}^N is N-dense in H'_{mpi}. In addition, there exist the sets $S_{mpi}^N \in \mathcal{S}_N(H)$ such that $c_{mpi}^N(s) = h_{mpi}^N(\tau s)$ for all $s \in S_{mpi}^N$. Consider the sets $H_m^N \equiv \bigcup\{H_{mpi}^N \mid p, i\}$ and $S_m^N \equiv H_m^N \cap \bigcap\{S_{mpi}^N \mid p, i\} \in \mathcal{S}_N(H)$.

Suppose that a takes its values in the interval $]-z, z[$. Introduce the numbers $x_{mpi} \equiv \inf\{a(s) \mid s \in H'_{mpi}\}$ and $y_{mpi} \equiv \sup\{a(s) \mid s \in H'_{mpi}\}$. Consider the functions $d_{mpik}^N \equiv -z\mathbf{1} + kc_{mpi}^N \wedge (x_{mpi} + z)\mathbf{1}$ and $e_{mpil}^N \equiv z\mathbf{1} - kc_{mpi}^N \wedge (z - y_{mpi})\mathbf{1}$. It is clear that $d_{mpik}^N \leqslant a \leqslant e_{mpik}^N$. Take the universally measurable functions $u_{mpik}^N \equiv -z\mathbf{1} + kh_{mpi}^N \wedge (x_{mpi} + z)\mathbf{1}$ and $v_{mpik}^N \equiv z\mathbf{1} - kh_{mpi}^N \wedge (z - y_{mpi})\mathbf{1}$. Let the subscripts be chosen in such a way that $X \equiv X_{mpi}^N \cap X_{nqj}^N \notin N$. Then there exists a closed set $F \notin N$ such that $F \subset X$. Denote by M the ideal $N(F)$. From the property of N-density it follows that there exists some point $s \in S_m^N \cap S_n^N \cap H_M$. Since $\tau s \in F$, we have $s \in H_{mpi}^N \cap H_{nqj}^N$. Therefore $x_{mpi} \vee x_{nqj} \leqslant a(s) \leqslant y_{mpi} \wedge y_{nqj}$ implies the inequality $d_{mpik}^N \leqslant x_{mpi}\mathbf{1} \leqslant y_{nqj}\mathbf{1} \leqslant e_{nqjl}^N$. Suppose now that the subscripts are such that $X \in N$. For $t \notin X_{mpi}^N$ the inequality $u_{mpik}^N(t) = -z \leqslant y_{nqj} \leqslant v_{nqjl}^N(t)$ holds. Similarly, $u_{mpik}^N(t) \leqslant x_{mpi} \leqslant z = v_{nqjl}^N(t)$ for all $t \notin X_{nqj}^N$. Consider the set $I_{mn}^N \equiv \bigcup\{X_{mpi}^N \cap X_{nqj}^N \mid X_{mpi}^N \cap X_{nqj}^N \in N\}$. Then $u_{mpik}^N(t) \leqslant v_{nqjl}^N(t)$ for all $t \notin I_{mn}^N$.

Now introduce the functions g_N and h_N in UM such that $g_N(t) \equiv \sup\{u_{mpik}^N(t)\}$, $h_N(t) \equiv \inf\{v_{mpik}^N(t)\}$, and the set $I_N \equiv \bigcup I_{mn}^N$. Let $t \notin I_N$. Then the above arguments imply that $g_N(t) \leqslant h_N(t)$. Since $t \in X_{mpi}^N$ for every m and for some p and i depending on m, there exists k such that $u_{mpik}^N(t) = x_{mpi}$ and $v_{mpik}^N(t) = y_{mpi}$. Consequently, $g_N(t) = h_N(t)$. Let $s \in S_m^N$. Then $s \in H_{mpi}^N \cap S_{mpi}^N$ for some values of the subscripts. Therefore $\tau s \in X_{mpi}^N$. So, one can find k such that $d_{mpik}^N(s) = x_{mpi} = u_{mpik}^N(\tau s)$ and $e_{mpik}^N(s) = y_{mpi} = v_{mpik}^N(\tau s)$. From these equalities it follows that $a(s) - g_N(\tau s) \leqslant e_{mpik}^N(s) - u_{mpik}^N(\tau s) < 1/m$ and $a(s) - h_N(\tau s) > d_{mpik}^N(s) - v_{mpik}^N(\tau s) \geqslant -1/m$. Consider the set $Y_N \equiv \bigcap S_m^N \setminus \tau^{-1} I_N$. Then $a(s) = g_N(\tau s)$ for every $s \in Y_N$, i.e., $a \sim g_N \circ \tau \bmod <\tau^{-1}N, \mathcal{F}_N(H)>$.

Let $N \subset M$ and $a(s) = g_M(\tau s)$ for all $s \in Y_M$. Suppose that $R \equiv \{t \in T \mid g_N(t) \neq g_M(t)\} \notin M$. Then there exists a closed set $F \subset R$ such that $F \notin M$. Denote by L the ideal $M(F)$. Due to the property of M-density there exists some

$s \in Y_N \cap Y_M \cap H_L$. Since $\tau s \in F$, we have $g_N(\tau s) \neq g_M(\tau s)$, which is impossible. Thus $g_N \sim g_M \bmod M$ and so $a \mapsto q \bmod \mathfrak{F}(H)$, where $q \equiv \overline{\{g_N\}}$.

Finally, let $a \in O_2$. Then there exists a sequence $\{\rho'_m\}$ of coverings in $C^c(\mathscr{K}_2)$ such that $\omega(a, Q) < 1/m$ for every $Q \in \rho'_m$. By definition, for each ρ'_m there exists a finite set of coverings $\{\rho'_{mk} \mid k \in K_m\}$ such that $\rho'_{mk} = \varkappa'_{mk} \cup \{R'_{mk}\}$, $R'_{mk} = \operatorname{cobod} \varkappa'_{mk}$, and $\rho'_m = \bigwedge \rho'_{mk}$. The family $\varkappa'_{mk}$ is an s-dense enclosure of some family $\varkappa_{mk} \in K(\mathscr{K}_2)$ complemented by some set $R_{mk} \in \mathscr{G}(\mathscr{K}_2)$. Consider the family $\rho_{mk} \equiv \varkappa_{mk} \cup \{R_{mk}\} \in K(\mathscr{K}_2)$. Then $\rho'_{mk} = s\rho_{mk}$ and $s \operatorname{bod} \rho_{mk} = H$. For the family $\rho_m \equiv \bigwedge \rho_{mk}$ we have $\rho'_m = s\rho_m$, $\rho_m \in K(\mathscr{K}_2)$, and $H = s \operatorname{bod} \rho_m$.

By definition, $\rho'_m = \{Q'_{mi} \mid i \in I_m\}$ and $\rho_m = \{Q_{mi} \mid i \in I_m\} \subset \mathscr{G}(\mathscr{K}_1)$. Therefore $Q_{mi} = \bigcup\{Q_{mi\alpha} \in \mathscr{K}_2 \mid \alpha\}$ and $Q_{mi\alpha} = \operatorname{coz} a_{mi\alpha}$ for some nonnegative functions $a_{mi\alpha} \in O_1$. According to what has been proved above, there exist nonnegative polyfunctions $\{f^N_{mi\alpha} \mid N\}$ and $Y^N_{mi\alpha}$ subject to $a_{mi\alpha}(s) = f^N_{mi\alpha}(\tau s)$ for all $s \in Y^N_{mi\alpha}$. Introduce $F^N_{mi\alpha} \equiv \operatorname{coz} f^N_{mi\alpha}$. One may suppose that no function $a_{mi\alpha}$ or $f_{mi\alpha}$ exceeds one.

Fix an ideal N. Then there exists some countable set of subscripts $\{\alpha_r \mid r \in \mathbb{N}\}$ such that $\bar{F}^N_{mi} = \sup\{\bar{F}^N_{mi\alpha} \mid \alpha\}$ in $\mathscr{U}_N$ where $F^N_{mi} \equiv \bigcup\{F^N_{mi\alpha_r} \mid r\}$. Again we shall write r instead of α_r in the subscript. Consider the open sets $Q^N_{mi} \equiv \bigcup\{Q^N_{mir} \mid r\}$. In the same way as before, we can verify that Q^N_{mi} is N-dense in Q_{mi}. Consider the functions $a^N_{mi} \equiv \sum a_{mir}/2^r$ and $f^N_{mi} \equiv \sum f^N_{mir}/2^r$. Then $Q^N_{mi} = \operatorname{coz} a^N_{mi}$, $F^N_{mi} = \operatorname{coz} f^N_{mi}$, and $a^N_{mi}(s) = f^N_{mi}(\tau s)$ for all $s \in Y^N_{mi} \equiv \bigcap Y^N_{mir}$. Suppose that $I^N_m \equiv T \setminus \bigcup F^N_{mi} \notin N$. Then there exists an open set $K \subset I^N_m$, $K \notin N$. Denote by M the ideal $N(K)$. Since we have started with a covering, $Q'_{mi} \cap H_M \neq \varnothing$ for some i. Therefore, $Q^N_{mi} \cap H_M \neq \varnothing$ implies $Q^N_{mi} \cap Y^N_{mi} \cap H_M \neq \varnothing$. Take a point s in the latter set. Then $\tau s \in F^N_{mi} \cap T_M$, which is impossible. So $I^N_m \in N$.

Suppose that a takes its values in the interval $]-z, z[$. Set $x_{mi} \equiv \inf\{a(s) \mid s \in Q'_{mi}\}$ and $y_{mi} \equiv \sup\{a(s) \mid s \in Q'_{mi}\}$. Introduce the functions $b^N_{mik} \equiv -z\mathbf{1} + ka^N_{mi} \wedge (x_{mi} + z)\mathbf{1}$ and $c^N_{mik} \equiv z\mathbf{1} - ka^N_{mi} \wedge (z - y_{mi})\mathbf{1}$. It is clear that $b^N_{mik} \leqslant a \leqslant c^N_{mik}$. Consider also $g^N_{mik} \equiv -z\mathbf{1} + kf^N_{mi} \wedge (x_{mi} + z)\mathbf{1}$ and $h^N_{mik} \equiv z\mathbf{1} - kf^N_{mi} \wedge (z - y_{mi})\mathbf{1}$ in UM. Suppose that the subscripts are such that $F = F^N_{mi} \cap F^N_{nj} \notin N$. Then there exists a closed set $K \notin N$ such that $K \subset F$. Denote by M the ideal $N(K)$. There must exist a point $s \in Y^N_{mi} \cap Y^N_{nj} \cap H_M$. Since $\tau s \in F$, we have $s \in Q^N_{mi} \cap Q^N_{nj}$. Therefore, $x_{mi} \vee x_{nj} \leqslant a(s) \leqslant y_{mi} \wedge y_{nj}$ implies $b^N_{mik} \leqslant x_{mi}\mathbf{1} \leqslant y_{ni}\mathbf{1} \leqslant c^N_{njl}$. Now let $F \in N$. Then $g^N_{mik}(t) \leqslant h^N_{njl}(t)$ for all $t \notin F$. Set $I^N_{mn} \equiv \bigcup\{F^N_{mi} \cap F^N_{nj} \mid F^N_{mi} \cap F^N_{nj} \in N\}$ and $I_N \equiv \bigcup I^N_m \cup \bigcup I^N_{mn} \in N$.

Consider the functions g_N and h_N in UM such that $g_N(t) \equiv \sup\{g^N_{mik}(t)\}$ and $h_N(t) \equiv \inf\{h^N_{mik}(t)\}$. Let $t \notin I_N$. Then from what has already been proved it follows that $g_N(t) \leqslant h_N(t)$. Since $t \in F^N_{mi}$ for every m and some $i = i(m)$, there exists k such that $g^N_{mik}(t) = x_{mi}$ and $h^N_{mik}(t) = y_{mi}$. Thus $g_N(t) = h_N(t)$. Define $Q^N_m \equiv \bigcup Q^N_{mi} \in \mathscr{U}_N(H)$ and $Y^N_m \equiv Q^N_m \cap \bigcap\{Y^N_{mi} \mid i\}$. Let $s \in Y^N_m$. Then $s \in Q^N_{mi} \cap Y^N_{mi}$ for some i. Therefore $\tau s \in F^N_{mi}$. Hence, for some k we have $b^N_{mik}(s) = x_{mi} = g^N_{mik}(\tau s)$ and $c^N_{mik}(s) = y_{mi} = h^N_{mik}(\tau s)$. So we have $a(s) - g_N(\tau s) \leqslant c^N_{mik}(s) - g^N_{mik}(\tau s) < 1/m$ and $a(s) - h_N(\tau s) \geqslant b^N_{mik}(s) - h^N_{mik}(\tau s) > -1/m$. These inequalities imply that $a(s) = g_N(\tau s)$ for all $s \in Y_N \equiv \bigcap Y^N_m \setminus \tau^{-1} I_N$. Since $g_N \sim g_M \bmod M$ for every $M \supset N$, one can consider the element $q \equiv \overline{\{g_N\}}$. For this element $a \mapsto q \bmod \mathfrak{F}(H)$ holds. The proposition is proved. □

Corollary 1. *Let* $\tau \colon T \twoheadleftarrow H \in \hat{\mathscr{P}}$. *Then for every function* $a \in O(H, \mathscr{H})$ *there exists* $p \in \mathbb{UM}$ *such that* $a \mapsto p \bmod \mathfrak{F}(H)$.

The proof of this corollary goes along the lines of the third part of the proof of Proposition 3.

COROLLARY 2. *Let* $\hat{\tau}\colon T \twoheadleftarrow \hat{H} \in \hat{\mathscr{P}}$. *Then there exists a topologically-lowering modulo* $\mathfrak{F}(\hat{H})$ *morphism* $\hat{\mathfrak{v}}$ *from the* cr*-extension* $\hat{u}\colon C \rightarrowtail \hat{O}$ *to the* cr*-extension* $u\colon C \rightarrowtail \mathbb{UM}$ *such that the former* cr*-extension is smaller then the latter with respect to this morphism.*

2.4. Preimages of $\tilde{\mathscr{P}}$ and corresponding extensions.

PROPOSITION 4. *Let* $\tau\colon T \twoheadleftarrow H \in \tilde{\mathscr{P}}$. *Then for every element* $p \in \mathbb{UM}$ *there exists a function* $a \in O(H, \mathscr{H})$ *such that* $p \mapsto a \bmod \mathfrak{R}(H)$.

PROOF. We shall use the notation from the previous proof. Let $F \in \mathscr{F}(\mathscr{T})$. Consider the set $G \equiv T \setminus F$. Then for the family $\varkappa \equiv \{\tau^{-1}G\} \in K(\mathscr{K}_0)$ there exists an s-dense enclosure $\varkappa' \in S(\mathscr{K}_0)$. Set $P \equiv \operatorname{bod}\varkappa'$ and $Q \equiv \operatorname{cobod}\varkappa'$. Since $P = \operatorname{cl}\tau^{-1}G$, we have $Q = \operatorname{int}\tau^{-1}F$ and thus $H_N \subset \tau^{-1}F$ implies $H_N \subset Q$. Since $\chi(Q) \in O_0$, $Q \in \mathscr{K}_1$.

Fix an ideal N. We shall prove that $H_N \in \mathscr{K}_2 \cap \text{co-}\mathscr{H}$. Take the sets S_{NM} and the sequences $\{F^i_{NM} \mid i\}$ from the proof of Lemma 13, and the sets $Q^i_{NM} \equiv \operatorname{int}\tau^{-1}F^i_{NM}$ and $Q_{NM} \equiv \bigcup Q^i_{NM} \in \mathscr{K}_1$. Choose a sequence of closed sets $\{K_j\}$ such that $\bigcup K_j \subset T \setminus \bigcup\{F^i_{NM} \mid i\}$ and $T \setminus \bigcup K_j \in N$. Denote by N_j the ideals $N(K_j)$. It follows from $N = \operatorname{top}\{N_j\}$ and $Q^i_{NM} \cap H_{N_j} = \varnothing$ that $Q_{NM} \cap H_N = \varnothing$.

Consider $Q_N \equiv \bigcup\{Q_{NM} \mid M\} \in \mathscr{G}(\mathscr{K}_1)$. Our conditions imply that for the single-element family $\varkappa_N \equiv \{Q_N\}$ there exists an s-dense enclosure $\varkappa'_N \in S(\mathscr{K}_1)$. Set $Q'_N \equiv \operatorname{bod}\varkappa'_N$. Since $\operatorname{cl}(Q'_N) \in O_1$, we have $Q'_N \in \mathscr{K}_2$. We shall verify now that Q_N is dense in $H \setminus H_N$. Let G be some open set in $H \setminus H_N$. Then $G \cap H_M \neq \varnothing$ for some $M \neq N$. Now $\mathscr{A}$ is a lattice with relative complements so there exist ideals M_N and M^*_N such that $M_N = M \wedge N$, $M_N \wedge M^*_N = \mathscr{P}$, and $M = M_N \vee M^*_N$. Then from $M = \operatorname{top}\{M_N, M^*_N\}$ it follows that $H_M = H_{M_N} \cup H_{M^*_N}$. So, the inclusion $M_N \supset N$ implies $G \cap H_{M^*_N} \neq \varnothing$. Consider the ideals $M_i \equiv M^*_N(F^i_{NM})$. Then $M^*_N = \operatorname{top}\{M_i\}$ implies $G \cap H_{M_i} \neq \varnothing$ for some i. But $H_{M_i} \subset Q^i_{NM}$; therefore $G \cap Q_{NM} \neq \varnothing$. From the density property it follows now that $H \setminus H_N \subset Q'_N$. Since $Q'_N \cap H_N = \varnothing$, $H \setminus H_N = Q'_N \in \mathscr{K}_2 \cap \text{co-}\mathscr{H}$.

Let N and M be arbitrary ideals. Consider $L \equiv N \wedge M$. Since $N = L \vee N^*_L$ and $M = L \vee M^*_L$, we have $N = \operatorname{top}\{L, N^*_L\}$ and $M = \operatorname{top}\{L, M^*_L\}$, according to the above reasoning. Therefore $H_N = H_L \cup H_{N^*_L}$ and $H_M = H_L \cup H_{M^*_L}$. Since $L \wedge N^*_L = \mathscr{P}$, there exists a universally measurable set $X \in N^*_L$ such that $T \setminus X \in L$. Choose a sequence $\{F_i\}$ of closed sets such that $\bigcup F_i \subset T_L \cap X$ and $T_L \setminus \bigcup F_i \in L$. Denote the ideals $L(F_i)$ by L_i. Then $L = \operatorname{top}\{L_i\}$ implies $H_L = \operatorname{cl}\cup H_{L_i}$. Then there exists a sequence $\{H_j\}$ of closed sets such that $\bigcup H_j \subset T_{N^*_L} \setminus X$ and $T_{N^*_L} \setminus \bigcup H_j \in N^*_L$. Consider the ideals $N_j \equiv N^*_L(H_j)$. Then from $N^*_L = \operatorname{top}\{N_j\}$ it follows that $H_{N^*_L} = \operatorname{cl}\cup H_{N_j}$. But we know that $\bigcup H_{L_j} \cap \bigcup H_{N_j} = \varnothing$. From the property of the elements of the subcover $\mathfrak{H}$ established in the previous paragraph it follows that $H_L \cap H_{N^*_L} = \varnothing$. Analogously, we have $H_L \cap H_{M^*_L} = \varnothing$ and $H_{N^*_L} \cap H_{M^*_L} = \varnothing$. Now the reasoning of the present paragraph implies that $H_N \cap H_M = H_L$ and $L \supset N \cup M$.

Let $\{X_N\}$ be some polyset. Consider the functions $x_N \equiv \chi(X_N)$ and the polyfunction $\{x_N\}$. Then for given N there exist closed sets Y^k_N and Z^l_N such that $\bigcup Y^k_N \subset X_N \subset T \setminus \bigcup Z^l_N$ and $T \setminus \bigcup Y^k_N \cup \bigcup Z^l_N \in N$. Take $Q^k_N \equiv \operatorname{int}\tau^{-1}Y^k_N$, $R^l_N \equiv \operatorname{int}\tau^{-1}Z^l_N$, $Q_N \equiv (\bigcup Q^k_N) \cap H_N$ and $R_N \equiv (\bigcup R^l_N) \cap H_N$. Then $Q_N \cap R_N = \varnothing$.

Denote by N_k the ideals $N(Y_N^k)$ and by N_l the ideals $N(Z_N^l)$. Then we have $H_{N_k} \subset Q_N^k$ and $H_{N_l} \subset R_N^l$ together with $N = \operatorname{top}\{N_k, N_l \mid k, l\}$. Therefore $H_N = \operatorname{cl} Q_N \cup \operatorname{cl} R_N$. Let $M \supset N$. Consider the closed sets $Y_{ik} \equiv Y_M^i \cap Y_N^k$ and $Z_{jl} \equiv Z_M^j \cap Z_N^l$. Then we have $T \subset \bigcup Y_{ik} \cup \bigcup Z_{jl} \in M$. Next, take the ideals $M_{ik} \equiv M(Y_{ik})$, $M_{jl} \equiv M(Z_{jl})$. Then $M = \operatorname{top}\{M_{ik}, M_{jl} \mid i, j, k, l\}$. So, the set $P \equiv \bigcup H_{M_{ik}} \cup \bigcup H_{M_{jl}}$ is dense in H_M. Besides this, we have $H_{M_{ik}} \cap H_{M_{jl}} = \varnothing$. Since $Q_N^k \cap (\bigcup H_{M_{jl}}) = \varnothing$ and $R_M^j \cap (\bigcup H_{M_{ik}}) = \varnothing$, $Q_N^k \cap R_M^j \cap P = \varnothing$. It follows now from the density of P that $Q_N^k \cap R_M^j \cap H_M = \varnothing$. Thus, $Q_N \cap R_M = \varnothing$ implies $Q_N \cap H_M \subset \operatorname{cl} Q_M$. Analogously, $R_N \cap H_M \subset \operatorname{cl} R_M$. Now let N and M be arbitrary ideals. Then it follows from $H_N \cap H_M = H_L$ and $L \supset N \cup M$ that $Q_N \cap R_M = (Q_N \cap H_L) \cap (R_M \cap H_L) \subset \operatorname{cl} Q_L \cap \operatorname{cl} R_L$. Since $Q_L \cap R_L = \varnothing$ we obtain $Q_N \cap R_M = \varnothing$.

The property just proved implies that the sets $Q \equiv \bigcup Q_N$ and $R \equiv \bigcup R_N$ are disjunct. Set $U \equiv Q \cup R$. Let $G \in \mathscr{H}$ and $G \cap U \cap H_N = \varnothing$. Then $G \cap (Q_N \cup R_N) \cap H_N = \varnothing$ implies $G \cap H_N = \varnothing$. Thus $H = sU$. Next, take $U_N \equiv Q_N \cup R_N$. Let $G \cap U_N \cap H_M = \varnothing$ for $M \supset N$. Since $H_N = \operatorname{cl} U_N$ and $H_M \in \mathscr{H}$, we have $G \cap H_M = \varnothing$. Consequently, U_N is N-dense in H. Consider the family $\varkappa = \{Q, R\} \in K^c(\mathscr{K}_2)$. By our conditions for $\varkappa$, there exists an s-dense enclosure $\varkappa' \equiv \{Q', R'\} \in S(\mathscr{K}_2)$. It is clear that $\operatorname{bod} \varkappa' = H$. Therefore $a \equiv \chi(Q') \in O_2$. If $s \in O_N$, then from $\tau s \in X_N$ it follows that $a(s) = 1 = x_N(\tau s)$. On the other hand, if $s \in R_N$, then from $\tau s \notin X_N$ it follows that $a(s) = 0 = x_N(\tau s)$.

Let $p \in \mathbb{UM}$ and $\{f_N\} \in p$. Then all these functions take their values in one and the same interval $]-z, z[$. Subdivide this interval by points z_{mi} in such a way that $z_{mi+1} - z_{mi} = 1/(2m)$. Consider the sets $X_N^{mi} \equiv f_N^{-1}(]z_{mi-1}, z_{mi+1}[)$. It is clear that $\{X_N^{mi} \mid N\}$ is a polyset. Set $x_N^{mi} \equiv \chi(X_N^{mi})$ and introduce the polyfunction $\{x_N^{mi} \mid N\}$. Due to what has already been proved, there exist sets $Q_{mi} \in \mathscr{H} \cap \text{co-}\mathscr{H}$ and $U_N^{mi} \in \mathscr{U}_N(H)$ and functions $a_{mi} \equiv \chi(Q_{mi})$ such that $a_{mi}(s) = x_N^{mi}(\tau s)$ for all $s \in U_N^{mi}$. Take b_{mi} and c_{mi} in O such that $b_{mi}(s) \equiv z_{mi-1}$ and $c_{mi}(s) \equiv z_{mi+1}$ for $s \in Q_{mi}$ and $b_{mi}(s) \equiv -z$, $c_{mi}(s) \equiv z$ for $s \notin Q_{mi}$. Set $U_N^m \equiv \bigcap U_N^{mi}$.

Suppose that $G \equiv H \setminus \bigcup\{Q_{mi} \mid i\} \neq \varnothing$. Then $G \cap H_N \neq \varnothing$ for some N. Therefore $G \cap U_N^m \cap H_N \neq \varnothing$. Take a point s in the last set. The latter relation would imply $\tau s \notin \bigcup\{X_N^{mi} \mid i\} = T$, which is impossible. So let $Q \equiv Q_{mi} \cap Q_{nj} \neq \varnothing$. Then $Q \cap H_N \neq \varnothing$ for some N. Therefore $Q \cap U_N^{mi} \cap U_N^{nj} \cap H_N \neq \varnothing$. Fix a point s in the latter set. For this point, $\tau s \in X_N^{mi} \cap X_N^{nj}$ must hold. Consequently, the inequality $z_{mi-1} \vee z_{nj-1} \leqslant f_N(\tau s) \leqslant z_{mi+1} \wedge z_{nj+1}$ implies $b_{mi}(s) \leqslant z_{mi-1}\mathbf{1} \leqslant z_{nj+1}\mathbf{1} \leqslant c_{nj}(s)$. On the other hand, if $Q = \varnothing$, then, obviously, $b_{mi} \leqslant c_{nj}$.

Introduce $b_m \equiv \max\{b_{mi}\}$ and $c_n \equiv \min\{c_{nj}\}$. If $s \in H$, then $s \in Q_{mi}$. Therefore $c_m(s) - b_m(s) \leqslant c_{mi}(s) - b_{mi}(s) < 1/m$. So, there exists a function $a \in A$ such that $b_m \leqslant a \leqslant c_n$. If $s \in U_N^m$, then $s \in Q_{mi}$ for some i. Thus, from $\tau s \in X_N^{mi}$ it follows that $a(s) - f_N(\tau s) \leqslant c_{mi}(s) - z_{mi-1} < 1/m$ and $a(s) - f_N(\tau s) \geqslant b_{mi}(s) - z_{mi+1} > -1/m$. Consequently, $a \sim f_N \circ \tau \bmod \mathscr{R}_N(H)$; this means that $p \mapsto a \bmod \mathfrak{R}(H)$, which proves the proposition. □

COROLLARY. *Let $\tilde{\tau} : T \twoheadleftarrow \tilde{H} \in \tilde{\mathscr{P}}$. Then there exists a topologically-lifting modulo $\mathfrak{R}(\tilde{H})$ morphism $\tilde{v}$ from the cr-extension $u \colon C \rightarrowtail \mathbb{UM}$ to the cr-extension $\tilde{u} \colon C \rightarrowtail \tilde{O}$ such that the former cr-extension is smaller then the latter one with respect to it.*

2.5. Description of the hyperstonean preimage and the Gordon preimage. Let $\tau \colon T \twoheadleftarrow H$ be the as-preimage constructed in §2.2.

PROPOSITION 5. *The as-preimage* $\tau\colon T \twoheadleftarrow H$ *realizes the Arens cr-extension* $u\colon C \rightarrowtail \mathbb{UM}$ *with respect to the topologically-lifting modulo* $\mathfrak{F}(H)$ *realizing isomorphism* r. *This means that* $\tau\colon T \twoheadleftarrow H$ *is the hyperstonean preimage* $\tau\colon T \twoheadleftarrow hT$.

PROOF. Since $\tau\colon T \twoheadleftarrow H \in \hat{\mathscr{P}} \cap \tilde{\mathscr{P}}$, the corollaries to Propositions 3 and 4 imply that there exists a topologically-lifting modulo $\mathfrak{F}(H)$ realizing isomorphism r between $u\colon C \rightarrowtail \mathbb{UM}$ and $u_\tau\colon C \rightarrowtail O$. The proposition is proved. □

THEOREM 1. *The hyperstonean preimage* $T \twoheadleftarrow hT$ *is an enclosable as-covering of* T *of the type* $ZZZ^c|^aZ$.

PROOF. Let $u\colon C \rightarrowtail O$ denote the cr-extension corresponding to the as-preimage $\tau\colon T \twoheadleftarrow hT$. Let $\hat{\tau}\colon T \twoheadleftarrow \hat{H} \in \hat{\mathscr{P}}$ and $\tilde{\tau}\colon T \twoheadleftarrow \tilde{H} \in \tilde{\mathscr{P}}$. The previous proposition and results of Subsection 2.3 imply the relation $\hat{u}\colon C \rightarrowtail \hat{O} \leqslant u\colon C \rightarrowtail O$ with respect to $r \circ \hat{v}$. It follows from 2.4 that $u\colon C \rightarrowtail O \leqslant \tilde{u}\colon C \rightarrowtail \tilde{O}$ with respect to $\tilde{v} \circ r^{-1}$. Hence $\hat{\tau}\colon T \twoheadleftarrow \hat{H} \leqslant \tau\colon T \twoheadleftarrow hT \leqslant \tilde{\tau}\colon T \twoheadleftarrow \tilde{H}$. The theorem is proved. □

Consider now the *universally measurable Kelley extension* $u\colon C \rightarrowtail UM$ studied by Kelley in **[21]** and also by the author in **[23]**. Introduce on the c-ring UM the saturated refinement $\mathfrak{A} : \mathscr{A} \rightarrowtail \mathscr{C}(UM)$ by setting $\mathfrak{A}(N) \equiv \{f \in UM \mid \forall n(T_N \cap \operatorname{coz}_n f \in N)\}$. Then $u\colon (C, \mathfrak{L}) \rightarrowtail (UM, \mathfrak{A})$ is a cr-extension. Its description was announced in **[22]**.

We call the realizing as-preimage $\tau\colon T \twoheadleftarrow gT$ of the Kaplan cr-extension $u\colon C \rightarrowtail UM$ "*the Gordon preimage of* T". It was introduced in **[24]** and studied in **[23]**. Here we give its complete description without proof (it can be found in our paper **[26]**).

THEOREM 2. *The Gordon preimage* $T \twoheadleftarrow gT$ *is an enclosable as-cover of* T *of the type* $ZZ^c|^aZ^c$.

References

1. K. Kuratowsky, *Topology*, Vol. I, Academic Press, New York and London, 1966.
2. K. Jacobs, *Measure and integral*, Academic Press, New York, 1978.
3. Z. Semadeni, *Banach spaces of continuous functions*, Vol. I, PWN, Warszawa, 1971.
4. N. Bourbaki, *Intégration*, Chaps. 3–5, 9, Actualités Sci. Indust., nos. 1175, 1244, 1343, Hermann, Paris, 1965, 1967, 1969.
5. R. F. Arens, *Operations induced in function classes*, Monatsh. Math. **55** (1951), 1–19.
6. A. D. Aleksandrov, *Additive functions in abstract spaces*. I, Mat. Sb. **6** (1940), 303–348; II **9** (1941), 563–628; III **13** (1943), 169–238. (Russian)
7. N. J. Fine, L. Gillman, and J. Lambek, *Rings of quotients of rings of functions*, McGill Univ. Press, Montreal, 1966.
8. V. K. Zakharov, *Functional characteristics of the absolute vector lattices of functions with the Baire property and of quasinormal functions, and modules of quotients of continuous functions*, Trudy Moskov. Mat. Obshch. **45** (1982), 68–104; English transl. in Trans. Moscow Math. Soc. **1984**, no. 1.
9. A. V. Arkhangelskiĭ and V. I. Ponomarev, *The fundamentals of general topology through problems and exercises*, "Nauka", Moscow, 1974; English transl., Reidel, Dordrecht, 1984.
10. V. K. Zakharov and A. V. Koldunov, *Sequential absolute and its characterization*, Dokl. Akad. Nauk SSSR **253** (1980), no. 2, 280–284; English transl. in Soviet Math. Dokl. **22** (1980).
11. F. Dashiell, A. Hager, and M. Henriksen, *Order-Cauchy complections of rings and vector lattices of continuous functions*, Canad. J. Math. **32** (1990), 657–685.
12. V. K. Zakharov, *On functions connected with sequential absolute, Cantor completion and classical ring of quotients*, Period. Math. Hungar. **19** (1988), no. 2, 113–133.
13. J. Flachsmeyer, *Topologization of Boolean algebras*, General Topology and its Relations to Modern Analysis and Algebra (Prague, 1976), Lecture Notes in Math., Vol. 609, Springer-Verlag, Berlin and New York, 1977, pp. 81–97.

14. D. H. Fremlin, *Review of the book "The bidual of $C(X)$, I" by S. Kaplan*, Bull. Amer. Math. Soc. **15** (1986), 98–101.
15. V. K. Zakharov, *Hyperstonean absolute of a completely regular space*, Dokl. Akad. Nauk SSSR **267** (1982), no. 2, 280–283; English transl. in Soviet Math. Dokl. **26** (1982).
16. ______, *Some perfect preimages connected with extensions of the family of continuous functions*, Colloq. Math. Soc. Janos Bolyai, Vol. 401, North-Holland, Amsterdam, 1983, pp. 703–728.
17. ______, *Hyperstonean cover and second dual extensions*, Acta Math. Hungar. **51** (1988), no. 1–2, 125–149.
18. ______, *Topological preimages corresponding to the classical extensions of the ring of continuous functions*, Vestnik Moskov. Univ. Ser. I Mat. Mekh. **1990**, no. 1, 44–47; English transl. in Moscow Univ. Math. Bull. **45** (1990).
19. ______, *CR-envelopes of rings of continuous functions*, Dokl. Akad. Nauk SSSR **294** (1987), no. 3, 531–534; English transl. in Soviet Math. Dokl. **35** (1987).
20. J.-P. Delfosse, *Caracterisations d'anneaux de fonctions continues*, Ann. Soc. Sci. Bruxelles Sér. I, **89** (1975), no. 3, 364–368.
21. J. L. Kelly, *Measures in Boolean algebras*, Pacific J. Math. **9** (1959), 1165–1177.
22. S. Kaplan, *The bidual of $C(X)$*, I, North-Holland, Amsterdam, 1985.
23. V. K. Zakharov, *Universally measurable extension and Arens extension of the Banach algebra of continuous functions*, Funktsional. Anal. i Prilozhen. **26** (1990); English transl. in Functional Anal. Appl. **26** (1990).
24. ______, *Lebesgue cover and Lebesguean extension*, Studia Sci. Math. Hungar. **23** (1988), 343–368.
25. H. Gordon, *The maximal ideal space of a ring of measurable functions*, Amer. J. Math. **88** (1966), 827–843.

St. Petersburg State University for Technology and Desing, 18, Bolshaya Morskaya, St. Petersburg, 191065, Russia

Translated by G. Rozenblum

Amer. Math. Soc. Transl.
(2) Vol. **159**. 1994

The Variational Principle and Completely Nonlinear Second-Order Equations

N. M. Ivochkina

§0. Introduction

Equations with a principal operator given in a divergence form belong to a special category in the theory of completely nonlinear elliptic second-order equations. The divergence structure excludes total ellipticity, that is, the ellipticity in C^2, if the principal operator in the equation is not linear or quasilinear [**1**].

In this paper we consider the class of completely nonlinear second-order Euler-Lagrange operators, described in Section 1. In this section a concept of a mixed operator measure with Euler-Lagrange operator as its density is put forward. The Monge–Ampère type operators are considered as examples. It is interesting to note that this class contains the action functional Lagrangian in General Relativity Theory.

It is known that the absence of total ellipticity for the Monge–Ampère type operators of order m is compensated by their ellipticity in a functional cone closely connected with structural peculiarities of the operators [**2–7**]. The investigation of solutions of the corresponding equations depends on properties of these cones. A description of these cones is given in Section 2. In Section 3 it is shown how the properties of the cones work in the proof of $C^{l+2+\alpha}(\bar{\Omega})$-solvability of the Dirichlet problem for the above equations. In Section 4 a concept of weak solutions of the equations with operators from Section 1 is discussed, that is, solutions without second derivatives, generally speaking. The notion of weak solutions for equations elliptic in a cone was introduced in [**8**]. In the same paper, solvability in the weak sense of the Dirichlet problem for equations connected with curvature, in particular, for the curvature equation of order M, was proved. In elliptic theory the loss of smoothness of solutions usually results from a decrease in smoothness of the data or degeneration of ellipticity. It seems that for the class of equations in question, besides the causes mentioned and regardless of these, some concrete structural peculiarities of the principal operators are responsible for a possible loss of smoothness of solutions. In Section 4 this point of view is illustrated by an example.

The Euler-Lagrange equations are naturally connected with variational problems. In Section 5 an analysis of some variational problems with completely nonlinear second-order Lagrangians is presented, and smooth solutions of the Dirichlet problem are interpreted in the variational sense. In fact, there we discuss the minimization of unbounded in C^2 and degenerated functionals.

1991 *Mathematics Subject Classification*. Primary 35D10, 35A15; Secondary 35A30, 35B65.

§1. Operators μ_m^v as variational derivatives

Let us consider a pair of Euclidean spaces $\mathbf{R}^n(x)$, $\mathbf{R}^n(p)$, $n \geqslant 2$, and functions $u(x)$, $x = (x^1, \dots, x^n)$, $v(p)$, $p = (p_1, \dots, p_n)$, u, $v \in C^2$. In the space $C^2(\Omega)$, where Ω is a domain in $\mathbf{R}^n(x)$, we define μ_m^v, $m = 1, \dots, n$, by

$$\mu_0^v[u] \equiv 1, \quad \mu_m^v[u] = \frac{1}{n-m+1} \frac{\delta}{\delta u} J_{m-1}^v(u),$$

$$J_{m-1}^v(u) = \int_\Omega v(u_x) \mu_{m-1}^v[u]\, dx,$$

where $\delta/\delta u$ is a variational derivative, i.e., the Euler-Lagrange operator

$$\frac{\delta}{\delta u} = -\left(\frac{d}{dx^i} \frac{\partial}{\partial u_i} - \frac{d^2}{dx^i dx^j} \frac{\partial}{\partial u_{ij}} \right).$$

In order to describe the structure of μ_m^v we introduce the matrices $u_{xx} = (u_{ij})$, $u_i = du/dx^i$, $v^{pp} = (v^{ij})$, $v^i = \partial v / \partial p_i$, $v^p_{,x} = (v^i_{,j})$, $v^i_{,j} = d(\partial v/\partial u_i)/dx^j$. Denote by $u_{\left[\begin{smallmatrix} i_1 \dots i_m \\ j_1 \dots j_m \end{smallmatrix}\right]}$, $v^{\left[\begin{smallmatrix} i_1 \dots i_m \\ j_1 \dots j_m \end{smallmatrix}\right]}$, $v^{[i_1 \dots i_m]}_{,[j_1 \dots j_m]}$ the minors of those matrices whose rows and columns are enumerated by the indices $i_1 < \dots < i_m$, $j_1 < \dots < j_m$, respectively.

THEOREM 1. *The following equality is valid:*[1]

$$\mu_m^v[u] = \frac{1}{\binom{n}{m}} v^{[i_1 \dots i_m]}_{,[j_1 \dots j_m]}. \tag{1}$$

PROOF. For $m = 1$ the assertion is evident. Supposing (1) to be valid for $m - 1$ we first derive two auxiliary identities for the inductive step to m:

$$\frac{d}{dx^j} \frac{\partial \mu_m^v}{\partial u_{ij}} = \frac{\partial \mu_m^v}{\partial u_i}, \quad \frac{d}{dx^i} \frac{\partial \mu_m^v}{\partial v^k_{,i}} = 0, \qquad k = 1, \dots, n. \tag{2}$$

Formula (2) follows from the relations proved in **[9]**:

$$\frac{\partial^2 u_{\left[\begin{smallmatrix} i_1 \dots i_m \\ j_1 \dots j_m \end{smallmatrix}\right]}}{\partial u_{ij} \partial u_{kl}} + \frac{\partial^2 u_{\left[\begin{smallmatrix} i_1 \dots i_m \\ j_1 \dots j_m \end{smallmatrix}\right]}}{\partial u_{ik} \partial u_{lj}} + \frac{\partial^2 u_{\left[\begin{smallmatrix} i_1 \dots i_m \\ j_1 \dots j_m \end{smallmatrix}\right]}}{\partial u_{il} \partial u_{jk}} \equiv 0, \qquad i, j, k, l = 1, \dots, n,$$

$$\sum_{i=i_1}^{i_m} \frac{d}{dx^i} \frac{\partial u_{\left[\begin{smallmatrix} i_1 \dots i_m \\ j_1 \dots j_m \end{smallmatrix}\right]}}{\partial u_{ij}} \equiv 0. \tag{3}$$

[1] Here and below the summation is taken over repeating indices from 1 to n, unless otherwise stated.

Indeed,

$$\frac{d}{dx^i}\frac{\partial \mu_m^v}{\partial u_{ij}} = \frac{\partial u_{\left[\substack{i_1\dots i_m\\ j_1\dots j_m}\right]}}{\partial u_{ij}}\frac{d}{dx^j}v^{\left[\substack{i_1\dots i_m\\ j_1\dots j_m}\right]} = u_{\left[\substack{i_1\dots k\dots i_m\\ j_1\dots j_m}\right]}\frac{\partial}{\partial u_k}v^{\left[\substack{i_1\dots i\dots i_m\\ j_1\dots j_m}\right]}$$

$$= \left(\frac{\partial}{\partial u_k}v^{ij}\frac{\partial v^{\left[\substack{i_1\dots i_m\\ j_1\dots j_m}\right]}}{\partial v^{kj}}\right)u_{\left[\substack{i_1\dots i_m\\ j_1\dots j_m}\right]} = \frac{\partial \mu_m^v}{\partial u_j},$$

$$\frac{d}{dx^i}\frac{\partial v^{[i_1\dots i_m]}_{,[i_1\dots i_m]}}{\partial v^k_{,i}} = \frac{d}{dx^i}\left(\frac{\partial v^{\left[\substack{i_1\dots i_m\\ j_1\dots j_m}\right]}}{\partial v^k_{,j}}\frac{\partial u_{\left[\substack{i_1\dots i_m\\ j_1\dots j_m}\right]}}{\partial u_{ij}}\right) = \frac{\partial^2 v^{\left[\substack{i_1\dots i_m\\ j_1\dots j_m}\right]}}{\partial v^{kl}\partial v^{st}}v^{lst}u_{\left[\substack{i_1\dots i_m\\ j_1\dots j_m}\right]} \equiv 0.$$

Now we calculate the Euler-Lagrange operator:

$$\left(\frac{\partial}{\partial u_i} - \frac{d}{dx^j}\frac{\partial}{\partial u_{ij}}\right)(v(u_x)\mu_{m-1}^v[u]) = \frac{\partial v}{\partial u_i}\mu_{m-1}^v - \frac{d}{dx^j}v\frac{\partial \mu_{m-1}^v}{\partial u_{ij}}$$

$$= v^i v^{[i_1\dots\bar{i}\dots i_{m-1}]}_{,[i_1\dots i_{m-1}]} - v^k v^{[ii_1\dots\bar{i}\dots i_{m-1}]}_{,[ki_1\dots i_{m-1}]} = v^k\frac{\partial v^{[i_1\dots i_m]}_{,[i_1\dots i_m]}}{\partial v^k_{,i}},$$

$$\frac{\delta J_{m-1}^v}{u} = -\frac{d}{dx^i}v^k\frac{\partial v^{[i_1\dots i_m]}_{,[i_1\dots i_m]}}{\partial v^k_{,i}} = -mv^{[i_1\dots i_m]}_{,[i_1\dots i_m]} = -(n-m+1)\mu_m^v[u]. \qquad \square$$

Let us present the divergent forms of μ_m^v. From (2) it follows that

$$\mu_m^v[u] = \frac{1}{m}\left(\frac{d}{dx^i}v^k\frac{\partial \mu_m^v}{\partial v^k_{,i}}\right)[u]. \tag{4}$$

Due to (3) written for $v(p)$, one has

$$\mu_m^v = \frac{1}{m}\frac{d}{dx^i}\frac{\partial}{\partial u_k}\left(v\frac{\partial \mu_m^v}{\partial v^k_{,i}}\right).$$

In terms of exterior forms, (4) may be written as

$$\omega_m^n[v;u] = d\omega_{m-1}^{n-1}[v;u], \tag{5}$$

where d is the exterior differential

$$\omega_m^n[v;u] = \mu_m^v[u]\,dx^1\wedge\cdots\wedge dx^n;$$

$$\omega_{n-1}^{m-1}[v;u] = \frac{1}{m}\sum_{k=1}^{m}\sigma(i,1,\dots\bar{i},\dots n)v^k(u_x)\frac{\partial \mu_m^v}{\partial v^k_{,i}}\wedge dx^1\cdots\wedge d\bar{x}^i\wedge\cdots\wedge dx^n;$$

$\sigma(i_1,\dots,i_n)$ is the sign of the permutation, while the notation $\bar{i}$ means that the corresponding member is omitted. As in **[10, 11]**, we show that $\omega_n^m[v;u]$ gives the mixed measure induced by u, v in the following sense. Associate to $\Omega\subset\mathbf{R}^n(x)$ the set

$\hat{\Omega} \subset \mathbf{R}^n(x)$ with the help of the mapping $H_{vu} = H_v \circ H_u$, where $H_u : \hat{p} = u_x \in \mathbf{R}^n(p)$, $H_v : \hat{x} = v^p \in \mathbf{R}^n(x)$,

$$H_{vu} : \Omega \to \hat{\Omega} = \{\hat{x} : x \in \Omega\}.$$

If u, $v \in C^2$, then

$$\omega_{m-1}^{n-1} = \frac{1}{m}\frac{1}{\binom{n}{m}} \sum_{i \notin \{i_{m+1} < \cdots < i_n\}} \sigma(i, i_1, \ldots, \bar{i}, \ldots, i_n)\hat{x}^i d\hat{x}^{i_1} \wedge \cdots \wedge d\bar{\hat{x}}^i \wedge \cdots \wedge dx^{i_n},$$

$$\omega_m^n = \frac{1}{\binom{n}{m}} \sum_{\{i_1 < \cdots < i_m ; i_{m+1} < \cdots < i_n\}} \sigma(i_1, \ldots, i_n) d\hat{x}^{i_1} \wedge \cdots \wedge d\hat{x}_{i_m} \wedge dx^{i_{m+1}} \wedge \cdots \wedge dx^{i_n}.$$

When one considers exterior forms or integrals in the role of volume in some sense, the question of invariance arises. In our situation the integral $J_{m-1}^v(u)$ may be called a volume. Since $v^i_{.j}$ is a mixed tensor, the type of invariance is determined by v. For example, if $v = (1+p^2)^{5/2}$, then J_{m-1}^v is invariant with respect to orthogonal transformations. Note also that (5) may be regarded as a Stokes formula and μ_m^v as a mixed measure density. It often turns out to be useful to abstract from the specific form of v in μ_m^v and deal with representations uniformized with respect to v instead of (1). If the eigenvalues of v^{pp} are positive, then μ_m^v admit the interpretation

(6) $$\mu_m^v = S_m(\lambda)/\binom{n}{m},$$

where $\lambda = (\lambda_1, \ldots, \lambda_n)$ is the vector of eigenvalues of $v^p_{.x}$ and S_m is the mth-order elementary symmetric function. Sometimes it is more convenient to use

(7) $$\mu_m^v[u] = \hat{u}_{\left[\begin{smallmatrix} i_1 \ldots i_m \\ j_1 \ldots j_m \end{smallmatrix}\right]}/\binom{n}{m} \equiv F_m[\hat{u}],$$

intermediate between (1) and (6), which differs from (6) in that the uniformization ends by finding the eigenvalues of v^{pp} and a subsequent dilation.

EXAMPLES. a) $m = 1$. The operators μ_1^v are quasilinear:

$$n\mu_1^v[u] = \frac{\partial^2 v(u_x)}{\partial u_i \partial u_j} u_{ij} = v^{ij}(u_x)u_{ij}.$$

By means of v^{pp} one finds out whether μ_1^v is elliptic.

b) $m = n$. The operators μ_n^v prove to be Monge–Ampère up to the multiplier $\det v_{pp}$.

c) $v = \frac{1}{2}p^2$. In this case we shall call μ_m^v Monge–Ampère type operators. For such operators, and only for them, (1) and (7) coincide. Denote these operators by $\mu_m^{(2)}$, $\mu_m^{(2)}[u] = F(u_{xx})$;

d) $v = \sqrt{1+p^2}$. Now we call μ_m^v curvature operators of order m and denote them by $\mu_m^{(1)}$. The eigenvalues of $v^i_{.j}$ have simple geometric meaning: they are the principal curvatures of the graph Γ of u. In particular, $\mu_1^{(1)}[u]$ is the mean curvature of Γ, $\mu_2^{(1)}[u]$ is the scalar curvature, and $\mu_n^{(1)}[u]$ is the Gaussian curvature. The geometric origin of $\mu_m^{(1)}[u]$ allows us to interpret $\mu_m[\Gamma]$ in a natural way.

The functional $J_2^{(1)}(\Gamma)$ admits a physical interpretation. Namely, let us consider the metric tensor $g_{\alpha\beta}$, α, $\beta = 1, \ldots, n$, of $\Gamma \subset \mathbf{R}^{n+1}$ induced by the Euclidean metric of $\mathbf{R}^{n+1}$. One has $J_2^{(1)} = \int_\Omega Rg dx$ with $g = \det g_{\alpha\beta}$, $R = \mu_2[\Gamma]$. Therefore

$J_2^{(1)}$ coincides up to a normalizing multiplier with the action functional S_g of the stationary gravitational field with metric $g_{\alpha\beta}$ in General Relativity Theory.

§2. Description of the cones generated by μ_m^v

Each operator μ_m^v can be regarded as measure density, and so it is worth describing the sets $\{u\}$ where these operators are positive. Besides, bearing in mind the control of J_{m-1}^v by means of its variational derivative, one ought to require a constant sign for $\mu_m^v[u]$, e.g., $\mu_m^v[u] > 0$. This is a reason to expect

(8) $$\{u \in C^2 : \mu_{m-1}^v[u] > 0,\ \mu_m^v[u] > 0\}$$

as the optimal set for J_{m-1}^v. The positivity set of μ_m^v for $m > 1$ consists of at least two connected components, which may be marked by some representatives $u^{(k)}$, $k = 0, 1, \dots$. From this point of view the set competitive with (8) is

(9) $$\{u \in C^2 : \mu_m^v[\Phi(\tau; u^{(0)}, u)] > 0,\ \tau \in [0,1],\ \Phi(0) = u^{(0)},\ \Phi(1) = u\},$$

where $\Phi(\tau)$ is a continuous function and $u^{(0)}$ can be chosen by the researcher.

From now on, we shall consider convex functions v only, $\det v^{pp} > 0$. The key role of the cones

(10) $$K_m^v = \{u \in C^2(\bar{\Omega});\ \mu_i^v[u] > 0,\ i = 1, \dots, m\}$$

in the study of the equations with the operators μ_m^v was first clarified in **[2]**. In **[8]** it was observed that the definition

(11) $$K_m^v = \{u \in C^2(\bar{\Omega});\ \mu_{m-1}^v[u] > 0,\ \mu_m^v[u] > 0\}$$

is equivalent to (10), i.e., we deal with the set (8) for convex v. In **[4, 6]** the definition of K_m^v as a component of the positivity set of μ_m^v containing $u^{(0)} = |x|^2$, i.e., as a set of the type (9) was used. In fact, the latter definition is equivalent to (10), (11), and for convex v the sets (8) and (9) with $u^{(0)} = x^2$ coincide.

REMARK. If $\partial\Omega \subset \mathbf{R}^n$ is a closed C^2 surface, then from $\mu_m[\partial\Omega] > 0$ it follows that $\mu_{m-1}[\partial\Omega] > 0$.

Let us uniformize K_m^v with respect to v in accordance with (6), (7). Namely, consider the number cones $\mathscr{K}_m$, K_m generated by polynomials $S_m(\lambda)$, $\lambda \in \mathbf{R}^n$, $F_m(T)$, $T \in \mathbf{R}^{n(n+1)/2}$:

$$\mathscr{K}_m = \{\lambda \in \mathbf{R}^n;\ S_i(\lambda) > 0,\ i = 1, \dots, m\},$$
$$K_m = \{T = (T^{ij}) \in \mathbf{R}^{n(n+1)/2};\ F_i(T) > 0,\ i = 1, \dots, m\}.$$

The S_m, F_m appear as examples of a-hyperbolic polynomials in **[12]**. According to the definition given by L. Gärding a homogeneous polynomial $P_m(x)$, $x \in \mathbf{R}^N$, is a-hyperbolic if the polynomial $\mathbf{R}^1 \ni t \mapsto P_m(ta + x)$ has only real roots for all $x \in \mathbf{R}^N$. S_m fits into this definition as an e-hyperbolic polynomial, $e = (1, \dots, 1)$, and F_m as a (δ_j^i)-hyperbolic one. In the paper mentioned above, to an a-hyperbolic polynomial is associated the cone $C(a, P_m)$ as a component of its positivity set containing the vector a. For the polynomials S_m and F_m the Gärding cone coincides with $\mathscr{K}_m$ and K_m, respectively. In **[12]** the following basic properties of $C(a, P_m)$ were proved.

a) If $b \in C(a)$, then P_m is b-hyperbolic, while $C(a, P_m) = C(b, P_m)$;

b) $C(a)$ is a convex set in $\mathbf{R}^N$, and $P_m^{1/m}$ is a convex function on $C(a)$. In particular, for all $b, c \in C(a)$

$$\frac{\partial P_m(b)}{\partial b^i} c^i \geqslant m P_m^{\frac{m-1}{m}}(b) P_m^{\frac{1}{m}}(c). \tag{12}$$

In K_m the inequality (12) takes the form

$$\frac{\partial F_m(T)}{\partial T^{ij}} S^{ij} \geqslant m F_m^{\frac{m-1}{m}}(T) F_m^{\frac{1}{m}}(S). \tag{13}$$

Let us give some other properties of $\mathscr{K}_m$ in terms of shortened symmetric functions $S_{l,i_1\dots i_{m-l}}(\lambda) = S_l(\lambda_{i_1\dots i_{m-l}})$, where for $k = 1, \dots, n$

$$\lambda^k_{i_1\dots i_{m-l}} = \begin{cases} \lambda^k, & k \notin \{i_1 \dots i_{m-l}\}, \\ 0, & k \in \{i_1 \dots i_{m-l}\}. \end{cases}$$

It was shown in **[2]** that

$$S_m \leqslant c S_{m-1,i_1} \leqslant \dots \leqslant c^{m-1} S_{1,i_1\dots i_{m-1}}, \qquad c = \max_{\{i_1\dots i_{m-1}\}} S_{1,i_1\dots i_{m-1}}$$

in $\mathscr{K}_m$ and, in particular,

$$S_{m-1,i} \geqslant S_m / \left(\tbinom{n}{m} S_1\right). \tag{14}$$

The situation of one-variable domination in S_m, e.g. λ^1: $S_{m-1} < 2S_{m-2,1}\lambda^1$, was considered in **[13]**. In this case

$$S_{m+1,1} \leqslant c(n) \sum_{i=2}^{n} S_{m-1,i} (\lambda^i)^2 \tag{15}$$

in $\mathscr{K}_m$.

Let us analyze (13), (14) from the functional point of view. From (4) it follows that

$$\frac{\partial \mu_m^v}{\partial u_{ij}} \xi^i \xi^j \geqslant \alpha(u_x) \frac{\mu_m^v}{\binom{n}{m} \mu_1^v} \xi^2 \tag{16}$$

in any cone K_m^v, where $v^{ij}(u_x)\xi^i\xi^j \geqslant \alpha(u_x)\xi^2$, $\alpha > 0$. As for an analogue of (13), one has

$$\frac{\partial \mu_m^{(2)}}{\partial u_{ij}} w_{ij} \geqslant m(\mu_m^{(2)}[u])^{\frac{m-1}{m}} (\mu_m^{(2)}[w])^{1/m} \tag{17}$$

in $K_m^{(2)}$. The inequality (17) contains, besides the convexity of $(\mu_m^{(2)})^{1/m}$, some information on the convexity in C^2 of the cone $K_m^{(2)}$. If $v \neq p^2$, then, generally, the cones

K_m^v are not convex in C^2. So, in general, one can expect the local variant of (17) only:

$$\frac{\partial^2(\mu_m^v)^{1/m}}{\partial u_{ij}\partial u_{kl}}\xi^{ij}\xi^{kl} \leqslant 0, \qquad u \in K_m^v. \tag{18}$$

The functional cones K_m^v are different for different v and the question of relations among them remains open. To some extent, the imbedding $K_m^{(2)} \subset K_{m-1}^{(1)}$ proved in **[14]** is the answer to this question. In the same paper the following analog of (17),

$$\frac{\partial \mu_{m-1}^{(1)}[u]}{\partial u_{ij}} w_{ij} \geqslant (m-1)(\mu_{m-1}^{(1)}[u])^{\frac{m-2}{m-1}}(\mu_{m-1}^{(2)}[w])^{\frac{1}{m-1}}/(1+u_x^2)^{\frac{m+1}{2(m-1)}}, \tag{19}$$

was derived with $u \in K_{m-1}^{(1)}$, $w \in K_m^{(2)}$. From (19) it follows that $K_{m-1}^{(1)}$ is star with respect to $K_m^{(2)}$.

Let us turn to the boundary of $\mathscr{K}_m$. It is determined by $S_m(\chi) = 0$. One has to restrict oneself to the solutions χ such that $(t\lambda + \chi) \in \mathscr{K}_m$ for all $t > 0$, $\lambda \in \mathscr{K}_m$. The description of $\partial\mathscr{K}_m$ originates from the definition of $C(a)$ in **[12]** as the set of vectors $b \in \mathbf{R}^N$ subject to $P_m(ta + b) > 0$ in the case of $P_m(a) > 0$, $t \geqslant 0$.

There is a simple observation for $\mathscr{K}_m$ in **[14]**, which proves to be valid in a general situation. It is the following one.

Set

$$P^{(k).a}(x) = \left.\frac{d^k P(ta + x)}{dt^k}\right|_{t=0}.$$

LEMMA 1. *Let $P(x)$ be an a-hyperbolic polynomial of degree m, a, $b \in C(P)$. Assume $t = 0$ to be a multiplicity-l root of $P(tb + y)$. Then 0 is a root of $P^{(1).a}(tb + y)$ of multiplicity $l - 1$.*

PROOF. It is easy to show that the cone $C(P)$ is completely determined by the inequalities $P^{(k).a}(x) > 0$, $k = 0, 1, \dots, m-1$. First let $l = 2$. If $P^{(1).a}(y) \neq 0$, then $y \in C(P^{(1).a})$ and there exists a small $t_0 > 0$ such that $y - t_0 b \in C(P)$, which is impossible because of the convexity of $C(P)$.

Suppose now that $l = s + 2$, $s \geqslant 1$. Differentiating the polynomials $P^{(k).a}(tb + y)$ s times with respect to t, $k = 0, 1, \dots$, we arrive at the variant considered for the auxiliary polynomials $(P^{(k).a}(x))^{(s).b}$, which are a-hyperbolic due to (12), $b \in C(P^{(s).b})$, $y \in \partial C(P^{(s).b})$. □

COROLLARY. *Let a, b, $c \in C(P)$, $P(a) > 0$, $y \in \partial C(P)$. Then*

$$0 < c_1 \leqslant \frac{P^{(1).a}(tb + y)}{P^{(1).c}(tb + y)} \leqslant c_2, \tag{20}$$

where c_1, c_2 depend on a, b, c, y, $0 < t \ll 1$.

Let us return to $\mathscr{K}_m$. Consider vectors $\lambda \in \mathbf{R}^n$ as functions dependent on $z = (z^1, \dots, z^k) \in Z$. Let $\mu(z) \in \mathscr{K}_m$ for all $z \in Z$. Denote by θ^ε the greatest root of $S_m(\tau\mu + \lambda) - \varepsilon$, $\varepsilon > 0$. Since $\mu \in \mathscr{K}_m$, θ^ε is a simple root, and its smoothness coincides with the smoothness of $\mu(z)$, $\lambda(z)$. Besides, we have the following assertion.

LEMMA 2. *On any compact* $\hat{Z} \subset Z$

$$|\theta^\varepsilon_z| \leqslant c(\|\mu, \lambda\|_{C^1(\hat{Z})}). \tag{21}$$

PROOF. Set $\chi = \theta^0\mu + \lambda$, $\tau = \theta^0 - \theta^\varepsilon$, where $\theta^0 = \lim \theta^\varepsilon$ as $\varepsilon \downarrow 0$, and rewrite S_m in the new notation: $S_m(\theta^\varepsilon\mu + \lambda) = S_m(\tau\mu + \chi)$. Lemma 1 may be applied to $S_m(\tau\mu + \chi)$; in particular, (20) holds with $P = S_m$, $b = c = \mu$, $a = e$, $y = \chi$. Therefore,

$$S_{m-1}(\tau\mu + \chi) = P^{(1),e}(\tau\mu + \chi) \leqslant c_2 S_m^{(1),\mu}(\tau\mu + \chi). \tag{22}$$

Note that $S_m^{(1),\mu}(\tau\mu + \chi)$ admits the following interpretation:

$$S_m^{(1),\mu}(\tau\mu + \chi) = \frac{d}{d\tau} S_m(\tau\mu + \chi) = \frac{\partial}{\partial\theta^\varepsilon} S_m(\theta^\varepsilon\mu + \lambda). \tag{23}$$

Bearing in mind $S_{m-1.i} > 0$, $i = 1, \dots, n$, in $\mathscr{K}_m$ along with $S_{m-1} = (\sum_1^n S_{m-1.i})/(n - m + 1)$, we derive

$$S_{m-1.i}(\tau\mu + \chi) \leqslant (n - m + 1) S_{m-1}(\tau\mu + \chi). \tag{24}$$

Calculating directly the derivative

$$\frac{\partial\theta^\varepsilon}{\partial z^i} = S_{m-1.i}(\tau\mu + \chi)(\theta^\varepsilon\mu_i + \lambda_i)/S_m^{(1),\mu}(\tau\mu + \chi)$$

and using (22)–(24) and the compactness of $\hat{Z}$ we obtain (22). □

COROLLARY. *If* $\mu(z) \in \mathscr{K}_m$, *then the greatest root* $\theta^0(z)$ *of* $S_m(\mu(z)t + \lambda(z))$ *belongs to* $C^{0,1}(\hat{Z})$, *while* $\|\theta^0\|_{C^{0,1}(\hat{Z})} \leqslant c(\|\mu, \lambda\|_{C^{0,1}(\hat{Z})})$.

The smoothness problem of all roots of hyperbolic polynomials, i.e., polynomials whose roots are all real, was solved in the general theory in **[15, 16]**. It was proved there that the smoothness of roots is not greater than $C^{0,1}$ and depends on the degree of the polynomial and the smoothness of the coefficients. The inequality in the corollary of Lemma 2 makes more precise the known results for $S_m(\mu t + \lambda)$ and its greatest root $\theta^0(z)$.

§3. Equations with the operators $\mu_m^{(1)}$, $\mu_m^{(2)}$

Let us consider the problem of finding the solutions of

$$\mu_m^v[u] = f(x) \tag{25}$$

in K_m^v with a strictly convex function v. By the definition of the cone , this problem may be well posed only if $f(x) > 0$. The inequalities (16) provide the ellipticity of (25) in K_m^v, while the vanishing of f leads, generally, to the degeneration of ellipticity. Thus, the equation (25) in K_m^v proves to be a completely nonlinear second-order elliptic equation.

The inequality (18) allows one to apply to (25) the known results by N. V. Krylov, M. V. Safonov, and L. Evans on smoothness of solutions u for completely nonlinear second-order elliptic equations under a priori boundedness of $\|u\|_{C^2}$. Therefore, to prove the solvability of (25) in $K_m^v \cap C^{l+2+\alpha}(\bar{\Omega})$ it suffices to obtain a priori estimates of u in $C^2(\bar{\Omega})$.

Consider in a bounded domain $\Omega \subset \mathbf{R}^n$ the Dirichlet problem for the equation (25),

$$u\big|_{\partial\Omega} = \Phi(x). \tag{26}$$

Let us formulate some results on its solvability in $C^{l+2+\alpha}(\bar{\Omega}) \cap K_m^v$, $l \geqslant 2$. We are interested in facts uniform with respect to m. So, many results, related to the mean curvature equations and Monge–Ampère equations, remain beyond our consideration.

The simplest equation among these of the type (25) is

$$\mu_m^{(2)}[u] = f(x). \tag{27}$$

THEOREM 2. *Let $\partial\Omega \subset C^{l+2+\alpha}$ be a closed surface in $\mathbf{R}^n$, $\mu_{m-1}[\partial\Omega] > 0$, $\Phi \in C^{l+2+\alpha}(\partial\Omega)$, $f > 0$ in $\bar{\Omega}$, $f \in C^{l+\alpha}(\bar{\Omega})$. Then there exists a solution $u \in K_m^{(2)} \cap C^{l+2+\alpha}(\bar{\Omega})$ of* (27), (26), *unique in $K_m^{(2)}$.*

The complete proof of Theorem 2 can be found in **[4]**. In the case of a convex domain Ω and $\Phi = 0$ this theorem was announced in **[17]**. The solvability of the problem (26), (27) in $\bar{K}_m^{(1)} \cap C^{1,1}(\bar{\Omega})$ for $f \geqslant 0$ and $\Phi = 0$ was proved in **[5]**.

The conditions of Theorem 2 are necessary because $\mu_{m-1}[\partial\Omega] > 0$ is a consequence of the ellipticity of (27) at $\Phi = 0$.

Now let $v = \sqrt{1+p^2}$, i.e., we deal with the curvature equation of order m:

$$\mu_m^{(1)}[u] = f(x). \tag{28}$$

THEOREM 3. *Let $\partial\Omega \in C^{l+2+\alpha}$, $\mu_m[\partial\Omega] > 0$, $\Phi \in C^{l+2+\alpha}(\partial\Omega)$, $f > 0$ in $\bar{\Omega}$, $f \in C^{l+\alpha}(\bar{\Omega})$. Suppose that there exists a number $\varepsilon > 0$ such that for any subdomain $E \subset \Omega$ with $\mu_{m-1}[\partial E] > 0$ the inequality*

$$\int_E F(x)\,dx \leqslant \frac{1-\varepsilon}{n} \int_{\partial E} \mu_{m-1}[\partial E]\,ds \tag{29}$$

holds and on $\partial\Omega$

$$f(x) \leqslant \frac{n-m}{n} \mu_m[\partial\Omega](x). \tag{30}$$

Then there exists a unique solution $u \in K_m^{(1)} \cap C^{l+2+\alpha}(\bar{\Omega})$ of the problem (28), (26).

The first results on solvability of the Dirichlet problem for the curvature equation of order m were obtained for convex Ω and the homogeneous boundary condition in **[6, 7, 13]** under more rigorous restrictions than (29), (30). In **[14]** Φ was admitted to be an arbitrary function and the strict variant of (30) was used. The role of (29), (30) in the theory of mth-order curvature equations was clarified in **[8, 18]**.

In **[18, 19]** the inequality (29) was employed to solve the problem (28), (29). It was shown in **[18]** that for any function $u \in K_m^{(1)}$ and subdomain $E \subset \Omega$ with $\mu_m[\partial E] > 0$, the inequality

$$\int_E \mu_m[u]\,dx < \int_{\partial E} \mu_{m-1}[\partial E]\,dx$$

holds. In this sense, the restriction (29) is necessary.

The inequality (30) is a natural extension to $m > 1$ of the well-known Serrin condition for the mean curvature equation **[20]**. It was shown in **[8]** that this condition is necessary for the problem (28), (26) to be solvable for an arbitrary boundary Φ.

We describe the role of the cones and the structure of the operators for estimating in $C^2(\Omega)$ solutions of the problems (26)–(28). The ellipticity of equations with operators μ_m^v in the cones K_m^v allows us to apply the method of comparison theorems and barrier functions. We present, as an example, one of the main theorems of this type **[4, 13]**.

THEOREM 4. *Let $w^{(1)} \in K_m^v$, $w^{(2)} \in C^2(\Omega)$, and let $(x,u) \mapsto f(x,u,p)$ be a continuous function, subject to a Lipshitz condition relative to p and nondecreasing with respect to u. Suppose*

$$\mu_m^v[w^{(1)}] \geqslant f(x, w^{(1)}, w_x^{(1)}),$$
$$(1 + \operatorname{sgn}(\mu_{m-1}^v[w^{(2)}](x)))\mu_m^v[w^{(2)}](x) \leqslant 2f(x, w^{(2)}(x), w_x^{(2)}).$$

Then

$$\max_{\Omega}(w^{(1)} - w^{(2)}) \leqslant \max_{\partial\Omega}(w^{(1)} - w^{(2)})^+.$$

The possibility of applying Theorem 4 depends on one's ability to construct appropriate barrier functions. In this situation the following property of the operators turns out to be useful.

LEMMA 3. *Let $v = (1+p^2)^{s/2}$, $s \geqslant 1$, $w = h(|x|)$. Then*

$$\mu_m^{(s)}[w] = \mu_m^v[w]$$
$$= (1+h'^2)^{\frac{m(s-2)}{2}} \left(\frac{h'}{|x|}\right)^{m-1} \left(\frac{m}{n}\frac{1+(s-1)h'^2}{1+h'^2}h'' + \frac{(n-m)}{n}\frac{h'}{|x|}\right). \tag{31}$$

PROOF. Calculate $\mu_m^{(s)}$ on functions of the form $h(u)$ in coordinates $(\tilde{x}, x^n)$, $\tilde{x} = (x^1, \dots, x^{n-1})$, choosing the basis vector $e_n = u_x/|u_x|$, if $|u_x| \neq 0$. One has

$$\mu_m^{(s)}[h(u)] = \frac{1}{\binom{n}{m}}(1+h'^2u_x^2)^{\frac{m(s-2)}{2}}(h')^{m-1}$$
$$\times \left(h'u_{\left[\begin{smallmatrix} i_1\dots i_m \\ i_1\dots i_m\end{smallmatrix}\right]} + h''u_x^2\frac{1+(s-1)h'^2u_x^2}{1+h'^2u_x^2}u_{\left[\begin{smallmatrix} s_1\dots s_{m-1} \\ s_1\dots s_{m-1}\end{smallmatrix}\right]}\right). \tag{32}$$

For $u = |x|$ the representation (32) is invariant with respect to orthogonal transformations and coincides with (31). □

The formula (31) clarifies the role of the eigenvalues of the matrix. The inequality (29) was used in **[18, 19]** to estimate solutions of (28), (26). It is shown in **[18]** that for any $u \in K_m^{(1)}$ and subdomain $E \subset \Omega$ such that $\mu_m[\partial E] > 0$, the inequality

$$\int_E \mu_m[u]\,dx < \int_{\partial E} \mu_{m-1}[\partial E]\,ds$$

holds; in this sense one can consider (29) to be necessary.

The inequality (30) is a natural extension for $m > 1$ of the well-known Serrin inequality **[20]** for the mean curvature equations; as was noted in **[8]**, (30) is necessary for solvability of (28), (26) with an arbitrary function Φ.

Let us describe the role of the cones and the operator structure in the derivation of a priori estimates in $C^2(\overline{\Omega})$ for the solution of (26)–(28). The ellipticity of equations with the operators μ_m^v and K_m^v allows one to use the method of comparison theorems and barrier functions for the qualitative analysis of solutions.

Evidently, for any m there is a qualitative difference between $\mu_m^{(s)}$, $s > 1$, and $\mu_m^{(1)}$ of the same type, as between uniform elliptic operators and mean curvature operators: in the second case h'' enters weakly because of the nonuniform behaviour at $|p| \gg 1$ of eigenvalues of $v^{pp} = ((\delta_j^i - \frac{p_i p_j}{1+p^2})/\sqrt{1+p^2})$. This difference leads to different requirements to $\partial\Omega$ in Theorems 2 and 3.

The differentiation of μ_m^v needs a special approach. In **[13]** one proposes to use the representation (6),

$$\binom{n}{m}\frac{d}{dx^i}\mu_m^v[u] = S_{m-1,\alpha}u_{\alpha\alpha,i}$$

with $u_{\alpha\alpha} = u_{ij}\tau_\alpha^i\tau_\alpha^j$, while $\tau_1, \dots, \tau_n$ are eigenvectors of the problem $u_{xx}\tau = \lambda A\tau$, $A^\alpha = (v^{pp})^{-1}$ and $(A\tau_\alpha, \tau_\beta) = \delta_\beta^\alpha$, $u_{\alpha\beta} = \delta_\beta^\alpha u_{\alpha\beta}$, α, $\beta = 1, \dots, n$. One has

$$u_{\alpha\alpha,i} = u_{i\alpha\alpha} - (A_i\tau_\alpha, \tau_\alpha)u_{\alpha\alpha},$$

$$\binom{n}{m}\frac{d}{dx^i}\mu_m^v[u] = S_{m-1,\alpha}u_{i\alpha\alpha} - S_{m-1,\alpha}u_{\alpha\alpha}(A_i\tau_\alpha, \tau_\alpha). \tag{33}$$

Considering (33) it is often convenient to deal with a normalized matrix A. The normalization corresponds to the multiplication of the equation by a function. For instance, in the case of the mth-order curvature equation it is easier from the point of view of calculations to treat $A = (\delta_j^i + p_i p_j) = (v^{pp})^{-1}/\sqrt{1+p^2}$. Then (33) for solutions of (28) can be written as

$$S_{m-1,\alpha}u_{i\alpha\alpha} = 2S_{m-1,\alpha}u_\alpha u_{\alpha\alpha}u_{i\alpha} + \binom{n}{m}((1+u_x^2)^{\frac{m}{2}}f)_i. \tag{34}$$

This allows one to introduce an auxiliary function and use the maximum principle in the standard way to estimate $|u_x|_\Omega$.

The cone and the structure of μ_m^v play an important role in deriving estimates on $\partial\Omega$ for second-order derivatives of solutions of the Dirichlet problem. When one deals with the mixed curvature $u_{nt}(x_0)$ where $x_0 \in \partial\Omega$, n and t are a normal and a tangent direction to $\partial\Omega$, respectively, one has to have a function $w(u_t, u_x)$ such that

a) in a neighbourhood of x_0

$$\frac{\partial\mu_m^{(1)}}{\partial u_{ij}}w_{ij} \leqslant c(f, \partial\Omega, \Phi)(1 + |w_x|); \tag{35}$$

b) $\partial w/\partial u_t \neq 0$ on $\partial\Omega$;

c) w can be controlled in a neighborhood of x_0 on $\partial\Omega$ by the data of (28), (26).

As was shown in [13], one can construct the function w starting with $W = u_t + u_n\omega_t - \sum_1^{n-1}(u_s - \varphi_s(x_0))^2$, where $x^n = \omega(\tilde{x})$ defines $\partial\Omega$ near x_0, $\varphi(\tilde{x}) = \Phi(\tilde{x}, \omega)$. From (34) it follows that

(36)

$$S_{m-1,\alpha} W_{\alpha\alpha} \leqslant c\left(1 + |W_x| + S_{m-1}(u) + S_{m-1,\alpha}|u_{\alpha\alpha}| + S_{m-1,\alpha}\left(2u_\alpha u_{\alpha\alpha} W_\alpha - u_{\alpha\alpha}^2 \sum_{i=1}^{n} (\eta_i^\alpha)^2\right)\right).$$

The arbitrariness in the choice of h, $w = h(W)$, allows one to use $S_{m-1,\alpha} W_\alpha^2$ to estimate the right-hand side of (35). It turns out (see [13]) that $S_{m-1,1} u_1 u_{11} W_1$ in (36) is uncontrollable at $S_{m-1} < 2S_{m-2,1} u_{11}$. In such a case the property of $\mathcal{K}_m$ expressed by (15) proves to be crucial for (35) to be valid.

In order to apply the comparison principle it is necessary to use besides (35) the inequality

$$\frac{\partial \mu_m^{(1)}[u]}{\partial u_{ij}} v_{ij} \geqslant c(f, \partial\Omega, \Phi)(1 + |v_x|)$$

for a barrier function v. The condition $\mu_m[\partial\Omega] > 0$ and the property (19) of $K_{m-1}^{(1)}$ are sufficient for the existence of the desired barrier v.

The above scheme leads to an estimate of $|u_{nt}|$ for the mth-order curvature equations. For the equation (27) this result may be obtained in a simpler way due to (17) and $A(p) = \mathrm{Id}$.

To evaluate $u_{nn}(x_0)$ one uses equation (28) as a rule. In the case $\Phi = 0$ it is easy to get a lower bound for the coefficient by $u_{nn}(x_0)$ in (27), (28). Let us demonstrate the scheme of evaluation of $u_{nn}(x_0)$ in the case of an arbitrary Φ, given in [4], considering, for example, equation (28) [14].

We fix $x_0 \in \partial\Omega$ and coordinates $\{(\tilde{x}, x^n); x_0 = 0\}$ so that the vector $(0, \dots, 0, 1)$ coincides with the inner normal to $\partial\Omega$ at x_0. Let $x^n = \omega(\tilde{x})$ be an equation of $\partial\Omega$ near x_0, $\varphi = \Phi(\tilde{x}, \omega)$. The coefficient of $u_{nn}(x_0)$ in (28) coincides up to a known positive factor with

$$P_{m-1}(\theta) = \sum_{\{1 \leqslant s_1 < \cdots < s_{m-1} \leqslant n-1\}} (\varphi + \theta\omega)^0 \begin{bmatrix} s_1 \dots s_{m-1} \\ s_1 \dots s_{m-1} \end{bmatrix} - \frac{\varphi_s^0 \varphi_t^0}{1 + (\varphi_x^0)^2} (\varphi + \theta\omega)^0 \begin{bmatrix} s_1 \dots s_{m-2} s \\ s_1 \dots s_{m-2} t \end{bmatrix}$$

at $\theta = (-u_n^0)$. Denote by θ^0 the greatest root of P_{m-1}. Since P_{m-1} is a w_{st}^0-hyperbolic polynomial, to majorize $u_{nn}(x_0)$ it suffices to find $\delta > 0$ such that $-u_n^0 \geqslant \theta^0 + \delta$. In the above situation it is natural to use Theorem 4 taking for $w^{(2)}$ an analogue of the Bernstein paraboloid. For example [14],

(37)

$$w^{(2)} = \varphi - (\theta^0 + \delta)(x^n - \omega) + a_s x^s (x^n - \omega) + a^{nn}(x^n - \omega + \varkappa \tilde{x}^2/2)^2, \quad 0 < \varkappa \ll 1.$$

The choice of a_s, $s = 1, \dots, n-1$ in (37) was crucial. A method for making the choice is given in [4]. According to [4], the linear part of $\mu_m^{(1)}[w^{(2)}]$ with respect to

$\theta^0 + \delta$ must vanish in a neighborhood of x_0. For (28) this means that one has to take for $a_1, \dots, a_{n-1}$ a solution of the algebraic linear systems

$$a_s \frac{d}{d\theta^0} \mu^{(1)}_{m-1,n}[w^{(2)}]\Big|_0 + \varkappa c^{st} a_t = \frac{\partial}{\partial x^s} \mu^{(1)}_{m-1,n}[w^{(2)}]\Big|_0,$$

with a positive definite matrix (c^{st}), $|c^{st}| \leqslant c(\|\varphi, \omega\|_{C^2})$, $s, t = 1, \dots, n-1$. The system is uniquely solvable for $\omega \in K^{(1)}_m$ and the solution is bounded uniformly as $\delta \downarrow 0$ by virtue of Lemma 2.

The above-mentioned choice of a_s, $s = 1, \dots, n-1$, allows one to satisfy all the conditions of Theorem 4 for the function (37) and $w^{(1)} = u$ at the expense of a^{nn} and $\varkappa, \delta$ small enough. This leads to an upper bound for $u_{nn}(x_0)$. A priori boundedness of u_{nn} from below follows from $\mu^v_1[u] \geqslant (\mu^v_m[u])^{1/m}$. Inside Ω one can estimate $|u_{xx}|$ due to special properties of v^{pp} with $v = p^2/2$ **[6]**, or $v = \sqrt{1+p^2}$ **[13]**.

§4. Equations with the operators $\mu^{(s)}_m$

There are no great possibilities for extending the results obtained for the equations (27), (28) to the equations with arbitrary operators μ^v_m. Let us analyze, for instance, the differentiation of (25) from the point of view of using auxiliary functions. Taking into account $((v^{pp})^{-1}\tau_\alpha, \tau_\alpha) = -(v^{pp}\eta^\alpha, \eta^\alpha)$, where $\eta^1, \dots, \eta^n$ is the dual vector field for $\tau_1, \dots, \tau_n$ and $(v^{pp}\eta^\alpha, \eta^\beta) = \delta^\alpha_\beta$, one has

$$u_{\alpha\alpha,i} = u_{i,\alpha\alpha} + u_{\alpha\alpha}(v^{pp}_{,i}\eta^\alpha, \eta^\alpha) = u_{i\alpha\alpha} + u_{\alpha\alpha} v^{\alpha\alpha\beta} u_{i\beta}, \qquad v^\alpha = v^i \eta^\alpha_i.$$

Set $w = h(u) W(u_x)$. It is easy to check that

$$\begin{aligned} S_{m-1,\alpha}(u) w_{\alpha\alpha} &= S_{m-1,\alpha}(h'' W u_\alpha^2 + 2h' W^\alpha u_\alpha u_{\alpha\alpha} + h W^{\alpha\alpha} u^2_{\alpha\alpha}) \\ &\quad - h S_{m-1,\alpha} v^{\alpha\alpha\beta} W^\beta u_{\alpha\alpha} u_{\beta\beta} + m h' W f + h W^\alpha f_\alpha. \end{aligned} \tag{38}$$

Due to $w_\alpha h' W u_\alpha + h W^\alpha u_{\alpha\alpha}$ the equality (38) takes the form

$$\begin{aligned} S_{m-1,\alpha}(u) w_{\alpha\alpha} &= (h'' - 2h'^2/(2h)) W S_{m-1,\alpha} u_\alpha^2 \\ &\quad + h S_{m-1,\alpha} W^{\alpha\alpha} u^2_{\alpha\alpha} + (2h'/h) S_{m-1,\alpha} w_\alpha \\ &\quad - S_{m-1,\alpha} v^{\alpha\alpha\beta} u_{\alpha\alpha}(w_\beta - h' W u_\beta) + m h' W f + h W^\alpha f_\alpha. \end{aligned} \tag{39}$$

Using (38), (39) one cannot do without additional assumptions on the dependence $v^{\alpha\alpha\beta}$ on u_γ, $\gamma = 1, \dots, n$, because μ^v_m is nonuniformly elliptic in K^v_m. We consider, for instance, $v = (1+p^2)^{s/2}/s$. In this case we shall write $\mu^{(s)}_m$ instead of μ^v_m. The derivatives v^{ijk} satisfy

$$v^{ijk} = \frac{s-2}{sv}(v^{ij} v^k + v^{ik} v^j + v^{jk} v^i) + (s-2)(s-1)/(v^2 v^i v^j v^k).$$

Since $(v^{pp}\eta^\alpha, \eta^\beta) = \delta^\alpha_\beta$, one obtains

$$v^{\alpha\beta\gamma} = \frac{s-2}{sv}(\delta^\alpha_\beta v^\gamma + \delta^\alpha_\gamma v^\beta + \delta^\gamma_\beta v^\alpha) + \frac{(s-2)(s-1)}{(vs)^2} v^\alpha v^\beta v^\gamma. \tag{40}$$

Considering w at a maximum point, it is natural to deal with (39) where $S_{m-1,\alpha} \times v^{\alpha\alpha\beta} u_\beta u_{\alpha\alpha}$ is not easily controlled. We check directly that

$$v^\alpha v^\beta + (s-2)\frac{u_x^2}{1+(s-1)u_x^2} v^\alpha u_\beta - \frac{1+u_x^2}{1+(s-1)u_x^2} u_\alpha u_\beta = 0, \qquad \alpha, \beta = 1, \dots, n.$$

So $(v^\alpha)^2 = c(s, |u_x|)u_\alpha^2$ with $0 < c_0(s) \leqslant c(s, |u_x|) \leqslant c_1(s)$. From (40) it follows that

$$\begin{aligned} S_{m-1,\alpha} v^{\alpha\alpha\beta} u_\beta u_{\alpha\alpha} &= \frac{(s-2)}{sv} S_{m-1,\alpha} u_{\alpha\alpha} \\ &\quad \times \left(2v^\alpha u_\alpha + (s-1)\frac{u_x^2}{1+u_x^2}(v^\alpha)^2\right) + m(s-2)\frac{u_x^2}{1+u_x^2} f \\ &= c(s, |u_x|) S_{m-1,\alpha} \frac{u_\alpha^2 u_{\alpha\alpha}}{(1+u_x^2)^{s/2}}(s-2) + m(s-2)\frac{u_x^2}{1+u_x^2} f. \end{aligned}$$

This expression allows us to get the boundedness of $|u_x|$ inside Ω by means of (39) for $s > 1$ and the maximum principle for a special function w **[3, 21]**.

To estimate the mixed derivatives on $\partial\Omega$ one can turn to (38) where the term $hS_{m-1,\alpha} v^{\alpha\alpha\beta} u_{\alpha\alpha} u_{\beta\beta} W^\beta$ causes basic difficulties. One has

$$\begin{aligned} &S_{m-1,\alpha} v^{\alpha\alpha\beta} W^\beta u_{\alpha\alpha} u_{\beta\beta} \\ &\quad = \frac{s-2}{sv}\left(S_{m-1,\alpha}\left(2W^\alpha v^\alpha u_{\alpha\alpha}^2 + \frac{s-1}{sv}(v^\alpha)^2 u_{\alpha\alpha} W^\beta u_{\beta\beta}\right) + mf W^\alpha v^\alpha\right). \end{aligned}$$

The term $(s-1)S_{m-1,\alpha}(v^\alpha)^2 W^\beta u_{\alpha\alpha} u_{\beta\beta}$ is not subject to the ellipticity condition (16) for $s \neq 1$. It seems there are no estimates of second-order derivatives for solutions of the equations with the operators $\mu_m^{(s)}$, $s \neq 1, 2$. Probably, we have solutions whose smoothness is not great enough for treating (25) in an ordinary way, no matter how smooth $\partial\Omega$, Φ, and f are.

Trudinger explained **[8]** how one could make sense of (25) on functions with small smoothness. He extended the notion of solution given in **[22, 23]** etc. to the curvature equations. We rephrase for (25) the corresponding definitions from **[8]**.

Definition 1. A function $u \in C^0(\Omega)$ is called K_m^v-admissible if $\mu_{m-1}^v[v](x_0) > 0$, $\mu_m^v[\varphi](x_0) > 0$ at each local maximum point x_0 of $u - \phi$, $\phi \in C^2(\Omega)$.

Definition 2. A function $u \in C^0(\Omega)$ is called a weak K_m^v-admissible solution of (25) if

a) for any $\varphi \in C^2(\Omega)$

$$\mu_m^v[\varphi](x_0) \geqslant f(x_0, \varphi(x_0), \varphi_x(x_0))$$

at each local maximum point $x_0 \in \Omega$ of $u - \varphi$;

b) for any $\varphi \in K_m^v$

$$\mu_m^v[\varphi](x_0) \leqslant f(x_0, \varphi(x_0), \varphi_x(x_0))$$

at each local minimum point of $u - \varphi$.

The choice of $v = \sqrt{1+p^2}$ is not essential in **[8]** if one is not dealing with a priori estimates. Taking this into account, the authors of **[19]** stated the following theorem on the existence of a K_m^v-admissible solution of the Dirichlet problem (25) for the equation

$$\mu_m^{(s)}[u] = f(x). \tag{41}$$

THEOREM 5 (see **[19]**). *Suppose that $s > 1$, $\partial\Omega$ being a closed surface of class C^4 in $\mathbf{R}^n$, $\mu_{m-1}[\partial\Omega] > 0$, $\Phi \in C^4(\partial\Omega)$, $f(x) > 0$ in Ω, $f \in C^2(\bar\Omega)$. Then there exists a unique K_m^v-admissible solution u of the problem* (41), (25), $u \in C^{0,1}(\bar\Omega)$.

A priori estimates in $C^{2+\alpha}(\bar\Omega)$ (uniform with respect to $C^1(\Omega)$) for solutions of the approximating family of problems from **[8]** are given in **[21]**.

§5. Variational problem

From Theorem 1 it follows that (25) is the Euler-Lagrange equation for

$$I_{m-1}^v(u) = \int_\Omega (v(u_x)\mu_{m-1}^v[u] + (n-m+1)fu)\,dx. \tag{42}$$

Being unbounded in $C^2(\bar\Omega)$, the functional (42) may have local extrema only. To analyze the situation we represent $\Delta I_{m-1}^v = I_{m-1}^v(w) - I_{m-1}^v(u)$, $u, w \in C^2$, in the form

$$\begin{aligned}\Delta I_{m-1}^v(u) = &-(n-m+1)\int_\Omega \eta(\mu_m^v[u] - f(x))\,dx\\ &+(n-m+1)\int_0^1(1-\tau)d\tau\int_\Omega \frac{\partial\mu_m^v}{\partial\tilde w_{ij}}\eta_i\eta_j dx\\ &+\int_0^1 d\tau\int_{\partial\Omega}\Big(\eta_j\frac{\partial\mu_{m-1}^v}{\partial\tilde w_{ij}}v(\tilde w_x) - \eta(v^j\frac{\partial\mu_m^v}{\partial v_{,i}^j})[\tilde w]\Big)\cos(nx^i)\,ds,\end{aligned} \tag{43}$$

where $\tilde w = u + \tau\eta$, $\eta = w - u$. Let us consider possibilities for minimizing (42).

LEMMA 4. *Let $f(x) > 0$ in $\bar\Omega$, $f \in C^0(\bar\Omega)$. Suppose that $u \in C^2(\bar\Omega)$ gives a local minimum to the functional* (42) *over $G_0 = \{w = u + \eta \in C^2(\bar\Omega), \operatorname{supp}\eta \subset \Omega\}$. Then $u \in K_m^v(\bar\Omega)$.*

PROOF. From (43) it follows that $\mu_m^v[u] = f > 0$ and for any $x \in \bar\Omega' \subset \Omega$

$$\frac{\partial\mu_m^v}{\partial u_{ij}}\xi^i\xi^j \geqslant 0, \qquad \xi \in \mathbf{R}^n. \tag{44}$$

This means that eigenvalues of the form (44) are nonnegative or, in terms of the representation (6), $S_{m-1,i}(u) \geqslant 0$, $i = 1, \dots, n$. Using $mS_m = S_{m-1,i}\lambda^i$ we obtain $\mu_m^v[u] > 0$, $\mu_{m-1}^v[u] > 0$ in $\bar\Omega'$ for any subdomain $\Omega' \subset \Omega$. Recalling the definition (11) we see that $u \in K_m^v(\bar\Omega')$ and $\mu_{m-1}^v[u] \geqslant (\mu_m^v[u])^{(m-1)/m}$. Since $f(x) \geqslant v > 0$ in $\bar\Omega$, one has $u \in K_m^v(\bar\Omega)$. □

Let us extend the set G_0 to $G_1 = \{w \in C^2(\bar\Omega), w|_{\partial\Omega} = \Phi\}$.

LEMMA 5. *Let $\partial\Omega \in C^3$, $\Phi \in C^3(\partial\Omega)$, $f(x) > 0$ in Ω, $f \in C^0(\bar\Omega)$. Then the functional* (42) *has no local minimum on G_1.*

PROOF. Suppose that the assertion is wrong, and there exists a function u, $u = \Phi$ on $\partial\Omega$, minimizing the functional (42) on G_1. Since $G_0 \subset G_1$, the conditions of Lemma 4 are valid for u and $u \in K_m^v$.

Let us give $\partial\Omega$ by $\psi(x) = 0$ taking for ψ a solution of the Poisson equation $\Delta\psi = 1$. It is well known that $\psi_n|_{\partial\Omega} \leqslant -\delta < 0$, with the inner normal n to $\partial\Omega$. Set $w^\varepsilon = u + \varepsilon\psi$, $\varepsilon > 0$. In view of (43) one has

$$\begin{aligned} 0 \leqslant (n-m+1)\varepsilon^2 \int_0^1 (1-\tau)d\tau \int_\Omega \frac{\partial\mu_m^v[\tilde{w}^\varepsilon]}{\partial\tilde{w}_{ij}^\varepsilon}\psi_i\psi_j\,dx \\ - \varepsilon\delta \int_0^1 d\tau \int_{\partial\Omega} \frac{\partial\mu_{m-1}^v[\tilde{w}^\varepsilon]}{\partial\tilde{w}_{nn}^\varepsilon}\,ds. \end{aligned} \tag{45}$$

The relation (45) is wrong for $0 < \varepsilon \ll 1$ because of $u \in K_m^v$; hence, the functional (42) has no local minimum on G_1. Analyzing (43), (45), it is not hard to show that there are no minimizing sequences in G_1, i.e., I_{m-1} has no minimum on G_1. □

Lemma 5 indicates that the Dirichlet condition alone proves to be insufficient to minimize (42) for $m > 1$. However, in view of Theorems 2, 3, we get generally speaking an unsolvable problem by adding some boundary condition. The degeneracy of (42) for $m > 1$ leads to this contradiction because the Euler-Lagrange equations (25) have second order only. To interpret optimal properties of solutions for problem (25), (26) let us consider a pair $(u; G)$, where G is a set in $C^2(\bar{\Omega})$, $u \in G$ a fixed function.

DEFINITION 3. A pair $(u; G)$ is called a minimizer of I_{m-1}^v if

$$I_{m-1}^v(u) \leqslant I_{m-1}^v(w) \tag{46}$$

whenever $w \in G$. If (46) is valid in $C^2(\bar{\Omega})$ locally only, we shall call the above pair a local minimizer for I_{m-1}^v.

Associate to each $u \in C^2(\bar{\Omega})$ the set $G_u = \{w \in C^2(\bar{\Omega}); (w-u)|_{\partial\Omega} = 0, (w_n - u_n)|_{\partial\Omega} \geqslant 0\}$.

THEOREM 6. *Let $\partial\Omega \subset C^2$, $\Phi \in C^2(\partial\Omega)$, $f(x) > 0$ in $\bar{\Omega}$, $f \in C^0(\bar{\Omega})$. A pair $(u; G_u)$, $u = \Phi$ on $\partial\Omega$, is a local minimizer for the functional* (42) *if and only if problem* (25), (26) *is solvable in K_m^v.*

This assertion is a simple consequence of Lemma 4 and the representation (43). If there exists a solution of (25), (26) in K_m^v, then, by virtue of Theorem 4, this solution is unique. Therefore one has

THEOREM 7. *Under the conditions of Theorem* 6 *there exists no more than one function $u \in C^2(\bar{\Omega})$, equal to Φ on $\partial\Omega$ such that the pair $(u; G_u)$ is a local minimizer of* (42).

Theorems 2, 3 give examples of minimizers in the sense of Definition 3.

THEOREM 8. *Under the conditions of Theorem* 2 *there exists a unique minimizer of the form $(u; G_u \cap C^{l+2+\alpha}(\bar{\Omega}) \cap K_m^{(2)})$ for*

$$I_{m-1}^{(2)}(u) = \int_\Omega \left(u_x^2/2\mu_{m-1}^{(2)}[u] + (n-m+1)f(x)u\right)\,dx.$$

THEOREM 9. *Under the conditions of Theorem* 3 *there exists a unique minimizer $(u; G_u \cap C^{l+2+\alpha}(\bar{\Omega}) \cap K_m^{(1)})$ for*

$$I_{m-1}^{(1)}(u) = \int_\Omega \left(\sqrt{1+u_x^2}\,\mu_{m-1}^{(1)}[u] + (n-m+1)f(x)u\right)\,dx.$$

In Theorems 8, 9 one asserts the existence of nonlocal minimizers. The proof in **[24]** of this fact is based on the special structure of $\mu_m^{(1)}$ and $\mu_m^{(2)}$.

In conclusion we give for $I_{m-1}^{(1)}$ another interpretation of the discussed properties of functionals.

THEOREM 10. *Suppose that* $0 < f(x) \leqslant g(x)$ *in* $\bar{\Omega}$, $f, g \in C^0(\bar{\Omega})$. *Let also* $u_1 \in K_m^{(1)}$ *be a solution of problem* (28), (26), $u_2 \in K_m^{(1)}$ *a solution of the equation* $\mu_m^{(1)}[u] = g$, *equal to* Φ *on* $\partial\Omega$. *Then*

$$I_{m-1}^{(1)}(u_1) \leqslant I_{m-1}^{(1)}(u_2).$$

Theorem 10 is a consequence of Theorem 5 and the nonlocal character of the minimizer $(u; G_u \cap K_m^{(1)})$ [24]. Its analogue is also valid for $I_{m-1}^{(2)}$.

References

1. N. M. Ivochkina, *Second order equations with d-elliptic operators*, Trudy Mat. Inst. Steklov. **147** (1980), 40–56; English transl. in Proc. Steklov Inst. Math. **1981**, no. 2.
2. ______, *Description of cones of stability generated by differential operators of Monge-Ampère type*, Mat. Sb. **122** (1983), no. 2, 265–275; English transl. in Math. USSR-Sb. **50** (1985).
3. ______, *Solution of the Dirichlet problem for certain equations of Monge-Ampère type*, Mat. Sb. **128** (1985), no. 3, 403–415; English transl. in Math. USSR-Sb. **56** (1987).
4. L. Caffarelli, L. Nirenberg, and J. Spruck, *The Dirichlet problem for nonlinear second order elliptic equations.* III. *Functions of the eigenvalues of the Hessian*, Acta Math. **155** (1985), 261–301.
5. N. V. Krylov, *On the first boundary value problem for nonlinear degenerating elliptic equations*, Izv. Akad. Nauk SSSR Ser. Mat. **51** (1987), 242–269; English transl. in Math. USSR-Izv. **30** (1988).
6. L. Caffarelli, L. Nirenberg, and J. Spruck, *Nonlinear second-order elliptic equations.* V. *The Dirichlet problem for Weingarten hypersurfaces*, Comm. Pure Appl. Math. **41** (1988), 47–70.
7. N. M. Ivochkina, *Solution of the Dirichlet problem for the curvature equation of order m*, Dokl. Akad. Nauk SSSR **299** (1988), no. 1, 35–38; English transl. in Soviet Math. Dokl. **37** (1988).
8. N. S. Trudinger, *The Dirichlet problem for the prescribed curvature equations*, Arch. Rational Mech. Anal. **111** (1990), 153–187.
9. N. M. Ivochkina, *Description of the algebraic structure of second order differential operators representable in divergence form*, Probl. Mat. Anal. **10**, Izdat. Leningrad. Univ., Leningrad, 1986, pp. 16–32. (Russian)
10. ______, *Variational problems connected to Monge-Ampère type operators*, Zap. Nauchn. Sem. Leningrad. Otdel. Mat. Inst. Steklov. (LOMI) **167** (1988), 186–189; English transl. in J. Soviet Math. **52** (1990).
11. ______, *Mixed volumes and connected variational problems*, Proc. of CMA **8** (1989), 957–965.
12. L. Gärding, *An inequality for hyperbolic polynomials*, J. Math. Mech. **8** (1959), 957–965.
13. N. M. Ivochkina, *Solution of the Dirichlet problem for equations of the mth order curvature*, Mat. Sb. **180** (1989), no. 7, 867–887; English transl. in Math. USSR-Sb. **66** (1990).
14. ______, *Solution of the Dirichlet problem for the curvature equation of order m*, Algebra i Analiz **2** (1990), no. 3, 192–217; English transl. in Leningrad Math. J. **2** (1991).
15. M. D. Bronstein, *Smoothness of roots of polynomials depending on parameters*, Sibirsk. Mat. Zh. **20** (1979), no. 3, 493–501; English transl. in Siberian Math. J. **20** (1979).
16. S. Wakabayashi, *Remarks on hyperbolic polynomials*, Tsukuba J. Math. **10** (1986), no. 1, 17–28.
17. N. M. Ivochkina, *Solution of the Dirichlet problem for some Monge-Ampère type equations*, Dokl. Akad. Nauk SSSR **279** (1984), no. 4, 796–798; English transl. in Soviet Math. Dokl. **30** (1984).
18. N. S. Trudinger, *A priori bounds and necessary conditions for solvability of prescribed curvature equations*, Manuscripta Math. **67** (1990), 99–112.
19. ______, *A priori bounds for graphs with prescribed curvature*, Festschrift for Jürgen Moser, 1989.
20. J. Serrin, *The problem of Dirichlet for quasilinear elliptic differential equations with many independent variables*, Philos. Trans. Roy. Soc. London Ser. A **264** (1969), 413–496.
21. N. M. Ivochkina, P. I. Prokofieva, and G. V. Yakunina, *On one class of Monge-Ampère type equations*, Probl. Mat. Anal. **13**, Izdat. Leningrad. Univ., Leningrad, pp. 89–106. (Russian)

22. M. G. Crandall and P.-L. Lions, *Viscosity solutions of Hamilton-Jacobi equations*, Trans. Amer. Math. Soc. **77** (1983), 1–42.
23. R. Jensen, *The maximum principle for viscosity solutions of fully nonlinear second order partial differential equations*, Arch. Rational Mech. Anal. **101** (1988), 1–27.
24. N. M. Ivochkina, *Minimization of functionals that generate curvature operators*, Zap. Nauchn. Sem. Leningrad. Otdel. Mat. Inst. Steklov (LOMI) **182** (1990), 29–36. (Russian)

St. Petersburg State University for Architecture and Civil Engineering, 4, 2nd Krasnoarmeĭskaya ul., 198005, St. Petersburg, Russia

Translated by B. A. PLAMENEVSKIĬ

Amer. Math. Soc. Transl.
(2) Vol. **159**, 1994

Regularity and Weak Regularity of Solutions of Quasilinear Elliptic Systems

A. I. Koshelev and S. I. Chelkak

§0. Introduction

Unlike the case of a single second-order equation, in the case of a system of equations or an equation of higher order with the number m of independent variables greater than two, the ellipticity and natural smoothness of the coefficients do not guarantee the regularity of solutions. For the existence of smooth solutions one has to impose more restrictive conditions, for example, various conditions limiting the spectrum dispersion of the matrix determining the ellipticity of the system. The first such condition was considered by H. O. Cordes in **[1]**. Then one of the authors of the present paper introduced another condition on dispersion of the spectrum ensuring continuity of generalized solutions of second-order systems. It was proved that this condition is precise. Note that one can find a review of the works on this subject in **[2, 3]**.

The present paper contains two parts. In the first part, we consider weighted Sobolev spaces and the spaces $H^{(l)}_{p,\alpha}$ and $H^{(l)}_{p,s,\alpha}$ used later for the analysis of the regularity of solutions of elliptic systems.

The second part of the paper is devoted to the study of weak regularity of generalized solutions of elliptic systems. The notion of "weak regularity of solutions" was introduced by one of the authors in **[4]**. We say that a generalized solution is weakly regular if some power of it is integrable over s-dimensional smooth manifolds (naturally, it is assumed that such integrability does not follow directly from imbedding theorems).

We consider in a bounded domain $\Omega \subset \mathbb{R}^m$ $(m \geqslant 2)$ with sufficiently smooth boundary $\partial\Omega$ the second-order system in divergence form:

$$L(u) \equiv D_j a_j(x, Du) - a_0(x, Du) = 0, \tag{0.1}$$

where $x = (x_1, \dots, x_m) \in \Omega$; D_j, $j = 1, \dots, m$, is the operator of differentiation with respect to x_j; $u(x)$ and $a_j(x, Du)$ are N-dimensional vector functions, $u = \{u^{(1)}(x), \dots, u^{(N)}(x)\}$; $a_i = \{a_i^{(1)}, \dots, a_i^{(N)}\}$, $i = 0, 1, \dots, m$. The notation $a_i(x, Du)$ means that a_i may depend on all $u^{(k)}$ and on all their first-order derivatives. As usual, in (0.1) and henceforth it is understood that we sum over repeating indices; if the bounds of the summations are not given explicitly, the summation over i, j, and t

---

1991 *Mathematics Subject Classification*. Primary 35D10, 35B65; Secondary 35J60.

goes from 1 to m, whereas the summation over l and k goes from 1 to N. Hereafter we shall use the following notation:

$$|D'u|^p = \sum_{k=1}^{N}\sum_{j=1}^{m}\left|\frac{\partial u^{(k)}}{\partial x_j}\right|^p, \qquad |Du|^p = |D'u|^p + |u|^p,$$

$$|D'^2 u|^p = \sum_{k=1}^{N}\sum_{i,j=1}^{m}\left|\frac{\partial^2 u^{(k)}}{\partial x_i \partial x_j}\right|^p, \qquad |D^2 u|^p = |D'^2 u|^p + |Du|^p.$$

It is assumed also that the functions $a_i(x,p)$ satisfy the following conditions:

1. $a_i(x, Du) \in L_q(\Omega')$ for all $u \in W_q^{(1)}(\Omega')$, where Ω' is any subdomain of Ω and $q > 1$ is arbitrary.

2. The functions $a_i(x,p)$ are measurable and for almost all $x \in \Omega$ they are continuously differentiable with respect to p. Moreover, for each collection of real N-dimensional vectors $\xi_i = (\xi_i^{(1)}, \dots, \xi_i^{(N)})$ the following inequalities, characterizing the ellipticity and the boundedness of the nonlinearity of the system,

$$\mu_0 \sum_{i=1}^{m} |\xi_i|^2 \leqslant \frac{\partial a_i^{(k)}}{\partial p_j^{(l)}} \xi_i^{(k)} \xi_j^{(l)} \leqslant \nu_0 \sum_{i=1}^{m} |\xi_i|^2, \qquad \left|\frac{\partial a_i^{(k)}}{\partial p_j^{(l)}}\right| \leqslant \nu_0, \tag{0.2}$$

hold, where μ_0 and ν_0 are some positive constants. Besides, let also the estimates

$$\left|\frac{\partial a_i}{\partial p_0}, \frac{\partial a_0}{\partial p_i}, \frac{\partial a_0}{\partial p_0}\right| \leqslant C, \qquad i = 1, \dots, m,$$

hold and let the domain Ω be sufficiently small.

In this paper we consider mostly systems with coefficients differentiable with respect to x. In this case, in addition to conditions 1, 2, we suppose the following condition to be fulfilled.

3. The functions $a_j(x,p)$, $j = 1, \dots, m$, are differentiable with respect to x and the inequalities

$$\left|\frac{\partial a_j}{\partial x_i}\right| \leqslant C|p| + \varphi(x), \qquad i, j = 1, \dots, m,$$

$$|a_i| \leqslant C|p| + \varphi(x), \qquad i = 0, 1, \dots, m,$$

hold with $\varphi(x) \in L_q(\Omega)$ for some q.

We introduce the $(L \times L)$-matrix, $L = mN$,

$$A = \left\{\frac{\partial a_i^{(k)}}{\partial p_j^{(l)}}\right\}, \qquad i, j = 1, \dots, m, \quad k, l = 1, \dots, N, \tag{0.3}$$

and let A^+ and A^- be its symmetric and skew-symmetric parts, respectively. Consider also the symmetric matrix $B = A^+A^- - A^-A^+ - (A^-)^2$ and let $\{\lambda_i\}$ and $\{\sigma_i\}$ be the collections of all the eigenvalues of A^+ and B, respectively. Set $\lambda = \inf_{i,x,p} \lambda_i$, $\Lambda = \sup_{i,x,p} \lambda_i$ and $\sigma = \sup_{i,x,p} \sigma_i$, where inf and sup are taken over all variables that λ_i and σ_i depend on. It follows from (0.2) that Λ and σ are finite and $\lambda > 0$.

We define a positive constant K by

$$K^2 = \begin{cases} \frac{\sigma}{\sigma+\lambda^2} & \text{for } \sigma \geqslant \frac{\lambda(\Lambda-\lambda)}{2}, \\ \frac{(\Lambda-\lambda)^2+4\sigma}{(\Lambda+\lambda)^2} & \text{for } \sigma \leqslant \frac{\lambda(\Lambda-\lambda)}{2}. \end{cases} \tag{0.4}$$

It is clear that $K < 1$. If A is Hermitian ($A^- = 0$) then $K = (\Lambda - \lambda)/(\Lambda + \lambda)$.

REMARK. The inequalities (0.2) and the matrix (0.3) contain derivatives $\partial a_i/\partial p_j$ for $i, j \neq 0$. In **[2]** and in other works by the authors, (0.2) and (0.3) were considered for $i, j = 0, 1 \ldots, m$. However, it is not hard to show that if the derivatives $\partial a_i/\partial p_0$, $\partial a_0/\partial p_i$, $i = 1, \ldots, m$, $\partial a_0/\partial p_0$ satisfy certain conditions, all the results of these works remain valid for the case considered. We are going to give here the most precise conditions on these derivatives. It is certainly sufficient that all of them be bounded (i.e., that condition 2 hold).

Along with the system (0.2) consider the boundary condition

$$u\big|_{\partial\Omega} = g, \tag{0.5}$$

where g is the trace of some function in $W_2^{(1)}(\Omega)$. Henceforth, we study the solution of the boundary value problem (0.1), (0.5). Since all the results of this paper concern the regularity of solutions inside Ω, it is evident that they also hold for any generalized solution of (0.1), because the latter may always be considered as a solution of some problem (0.1), (0.5) (moreover, in a sufficiently small Ω).

For solving (0.1), (0.5) in **[2]** we used the iterative procedure

$$\Delta u_{n+1} - u_{n+1} = \Delta u_n - u_n - \varepsilon L(u_n), \tag{0.6}$$

$$u_{n+1}\big|_{\partial\Omega} = g, \tag{0.7}$$

where ε is a positive constant determined by

$$\varepsilon = \begin{cases} (\sigma + \lambda^2)^{-1} & \text{for } \sigma \geqslant \frac{\lambda(\Lambda-\lambda)}{2}, \\ 2(\Lambda + \lambda)^{-1} & \text{for } \sigma \leqslant \frac{\lambda(\Lambda-\lambda)}{2}. \end{cases}$$

The convergence of (0.6), (0.7) in the energy metric of $W_2^{(1)}(\Omega)$ to the obviously existing unique generalized solution of (0.1), (0.5) was proved by one of the authors in 1965 (see, e.g., **[4,** Theorem 2.3]). In the present paper we consider the convergence of the procedure (0.6), (0.7) in the norms of the spaces $H^{(l)}_{p,s,\alpha}(\omega)$.

In the sequel we denote by $B_\rho = B_\rho(x^0) \in \mathbb{R}^m$ the ball of radius ρ centered at x^0 and by $\zeta(x)$ a cut-off function for B_ρ, i.e., a smooth function vanishing outside B_{ρ_2} and equal to 1 in B_{ρ_1} for some ρ_1, ρ_2, $0 < \rho_1 < \rho_2 < \rho$.

Besides that, if $\Pi = \{y \in \mathbb{R}^m \mid y_i = x_i^0, i = 1, \ldots, m-s\}$ is an s-dimensional linear manifold containing x^0, we denote by $\Pi^\perp = \{y \in \mathbb{R}^m \mid y_i = 0, i = m-s+1, \ldots, m\}$ the subspace orthogonal to Π. Here $x = (\hat{x}, \tilde{x})$ where $\hat{x} = (x_1, \ldots, x_{m-s})$ are the coordinates in $\Pi^\perp$ and $\tilde{x} = (x_{m-s+1}, \ldots, x_m)$ are the coordinates in Π. The ball in $\Pi^\perp$ of radius ρ centered at $\hat{x}^0$ will be denoted by $\hat{B}_\rho = \hat{B}_\rho(\hat{x}^0)$ and the cut-off function for $\hat{B}_\rho$ by $\hat{\zeta}(\hat{x})$. The gradient with respect to $\hat{x}$ and $\tilde{x}$ is denoted by $\hat{\nabla}$ and $\tilde{\nabla}$, respectively, $\nabla = (\hat{\nabla}, \tilde{\nabla})$, and the Laplace operator is denoted by $\hat{\Delta}$ and $\tilde{\Delta}$, respectively.

Finally, let $Q_\rho = \Pi \times \hat{B}_\rho$ be the cylinder in $\mathbb{R}^m$, $\omega \Subset \omega' \subset \mathbb{R}^m$ and $\Pi \cap \omega \neq \varnothing$. In this case denote by $\xi(x)$ the cut-off function "in the direction of Π" for the domain

$K_\rho \equiv Q_\rho \cap \omega'$. This means that ξ is a smooth function such that $\xi \equiv 1$ for $x \in Q_\rho \cap \omega$ and $\xi \equiv 0$ for $x \in Q_\rho \setminus K_\rho$. It is clear that for ρ small enough one can take ξ not depending on $\hat{x}$, i.e., $\xi = \tilde{\xi}(\tilde{x})$; then $\hat{\nabla}\tilde{\xi} \equiv 0$.

It is known that one can assume the estimates

$$|D'\tilde{\xi}| \leqslant C\tilde{\xi}^b, \quad |D'^2\tilde{\xi}| \leqslant C\tilde{\xi}^{2b-1}, \qquad C = C(b), \tag{0.8}$$

to hold for all $b \in [\frac{1}{2}, 1)$. So, for example, these inequalities are valid if the boundary of ω' is flat and $\tilde{\xi}(x) = [\operatorname{dist}(x, \partial\omega')]^{1/(1-b)}$ near $\partial\omega'$.

The letter C stands for various nonessential constants.

§1. The spaces $W^{(l)}_{p,\alpha}(\omega, x^0)$, $H^{(l)}_{p,\alpha}(\omega)$, $W^{(l)}_{p,s,\alpha}(\omega, \Pi)$, and $H^{(l)}_{p,s,\alpha}(\omega)$

The study of the regularity of solutions of elliptic systems is connected with using, besides Sobolev and Hölder spaces, some other function spaces. In particular, many results in this field were obtained on the basis of properties of Morrey and Campanato spaces. The definitions of these spaces and corresponding regularity results for solutions of elliptic systems are contained, for example, in the book by Giaquinto [5].

On the other hand, it is convenient to study the regularity of a generalized solution of (0.1), (0.5) by analyzing the convergence of the iterative procedure (0.6), (0.7) in Sobolev weighted spaces and in the spaces $H^{(l)}_{p,\alpha}$ and $H^{(l)}_{p,s,\alpha}$. Here we give the definitions of these spaces and establish some of their properties.

Let $\omega \in \mathbb{R}^m$ be a bounded domain, x, $x^0 \in \omega$, and $r = |x - x^0|$. For $\alpha > -m$ and $p \geqslant 1$ we define the weighted Sobolev space $W^{(l)}_{p,\alpha}(\omega, x^0)$, $l = 1, 2$, as the set of all measurable (vector) functions possessing all generalized derivatives in ω up to order l and having finite norm

$$\|u\|_{W^{(l)}_{p,\alpha}(\omega,x^0)} = \left(\int_\omega r^\alpha |D^l u|^p \, dx\right)^{1/p}.$$

In the same way one can define the spaces $W^{(l)}_{p,\alpha}$ for $l > 2$, but in the present paper we do not use $W^{(l)}_{p,\alpha}$, $l > 2$.

It is clear that $W^{(l)}_{p,0}(\omega, x^0)$ coincides with the usual Sobolev space $W^{(l)}_p(\omega)$ and $W^{(l)}_{p,\alpha}(\omega, x^0) \subset W^{(l)}_p(\omega)$ for $\alpha \in (-m, 0]$.

In a similar way the weighted Lebesgue space $L_{p,\alpha}(\omega, x^0)$ is defined with norm

$$\|u\|_{L_{p,\alpha}(\omega,x^0)} = \left(\int_\omega r^\alpha |u|^p \, dx\right)^{1/p}.$$

The spaces $W^{(l)}_{p,\alpha}(\omega, x^0)$ depend on x^0. Now denote by $H^{(l)}_{p,\alpha}(\omega)$ the set of functions $u(x)$ belonging to $W^{(l)}_{p,\alpha}(\omega, x^0)$ for all $x^0 \in \omega$ and having finite norm

$$\|u\|_{H^{(l)}_{p,\alpha}(\omega)} = \sup_{x^0 \in \omega} \|u\|_{W^{(l)}_{p,\alpha}(\omega,x^0)}.$$

The spaces $W^{(l)}_{p,\alpha}(\omega, x^0)$, $L_{p,\alpha}(\omega, x^0)$, and $W^{(l)}_{p,\alpha}(\omega)$ are Banach spaces; for $p = 2$, $W^{(l)}_{p,\alpha}(\omega, x^0)$ and $L_{p,\alpha}(\omega, x^0)$ are Hilbert spaces. The class C^∞ of functions that are smooth in ω is dense in all these spaces. As usual, we denote by $\overset{\circ}{W}{}^{(l)}_{p,\alpha}(\omega, x^0)$, $\overset{\circ}{H}{}^{(l)}_{p,\alpha}(\omega)$ the closures of the set $\overset{\circ}{C}{}^\infty(\omega)$ of smooth compactly supported in ω functions in the corresponding norms.

The spaces $H^{(1)}_{p.\alpha}(\omega)$ with $p = 2$ were used for the analysis of elliptic equations by L. Nirenberg **[6]** (in the case $m = 2$) and by H. O. Cordes **[1]**. For the study of elliptic systems the spaces $H^{(l)}_{p.\alpha}(\omega, x^0)$ were applied in **[2, 3]**.

Now we give some properties of the spaces under consideration.

LEMMA 1.1. *Let* $m \geqslant 2$.

1) *If* $\alpha, \beta > -m$ *and* $u \in W^{(1)}_{2.\alpha}(B_\rho, x^0) \cap L_{2.\beta}(B_\rho, x^0)$, *then*

$$\|u\|_{L_{2.\alpha}(B_\rho,x^0)} \leqslant C\rho\|D'u\|_{L_{2.\alpha}(B_\rho,x^0)} + C(\rho)\|u\|_{L_{2.\beta}(B_\rho,x^0)}; \tag{1.1}$$

additionally, if supp $u \subset B_\rho$, *then one can omit the second term in* (1.1), *i.e.*,

$$\|u\|_{L_{2.\alpha}(B_\rho,x^0)} \leqslant C\rho\|D'u\|_{L_{2.\alpha}(B_\rho,x^0)}; \tag{1.2}$$

2) *if* $\alpha \in (-m, -m+2)$ *and* $u \in W^{(1)}_{2.\alpha}(\omega, x^0)$, *then* $u(x^0)$ *is determined, where*

$$|u(x^0)| \leqslant C\rho^{(2-m-\alpha)/2}\|D'u\|_{L_{2.\alpha}(B_\rho,x_0)} + C\|u\|_{L_2(B_\rho)};$$

3) *if* $\alpha \in (-m, -m+2)$ *and* $u \in W^{(2)}_{2.\alpha}(B_\rho, x^0)$, *then* $u(x^0)$ *and* $\nabla u(x^0)$ *are determined and subject to the estimates*

$$\begin{aligned} |u(x^0)| &\leqslant C\rho^{(4-m-\alpha)/2}\|D'^2u\|_{L_{2.\alpha}(B_\rho,x^0)} + C\|u\|_{W^{(1)}_2(B_\rho)}, \\ |\nabla u(x^0)| &\leqslant C\rho^{(2-m-\alpha)/2}\|D'^2u\|_{L_{2.\alpha}(B_\rho,x^0)} + C\|u\|_{W^{(1)}_2(B_\rho)}; \end{aligned} \tag{1.3}$$

4) *if* $m \geqslant 4$, $\alpha \in (-m, -m+4)$ *and* $u \in W^{(2)}_{2.\alpha}(B_\rho, x^0)$, *then* $u(x^0)$ *is determined and subject to* (1.3).

PROOF. The proof of (1.2) was given in **[7]**; the inequality (1.1) follows immediately from (1.2) applied to the function $u\zeta$. Part 2 of the lemma was proved in **[8]** and parts 3, 4 are simple corollaries of Part 2. □

It follows from Lemma 1.1 that for $u \in W^{(2)}_{2.\alpha}(B_\rho, x^0)$ a function $\mathring{u}(x)$ can be well defined in the following way:

$$\mathring{u}(x) = \begin{cases} u(x) - T_1(x^0, x)\zeta, & \alpha \in (-m, 2-m), m \geqslant 2, \\ u(x) - u(x^0)\zeta, & \alpha \in (2-m, 4-m), m \geqslant 4, \\ u(x), & \alpha > 4-m, \end{cases} \tag{1.4}$$

where $T_1(x^0, x)$ is the linear part of the Taylor expansion of $u(x)$ at the point x^0 and ζ is a cut-off function for B_ρ. In addition, if $\alpha \leqslant 0$, the estimate

$$\|\mathring{u}\|_{W^{(2)}_{2.\alpha}(B_\rho,x^0)} \leqslant (1 + C\rho^\delta)\|u\|_{W^{(2)}_{2.\alpha}(B_\rho,x^0)} + C\|u\|_{W^{(1)}_2(B_\rho)} \tag{1.5}$$

holds with some $\delta > 0$.

LEMMA 1.2. *Let $m \geqslant 4$ and $u \in \overset{\circ}{W}{}^{(2)}_{2,\alpha}(B_\rho, x^0)$. Then*
1) *for $\alpha \in (-m+3, m)$ the inequality*

$$\|D'^2 u\|_{L_{2,\alpha}(B_\rho,x^0)} \leqslant C_{\alpha,m}\|\Delta u\|_{L_{2,\alpha}(B_\rho,x^0)} \tag{1.6}$$

is valid, where

$$C^2_{\alpha,m} = \begin{cases} 1, & \alpha \in (3-m, 0], \\ 1 + \frac{4\alpha(m-1)}{(m-\alpha)^2}, & \alpha \in [0, m); \end{cases} \tag{1.7}$$

2) *for $\alpha \in (-m+2, m-2)$ additionally there is the inequality*

$$\|D'^2 u\|^2_{L_{2,\alpha}(B_\rho,x^0)} \leqslant -M^2_{\alpha,m}\int_{B_\rho} u\Delta u r^\alpha dx, \tag{1.8}$$

where

$$M^2_{\alpha,m} = \begin{cases} 1, & \alpha \in (-m+2, 0], \\ 1 + \frac{2\alpha}{-\alpha+m-2}, & \alpha \in [0, m-2). \end{cases} \tag{1.9}$$

PROOF. The proof of part 1 of the lemma was given in [2, 3]. The proof of the second part goes along the same lines; moreover, it is sufficient to consider only smooth compactly supported functions.

Let (r, θ) be spherical coordinates centered at x^0, and $\{Y_{n,k}(\theta)\}$, $n = 0, 1, \dots$, $k = 1, \dots, k_n$, a complete system of spherical functions, where k_n is the multiplicity of the eigenvalue $\lambda_n = n(n+m-2)$ of the Beltrami operator. Then we have

$$I_1 \equiv \int_{B_\rho} |D'u|^2 r^\alpha dx = \sum_{n,k}\int_0^\rho \left(|u'_{n,k}(r)|^2 + \lambda_n \frac{|u_{n,k}(r)|^2}{r^2}\right) r^{\alpha+m-1} dr,$$

$$I_2 \equiv -\int_{B_\rho} u\Delta u r^\alpha dx$$
$$= \sum_{n,k}\int_0^\rho \left[|u'_{n,k}(r)|^2 + \left(\lambda_n - \frac{\alpha(\alpha+m-2)}{2}\right)\frac{|u_{n,k}(r)|^2}{r^2}\right] r^{\alpha+m-1} dr,$$

where $u_{n,k} = u_{n,k}(r)$ are the Fourier coefficients of $u(x)$, the sum is extended over all n and k, and the prime means differentiation with respect to r.

When $\alpha \in (-m+2, 0]$ these inequalities imply (1.8) immediately, since $M_{\alpha,m} = 1$ for such α.

We show now that (1.8) holds for $\alpha \in (0, m-2)$ too. Note that the Hardy inequality [9, p. 296] implies the estimate

$$\int_0^\rho |u_{n,k}|^2 r^{\alpha+m-3} dr \leqslant \frac{4}{(\alpha+m-2)^2}\int_0^\rho |u'_{n,k}|^2 r^{\alpha+m-1} dr.$$

Therefore, for $\alpha \in (0, m-2)$ we have

$$I_1 = I_2 + \frac{\alpha(\alpha+m-2)}{2}\sum_{n,k}\int_0^\rho |u_{n,k}|^2 r^{\alpha+m-3} dr$$
$$\leqslant I_2 + \frac{2\alpha}{\alpha+m-2}\sum_{n,k}\int_0^\rho |u'_{n,k}|^2 r^{\alpha+m-1} dr \leqslant I_2 + \frac{2\alpha}{\alpha+m-2} I_1;$$

thus,

$$I_1 \leqslant \Big(1 + \frac{2\alpha}{-\alpha + m - 2}\Big) I_2 = M^2_{\alpha,m} I_2.$$

So we have established (1.8) for all α, $|\alpha| < m - 2$. □

COROLLARY OF LEMMA 1.2. *Let $u \in W^{(2)}_{2,\alpha}(B_\rho, x^0)$ and $m \geqslant 4$, $\alpha \in (3 - m, 0]$. Then*

$$\|u\|^2_{W^{(2)}_{2,\alpha}(B_\rho,x^0)} \leqslant (1 + C\rho^2)\|\Delta u\|^2_{L_{2,\alpha}(B_\rho,x^0)} + C\|u\|^2_{W^{(2)}_2(B_\rho)}.$$

PROOF. Let us apply (1.6) to the function $u\zeta$. Taking into account the boundedness of r^α in the domain where $\zeta \not\equiv 1$ and (1.1), (1.6) we obtain the desired estimate. □

LEMMA 1.3. *For some $\gamma > 0$ the following imbeddings hold:*

(1.10) $$H^{(1)}_{2,\alpha}(\omega) \subset C^{(\gamma)}(\omega), \qquad \alpha \in (-m, 2 - m), \quad m \geqslant 2,$$

(1.11) $$H^{(2)}_{2,\alpha}(\omega) \subset C^{(1,\gamma)}(\omega), \qquad \alpha \in (-m, 2 - m), \quad m \geqslant 2,$$

(1.12) $$H^{(2)}_{2,\alpha}(\omega) \subset C^{(\gamma)}(\omega), \qquad \alpha \in (2 - m, 4 - m), \quad m \geqslant 4,$$

where $C^{(l,\gamma)}(\omega)$ is the set of functions whose derivatives of order l are Hölder with the exponent $\gamma > 0$ in ω and $C^{(\gamma)}(\omega) = C^{(0,\gamma)}(\omega)$.

The imbedding (1.10) was proven in **[1]** (see also **[2, 3]**) and (1.11), (1.12) follow from (1.10).

Now we turn to the spaces $W^{(l)}_{p,s,\alpha}(\omega, \Pi)$ and $H^{(l)}_{p,s,\alpha}(\omega)$. For the case $l = 1$ these spaces were introduced in **[4]**.

Let $s \in [1, m-1]$ be an integer and $\Pi \subset \mathbb{R}^m$ some s-dimensional linear manifold. For $x \in \mathbb{R}^m$ denote $r_s = \operatorname{dist}\{x, \Pi\}$.

For a bounded domain $\omega \subset \mathbb{R}^m$, $p \geqslant 1$ and $\alpha > s - m$ we define the weight Sobolev spaces $W^{(l)}_{p,s,\alpha}(\omega, \Pi)$, $l = 1, 2$, as the collections of all measurable (vector) functions $u(x)$ having generalized derivatives up to order l in ω with finite norm

(1.13) $$\|u\|_{W^{(l)}_{p,s,\alpha}(\omega,\Pi)} = \Big[\int_\omega r_s^\alpha |D^l u|^p dx\Big]^{1/p}.$$

It is clear that $W^{(l)}_{p,s,0}(\omega, \Pi)$ coincides with $W^{(l)}_p(\omega)$ and $W^{(l)}_{p,s,\alpha}(\omega, \Pi) \subset W^{(l)}_p(\omega)$ for $\alpha \in (s - m, 0]$. Note also that only the case $\Pi \cap \bar\omega \neq \varnothing$ is of interest, since if $\Pi \cap \bar\omega = \varnothing$, then $W^{(l)}_{p,s,\alpha}(\omega, \Pi)$ coincides with $W^{(l)}_p(\omega)$.

The norm (1.13) for $p = 2$ is invariant under orthogonal transformations of coordinates and in the case $p \neq 2$ orthogonal transformations of coordinates produce norms equivalent to (1.13). In this connection hereafter we assume that the coordinate system agrees with Π in such a way that $\Pi = \{y \in \mathbb{R}^m \mid y_i = x_i^0, \quad i = 1, \dots, m - s\}$, where x^0 is some point in Π. Here

$$r_s^2 = \sum_{i=1}^{m-s} (x_i - x_i^0)^2.$$

Similarly to $W_{p.s.\alpha}^{(l)}(\omega, \Pi)$ we define the weighted Lebesgue spaces $L_{p.s.\alpha}(\omega, \Pi)$. The spaces $W_{p.s.\alpha}^{(l)}(\omega, \Pi)$ are Banach and for $p = 2$ they are Hilbert spaces.

The space $W_{p.s.\alpha}^{(l)}(\omega, \Pi)$ depends on the manifold Π. Denote by $H_{p.s.\alpha}^{(l)}(\omega)$ the set of functions $u(x)$, belonging to $W_{p.s.\alpha}^{(l)}(\omega, \Pi)$ for each Π such that $\Pi \cap \bar{\omega} \neq \emptyset$, with finite norm

$$\|u\|_{H_{p.s.\alpha}^{(l)}(\omega)} = \sup_{\Pi} \|u\|_{W_{p.s.\alpha}^{(l)}(\omega, \Pi)},$$

where the supremum is taken over all Π intersecting $\bar{\omega}$.

The space $H_{p.s.\alpha}^{(l)}(\omega)$ is a Banach space, and the set $C^\infty(\omega)$ is dense in $H_{p.s.\alpha}^{(l)}(\omega)$. Denote by $\overset{\circ}{W}_{p.s.\alpha}^{(l)}(\omega)$ and $\overset{\circ}{H}_{p.s.\alpha}^{(l)}(\omega)$ the closures of $\overset{\circ}{C}^\infty(\omega)$ in the corresponding norms. The spaces $W_{p.\alpha}^{(l)}(\omega, x^0)$ and $H_{p.\alpha}^{(l)}(\omega)$ may be considered as $W_{p.s.\alpha}^{(l)}(\omega, \Pi)$ and $H_{p.s.\alpha}^{(l)}(\omega)$ for the case $s = 0$.

It is clear that

$$(1.14) \quad \|u\|_{L_{p.s.\alpha}(\omega, \Pi)} = \big\| \|u\|_{L_p(M^\| \cap \omega)} \big\|_{L_{p.\alpha}(\hat{\omega}, \hat{x}^0)} = \big\| \|u\|_{L_{p.\alpha}(M^\perp \cap \omega, \hat{x}^0)} \big\|_{L_p(\tilde{\omega})},$$

where $x^0 \in \Pi$, $M^\| = M^\|(\hat{x})$ is the s-dimensional linear manifold parallel to Π and passing through $x = (\hat{x}, \tilde{x})$; $M^\perp = M^\perp(\tilde{x})$ is the $(m - s)$-dimensional manifold orthogonal to Π and passing through x; $\hat{\omega}$ is the projection of ω onto $\Pi^\perp$ and $\tilde{\omega}$ is the projection of ω onto Π.

From (1.14) there follow assertions for the spaces $W_{p.s.\alpha}^{(l)}(\omega, \Pi)$ similar to those in Lemma 1.1. In particular, if $m \geq 3$, $s \in [1, m-2]$, $\alpha \in (s-m, s-m+2)$, and $u \in W_{2.s.\alpha}^{(2)}(Q_\rho, \Pi)$, then for almost all $\tilde{x}$ (with respect to Lebesgue measure in Π) the norm $\|u\|_{W_{2.\alpha}^{(2)}(M^\perp \cap Q_\rho, \hat{x}^0)}$ is finite. Thus, according to Lemma 1.1, for almost all $\tilde{x}$ the values $u(\hat{x}^0, \tilde{x})$ and $\hat{\nabla} u(\hat{x}^0, \tilde{x})$ are defined and they are subject to the estimates

$$(1.15) \quad \begin{aligned} |u(\hat{x}^0, \tilde{x})| &\leq C\rho^{(4-m+s-\alpha)/2} \|D'^2 u\|_{L_{2.\alpha}(M^\perp \cap Q_\rho, \hat{x}^0)} + C\|u\|_{W_2^{(1)}(M^\perp \cap Q_\rho)}, \\ |\hat{\nabla} u(\hat{x}^0, \tilde{x})| &\leq C\rho^{(2-m+s-\alpha)/2} \|D'^2 u\|_{L_{2.\alpha}(M^\perp \cap Q_\rho, \hat{x}^0)} + C\|u\|_{W_2^{(1)}(M^\perp \cap Q_\rho)}. \end{aligned}$$

Moreover, if $u \subset W_{2.s.\alpha}^{(2)}(Q_\rho, \Pi)$ for $m \geq 5$, $s \in [1, m-4]$, and $\alpha \in (s-m+2, s-m+4)$, then the value $u(\hat{x}^0, \tilde{x})$ is defined for almost all $\tilde{x}$ and (1.15) holds as well.

Thus for almost all $\tilde{x}$ for $u \in W_{2.s.\alpha}^{(2)}(Q_\rho, \Pi)$ the function $\overset{\circ}{u}_s$ is well defined by

$$(1.16) \quad \overset{\circ}{u}_s(x) = \begin{cases} u(x) - \hat{T}_1(\hat{x}^0, \tilde{x})\hat{\zeta}, & \alpha \in (s-m, s-m+2), \quad m-s \geq 2, \\ u(x) - \dot{u}(\hat{x}^0, \tilde{x})\hat{\zeta}, & \alpha \in (s-m+2, s-m+4), \quad m-s \geq 4, \\ u(x), & \alpha > s-m+4, \end{cases}$$

where $\hat{T}_1(\hat{x}^0, \hat{x})$ is the linear part of the Taylor expansion of the function $v(\hat{x}) = u(\hat{x}, \tilde{x})$ at $\hat{x}^0$.

Note that for a function u compactly supported in $\mathbb{R}^m$ the estimate

$$(1.17) \quad \|u\|_{L_{p.\alpha}(\mathbb{R}^m, x^0)} \leq C\|D'u\|_{L_{p.\alpha+p}(\mathbb{R}^m, x^0)},$$

proved in [10, Corollary 2.1.6/2] holds for $\alpha > -m$.

If, however, $u \in W_{p.\alpha+p}^{(1)}(B_\rho, x^0)$ and it is not compactly supported or $\alpha \in (-m-p, -m)$, then instead of (1.12) the inequality

(1.18) $$\|\overset{\circ}{u}\|_{L_{p,\alpha}(B_\rho,x^0)} \leqslant C\|D'^l u\|_{L_{p,\alpha+p}(B_\rho,x^0)} + C\|u\|_{L_{p,\beta}(B_\rho,x^0)}$$

is valid for all $\beta > -m$, where $\overset{\circ}{u}$ is defined by (1.4).

It is not hard to prove, using (1.2), that it is possible to take $\|D'^l u\|_{L_{p,s,\alpha}(\omega,\Pi)}$ and $\|D'^l u\|_{H_{p,s,\alpha}(\omega)}$ as equivalent norms in the spaces $\overset{\circ}{W}{}^{(l)}_{p,s,\alpha}(\omega,\Pi)$ and $\overset{\circ}{H}{}^{(l)}_{p,s,\alpha}(\omega)$, respectively, and to derive from (1.17) the inequality

(1.19) $$\|u\|_{L_{p,s,\alpha}(Q_\rho,\Pi)} \leqslant C\|D'u\|_{L_{p,s,\alpha+p}(Q_\rho,\Pi)},$$

valid for all $u \in \overset{\circ}{W}{}^{(1)}_{p,s,\alpha}(Q_\rho,\Pi)$ for $\alpha > -m+s$. If, however, $\alpha \in (-m+s-p, -m+s)$ or $u(x)$ is not compactly supported, $u \in W^{(1)}_{p,s,\alpha}(Q_\rho,\Pi) \cap L_{p,s,\beta}(Q_\rho,\Pi)$, then the following estimate, analogous to (1.18),

(1.20) $$\|\overset{\circ}{u}_s\|_{L_{p,s,\alpha}(Q_\rho,\Pi)} \leqslant C\|D'u\|_{L_{p,s,\alpha+p}(Q_\rho,\Pi)} + \|u\|_{L_{p,s,\beta}(Q_\rho,\Pi)}$$

replaces (1.19) for all $\beta > -m+s$, where $\overset{\circ}{u}_s$ is defined by (1.16).

The following inequality (similar to (1.5)) also holds for the function $\overset{\circ}{u}_s(x)$ in the case of $\alpha \leqslant 0$:

(1.21) $$\|\overset{\circ}{u}_s\|_{W^{(2)}_{2,s,\alpha}(Q_\rho,\Pi)} \leqslant (1+C\rho^\delta)\|u\|_{W^{(2)}_{2,s,\alpha}(Q_\rho,\Pi)} + C\|u\|_{W^{(1)}_2(Q_\rho)}.$$

LEMMA 1.4. *Let $m \geqslant 3$, let $s \in [1, m-2]$ be an integer, $\alpha \in (s-m, m-s)$.*

1) *If $u \in \overset{\circ}{W}{}^{(1)}_{p,s,\alpha}(\omega,\Pi)$, and Π is some s-dimensional manifold, $\Pi \cap \bar{\omega} \neq \varnothing$, $x^0 \in \Pi$, then*

(1.22) $$\|u\|_{L_p(\Pi\cap\omega)} \in W^{(1)}_{q,\beta}(\hat{\omega}, \hat{x}^0).$$

In (1.22) $\hat{\omega}$ denotes the projection of ω onto $\Pi^\perp$, and q and β are parameters,

$$1 < q \leqslant p, \quad \beta > \frac{(\alpha q - (p-q)(m-s))}{p}$$

(if $q = p$ one must take $\beta \geqslant \alpha$);

2) *if $u \in W^{(1)}_{p,s,\alpha}(\omega,\Pi)$, then $\|u\|_{L_p(\Pi\cap\omega')} \in W^{(1)}_{q,\beta}(\hat{\omega}, \hat{x}^0)$ for arbitrary $\omega' \Subset \omega$.*

PROOF. In order to prove the first part of the lemma it is sufficient to establish the required estimate for $u(x) \in \overset{\circ}{C}{}^\infty(\omega)$. Choose a coordinate system consistent with Π and denote

$$v(\hat{x}) = \|u\|_{L_p(\Pi\cap\omega)} = \left(\int_{\Pi\cap\omega} |u(\hat{x},\tilde{x})|^p d\tilde{x}\right)^{1/p}.$$

The function $v(\hat{x})$ has a compact support in $\hat{\omega}$ since $u(x)$ is compactly supported in ω. Let $E \subset \hat{\omega}$ be the set of all $\hat{x}$ such that $v(\hat{x}) = 0$. This set is closed and $v = \nabla v \equiv 0$ for $\hat{x} \in \text{int}(E)$. If $\hat{x} \in \hat{\omega} \setminus E$, it is clear that for $i = 1, \ldots, m-s$ we have

$$\frac{\partial v}{\partial x_i} = \left(\int_{\Pi\cap\omega} |u|^p d\tilde{x}\right)^{(1-p)/p} \int_{\Pi\cap\omega} \operatorname{sgn} u |u|^{p-1} \frac{\partial u}{\partial x_i} d\tilde{x}.$$

Consequently,

$$\|v\|^q_{W^{(1)}_{q,\beta}(\hat\omega,\hat x^0)} = \int_{\hat\omega\setminus E} r_s^\beta \sum_{i=1}^{m-s}\left|\frac{\partial v}{\partial x_i}\right|^q d\hat x$$
$$\leqslant \int_{\hat\omega\setminus E} r_s^\beta \Big(\int_{\Pi\cap\omega}|u|^p d\tilde x\Big)^{(1-p)q}\sum_{i=1}^{m-s}\Big(\int_{\Pi\cap\omega}|u|^{p-1}\left|\frac{\partial u}{\partial x_i}\right| d\tilde x\Big)^q d\hat x.$$

Applying the Hölder inequality with exponents p and $p' = p/(p-1)$ we obtain

$$\|v\|^q_{W^{(1)}_{q,\beta}(\hat\omega,\hat x^0)} \leqslant \int_{\hat\omega\setminus E} r_s^\beta \Big(\int_{\Pi\cap\omega}|u|^p d\tilde x\Big)^{(1-p)q/p}\Big(\int_{\Pi\cap\omega}|u|^p d\tilde x\Big)^{(p-1)q/p}$$
$$\times\sum_{i=1}^{m-s}\Big(\int_{\Pi\cap\omega}\left|\frac{\partial u}{\partial x_i}\right|^p d\tilde x\Big)^{q/p} d\hat x$$
$$\leqslant \Big(\int_{\hat\omega} r_s^\beta\sum_{i=1}^{m-s}\Big(\int_{\Pi\cap\omega}\left|\frac{\partial u}{\partial x_i}\right|^p d\tilde x\Big)^{q/p} d\hat x.$$

This estimate immediately implies (1.22) for $q = p$. If $q < p$ apply again the Hölder inequality with exponents p/q and $p/(p-q)$ to the right-hand side of the last estimate, first to the sums and then to the integral. We obtain

$$\|v\|^q_{W^{(1)}_{q,\beta}(\hat\omega,\hat x^0)} \leqslant C\int_{\hat\omega} r_s^\beta\Big(\sum_{i=1}^{m-s}\int_{\Pi\cap\omega}\left|\frac{\partial u}{\partial x_i}\right|^p d\tilde x\Big)^{q/p} d\hat x$$
$$= C\int_{\hat\omega} r_s^{\beta-\alpha q/p}\Big(r_s^\alpha\sum_{i=1}^{m-s}\int_{\Pi\cap\omega}\left|\frac{\partial u}{\partial x_i}\right|^p d\tilde x\Big)^{q/p} d\hat x$$
$$\leqslant C\Big(\int_{\hat\omega} r_s^{\frac{\beta p-\alpha q}{p-q}} d\hat x\Big)^{(p-q)/p}\Big(\int_{\hat\omega} r_s^\alpha\sum_{i=1}^{m-s}\int_{\Pi\cap\omega}\left|\frac{\partial u}{\partial x_i}\right|^p d\tilde x\, d\hat x\Big)^{q/p}.$$

Under our conditions we have $\frac{\beta p-\alpha q}{p-q} > s-m$; therefore the integral

$$\int_{\hat\omega} r_s^{(\beta p-\alpha q)/(p-q)} d\hat x \tag{1.23}$$

is finite. Thus,

$$\|v\|^q_{W^{(1)}_{q,\beta}(\hat\omega,\hat x^0)} \leqslant C\Big(\int_{\hat\omega}\int_{\Pi\cap\omega} r_s^\alpha\sum_{i=1}^{m-s}\left|\frac{\partial u}{\partial x_i}\right|^p d\tilde x\, d\hat x\Big)^{q/p} \leqslant C\|u\|^q_{W^{(1)}_{p,s,\alpha}(\omega,\Pi)},$$

which means that (1.22) is valid.

If $u \in W^{(1)}_{p,s,\alpha}(\omega,\Pi)$, $\omega' \Subset \omega$, and $\varphi(x) \in \overset{\circ}{C}{}^\infty(\omega)$, $\varphi(x) \equiv 1$ in ω', then $u\varphi \in \overset{\circ}{W}{}^{(1)}_{p,s,\alpha}(\omega,\Pi)$. Besides this, $\|u\|_{L_p(\Pi\cap\omega')} \leqslant \|u\varphi\|_{L_p(\Pi\cap\omega)}$. Now part 1 implies $\|u\|_{L_p(\Pi\cap\omega')} \in W^{(1)}_{q,\beta}(\hat\omega,\hat x^0)$ and this completes the proof of the lemma. □

As corollaries we get the following assertions.

LEMMA 1.5. *Lemma* 1.4 *remains valid if one replaces everywhere in its formulation the space* $W^{(1)}_{p,s,\alpha}(\omega,\Pi)$ *by the space* $H^{(1)}_{p,s,\alpha}(\omega)$, *and* $W^{(1)}_{q,\beta}(\hat\omega,\hat x^0)$ *by* $H^{(1)}_{q,\beta}(\hat\omega)$. *Here* (1.22) *holds for any s-dimensional manifold* Π.

The proof of the lemma follows from the definition of the norms in $H^{(1)}_{p,s,\alpha}$ and $H^{(1)}_{q,\beta}$ and from the uniform boundedness of the integral (1.23) with respect to $\hat{x}^0 \in \hat{\omega}$.

LEMMA 1.6. *Let $m \geqslant 3$, let $s \in [1, m-2]$ be an integer, $\alpha \in (s-m, s-m+2)$ and $u \in \overset{\circ}{H}{}^{(1)}_{2,s,\alpha}(\omega)$. Then $u \in L_2(\Pi \cap \omega)$ for any s-dimensional manifold Π and the norm $\|u\|_{L_2(\Pi\cap\omega)}$ is a Hölder function of the coordinates orthogonal to Π.*

If $u \in H^{(1)}_{2,s,\alpha}(\omega)$, then $u \in L_2(\Pi \cap \omega')$ for all $\omega' \Subset \omega$.

Proof of this assertion follows immediately from Lemma 1.5 and the imbedding (1.10). Lemma 1.6 strengthens Lemma 2 in **[4]**.

LEMMA 1.7. *Let $u \in \overset{\circ}{H}{}^{(2)}_{2,s,\alpha}(\omega)$. Then*

1) if $m \geqslant 3$, $s \in [1, m-2]$ is an integer and $\alpha \in (s-m, s-m+2)$, then $\nabla u \in L_2(\Pi \cap \omega)$, $u \in L_p(\Pi \cap \omega)$, $p = 2s(s-2)^{-1}$ for $s > 2$, and $p > 1$ is arbitrary for $s = 2$; here Π is an arbitrary smooth s-dimensional manifold, $\Pi \cap \omega \neq \varnothing$. Besides this, $u \in L_q(P \cap \omega)$, $q = 2(s-1)(s-2)^{-1}$ for $s > 2$, and $q > 1$ for $s = 2$, where P is an arbitrary $(s-1)$-dimensional manifold, $P \cap \omega \neq \varnothing$. Finally, $u \in C^{(1/2)}(\omega)$ in the case of $s = 1$;

2) if $m \geqslant 5$, $s \in [1, m-4]$ is an integer, $\alpha \in (s-m+2, s-m+4)$, then $u \in L_2(\Pi \cap \omega)$ for an s-dimensional manifold Π.

All the norms in the Lebesgue spaces mentioned here are Hölder functions of the coordinates orthogonal to the manifolds Π and P.

If $u \in H^{(2)}_{2,s,\alpha}(\omega)$ all the statements remain valid with ω replaced by ω', $\omega' \Subset \omega$.

PROOF. The proof of the first part of the lemma follows from Lemma 1.5, the imbedding (1.10), and the imbedding theorems for Sobolev spaces. The second part follows immediately from (1.19) and Lemma 1.5. □

In studying the weak regularity of solutions of elliptic systems we shall need, besides Lemmas 1.6 and 1.7, some estimates of the integrals of potential type in the norms of the weighted spaces introduced here.

Consider the operator

$$A_\lambda f = \int_\omega \frac{f(x)}{r^\lambda(x,y)}\,dx,$$

where $r(x,y) = |x-y|$, $\lambda > -m$, and ω is a bounded domain in $\mathbb{R}^m$.

The boundedness of the operator A_λ acting from $L_{2,\alpha}(\omega, x^0)$ to $L_{2,\alpha}(\omega, x^0)$ was proved by E. Stein **[11]**. Some estimates sufficient for the case studied here may be found also in **[12]**. Now we consider the operator A_λ in the spaces $L_{2,s,\alpha}(\omega, \Pi)$. We prove two lemmas (cf. **[12]**). Let, as before,

$$r_s^2(x,y) = \sum_{i=1}^{m-s} (x_i - y_i)^2.$$

LEMMA 1.8. *The function*

$$h(x^0, y) = \int_\omega r^{\alpha_1}(x,y)\, r_s^{\alpha_2}(x^0, x)\, r_s^{\alpha_3}(x^0, y)\,dx \tag{1.24}$$

is uniformly bounded with respect to x^0, $y \in \omega$ if ω is a bounded domain and

$$\alpha_1 > -m, \qquad \alpha_2 > s-m, \qquad \alpha_3 > 0, \qquad \alpha_1 + \alpha_2 + \alpha_3 = s - m.$$

PROOF. The proofs of this lemma and of the next one follow **[12]** in the main. Passing to new coordinates in (1.24),

$$x = \rho x', \qquad x^0 = \rho x^{0'}, \qquad y = \rho y', \qquad \rho = r_s(x^0, y), \tag{1.25}$$

and here

$$h(x^0, y) = \rho^s \int_{\omega'} r^{\alpha_1}(x', y') r_s^{\alpha_2}(x^{0'}, x')\, dx',$$

where ω' is the image of ω under the transformation of coordinates (1.25). It is clear that $\operatorname{diam} \omega' = \rho^{-1} \operatorname{diam} \omega$. In addition, one can assume that

$$x_i^{0'} = 0, \qquad i = 1, \dots m - s,$$
$$y_1{}' = 1, \quad y_i{}' = 0, \qquad i = 2, \dots, m.$$

Put $E_{m-s} = \{x' \in \mathbb{R}^m \mid |x_i'| < \operatorname{diam} \omega', i = m - s + 1, \dots, m\}$; this set is an $(m-s)$-dimensional layer in $\mathbb{R}^m$. Then

$$h(x^0, y) \leqslant \rho^s \int_{E_{m-s}} \left((x_1' - 1)^2 + x_2'^2 + \dots + x_m'^2\right)^{\alpha_1/2} \left(x_1'^2 + \dots + x_{m-s}'^2\right)^{\alpha_2/2} dx'.$$

It is not hard to show that under the conditions of the lemma this integral converges at both poles and at infinity. Together with the estimate for $\operatorname{diam} \omega'$ it implies the uniform boundedness of $h(x^0, y)$. □

LEMMA 1.9. *The operator A_λ acts boundedly from the space $L_{2,s,\beta-2s+2\varepsilon}(\omega, \Pi)$ to the space $L_{2,s,\beta}(\omega, \Pi)$, $\varepsilon = m - \lambda$, if $m \geqslant 2$, $s \in [1, m-1]$ is an integer, Π is an s-dimensional linear manifold, and*

$$s - m < \beta < m + s - 2\varepsilon, \qquad 0 \leqslant \lambda < m.$$

PROOF. Since $\varepsilon < m$, we have

$$\alpha_1 = -\frac{m}{s} + 2 + \frac{\beta}{s} + \frac{2\varepsilon}{s} < \frac{m}{s} + 2 + \frac{\beta}{s} = \alpha_2,$$
$$\alpha_3 = -\frac{m}{s} + 4 - \frac{\beta}{s} < \frac{m}{s} + 4 - \frac{\beta}{s} - \frac{2\varepsilon}{s} = \alpha_4.$$

Consider the intervals $E_1 = (\alpha_1, \alpha_2)$ and $E_2 = (\alpha_3, \alpha_4)$; we show that under our conditions $E \equiv E_1 \cap E_2 \neq \varnothing$. In fact, E_1 and E_2 can be disjoint only if $\alpha_3 \geqslant \alpha_2$ or $\alpha_1 \geqslant \alpha_4$. The condition $\alpha_3 \geqslant \alpha_2$ is equivalent to the inequality $\beta \leqslant s - m$ and $\alpha_1 \geqslant \alpha_4$ to the inequality $\beta \geqslant m + s - 2\varepsilon$. Both cases are excluded by the conditions of the lemma. Thus $E \neq \varnothing$.

Fix $\varkappa \in E$ and denote $\beta = 2\gamma$. Then

$$\begin{aligned} N^2 = \|A_\lambda f\|^2_{L_{2,s,\beta}(\omega,\Pi)} &= \int_\omega \left(\int_\omega \frac{f(x)}{r^\lambda(x,y)}\, dx \right)^2 r_s^{2\gamma}(x^0, y)\, dy \\ &= \int_\omega \left(\int_\omega \varphi_1(x^0, x, y) \varphi_2(x^0, x, y)\, dx \right)^2 r_s^{(\varkappa - 3)s/2}(x^0, y)\, dy, \end{aligned}$$

where

$$\varphi_2(x^0, x, y) = f(x) r^{(-m+\varepsilon)/2}(x, y) r_s^{(m - \varkappa s + 2\gamma + 2\varepsilon)/4}(x^0, x) r_s^{(-m+s+2\gamma)/4}(x^0, y),$$
$$\varphi_1(x^0, x, y) = r^{(-m+\varepsilon)/2}(x, y) r_s^{(-m + \varkappa s - 2\gamma - 2\varepsilon)/4}(x^0, x) r_s^{(m - s + 2\gamma + (3 - \varkappa)s)/4}(x^0, y).$$

Applying the Cauchy-Bunyakovskiĭ inequality to the inner integral we obtain

$$N^2 \leqslant \int_\omega r_s^{(\varkappa-3)s/2}(x^0,y)\Big(\int_\omega f^2(x) r^{-m+\varepsilon}(x,y) r_s^{(m-\varkappa s+2\gamma+2\varepsilon)/2}(x^0,x) \times r_s^{(-m+s+2\gamma)/2}(x^0,y)\,dx\Big) h_1(x^0,y)\,dy,$$

where

$$h_1(x^0,y) = \int_\omega r^{-m+\varepsilon}(x,y) r_s^{(-m+\varkappa s-2\gamma-2\varepsilon)/2}(x^0,x) r_s^{(m-s+2\gamma+(3-\varkappa)s)/2}(x^0,y)\,dx.$$

One can check easily that for $\varkappa \in E$ the function $h_1(x^0,y)$ is subject to the conditions of Lemma 1.8. Therefore $h_1(x^0,y)$ is bounded and

$$N^2 \leqslant C \int_\omega r_s^{((\varkappa-2)s-m+2\gamma)/2}(x^0,y) \times \Big(\int_\omega f^2(x) r^{-m+s}(x,y) r_s^{(m-\varkappa s+2\gamma+2\varepsilon)/2}(x^0,x)\,dx\Big)\,dy.$$

Next, we change the order of integration and get

$$N^2 \leqslant C \int_\omega f^2(x) r_s^{\beta-2s+2\varepsilon}(x^0,x) h_2(x^0,x)\,dx,$$

where

$$h_2(x^0,x) = \int_\omega r^{-m+\varepsilon}(x,y) r_s^{((\varkappa-2)s-m+2\gamma)/2}(x^0,y) r_s^{(m-\varkappa s-2\gamma-2\varepsilon+4s)/2}(x^0,x)\,dy.$$

The function $h_2(x^0,x)$ is also bounded, since it is subject to the conditions of Lemma 1.8. Thus,

$$N^2 \leqslant C \int_\omega f^2(x) r_s^{\beta-2s+2\varepsilon}(x^0,x)\,dx = C\|f\|^2_{L_{2,s,\beta-2s+2\varepsilon}(\omega,\Pi)},$$

which gives us the required boundedness of A_λ. □

Finally, in the next two lemmas we obtain for the spaces $W^{(2)}_{2,s,\alpha}(\omega,\Pi)$ some estimates similar to (1.6).

LEMMA 1.10. *If* $u \subset W^{(2)}_{2,s,\alpha}(Q_\rho,\Pi)$, $s = 1,\dots,m-4$, $m \geqslant 5$, $\alpha \in (-m+s+3, m-s-2)$, *then*

$$\|D'^2 u\|_{L_{2,s,\alpha}(Q_\rho,\Pi)} \leqslant C'_{\alpha,m-s}\|\Delta u\|_{L_{2,s,\alpha}(Q_\rho,\Pi)},$$

where $C'_{\alpha,m} = \max\{C_{\alpha,m}, M_{\alpha,m}\}$ *and* $C_{\alpha,m}$, $M_{\alpha,m}$ *are defined in* (1.7) *and* (1.9).

PROOF. It is sufficient to prove our lemma for smooth compactly supported functions. There is an obvious representation

$$\begin{aligned}\|D'^2 u\|^2_{L_{2,s,\alpha}(Q_\rho,\Pi)} &= \int_\Pi d\tilde{x} \int_{\hat{B}_\rho} \sum_{i,j=1}^{m-s} |D_{ij}u|^2 r_s^\alpha\,d\hat{x} \\ &+ 2\int_\Pi d\tilde{x} \int_{\hat{B}_\rho} \sum_{i=1}^{m-s} \sum_{j=m-s+1}^{m} |D_{ij}u|^2 r_s^\alpha\,d\hat{x} \qquad (1.26)\\ &+ \int_\Pi d\tilde{x} \int_{\hat{B}_\rho} \sum_{i,j=m-s+1}^{m} |D_{ij}u|^2 r_s^\alpha\,d\hat{x}.\end{aligned}$$

For the first term on the right-hand side of (1.26), according to (1.6), we have

$$\int_{\Pi} d\tilde{x} \int_{\hat{B}_\rho} \sum_{i,j=1}^{m-s} |D_{ij}u|^2 r_s^\alpha d\hat{x} \leqslant C^2_{\alpha,m-s} \|\hat{\Delta} u\|^2_{L_{2,s,\alpha}(Q_\rho,\Pi)}. \tag{1.27}$$

In the third term integrate twice by parts with respect to $\tilde{x}$, taking into account that $u(x)$ is compactly supported and that r_s^α does not depend on $\tilde{x}$, and obtain

$$\int_{\Pi} d\tilde{x} \int_{\hat{B}_\rho} \sum_{i,j=m-s+1}^{m} |D_{ij}u|^2 r_s^\alpha d\hat{x} = \|\tilde{\Delta} u\|^2_{L_{2,s,\alpha}(Q_\rho,\Pi)}. \tag{1.28}$$

Consider the middle term on the right-hand side of (1.26). Apply (1.8) and integrate by parts over $\tilde{x}$ to get

$$\begin{aligned}
&\int_{\Pi} d\tilde{x} \int_{\hat{B}_\rho} \sum_{i=1}^{m-s} \sum_{j=m-s+1}^{m} |D_{ij}u|^2 r_s^\alpha d\hat{x} \\
&\leqslant -M^2_{\alpha,m-s} \int_{\Pi} d\tilde{x} \int_{\hat{B}_\rho} \sum_{j=m-s+1}^{m} D_j u \hat{\Delta} D_j u r_s^\alpha d\hat{x} \\
&= M^2_{\alpha,m-s} \int_{Q_\rho} \tilde{\Delta} u \hat{\Delta} u r_s^\alpha dx.
\end{aligned} \tag{1.29}$$

Now the inequality (1.26) and the relations (1.27), (1.28), (1.29) lead us, for ρ small enough, to the required estimate. □

REMARK. It is not hard to see that $C'_{\alpha,m} = C_{\alpha,m}$ for $\alpha \in (-m+3, \alpha^*]$ and $C'_{\alpha,m} = M_{\alpha,m}$ for $\alpha \in [\alpha^*, m-2)$, where $\alpha^* = 1 + \sqrt{(m-1)(m-5)}$.

In the sequel we shall need the so-called "Cauchy inequality with η":

$$ab \leqslant \eta a^2 + C(\eta) b^2. \tag{1.30}$$

LEMMA 1.11. *Let $m \geqslant 5$, $s = 1, \dots, m-4$, $u \in W^{(2)}_{2,s,\alpha}(Q_\rho, \Pi)$, $\alpha \in (-m+s+3, m-s-2)$; $\tilde{\xi}$ and $\hat{\zeta}$ are the cut-off functions defined earlier and $\tilde{\xi}$ is subject to* (0.8).

1) *If $u(x)$ vanishes near the lateral area of the cylinder Q_ρ, then*

$$\|\tilde{\xi} D'^2 u\|_{L_{2,s,\alpha}(Q_\rho,\Pi)} \leqslant C'_{\alpha,m-s}(1+\eta)\|\tilde{\xi}\Delta u\|_{L_{2,s,\alpha}(Q_\rho,\Pi)} + C\|u\|_{L_{2,s,\beta}(Q_\rho,\Pi)}. \tag{1.31}$$

2) *If $u(x)$ is not compactly supported in a neighborhood of the lateral area of Q_ρ, then*

$$\|\tilde{\xi} D'^2 u\|_{L_{2,s,\alpha}(Q_\rho,\Pi)} \leqslant C'_{\alpha,m-s}(1+\eta)\|\tilde{\xi}\Delta u\|_{L_{2,s,\alpha}(Q_\rho,\Pi)} + C\|u\|_{W^{(2)}_{2,s,\beta}(Q_\rho,\Pi)}, \tag{1.32}$$

$$\|\tilde{\xi} D'^2 (\hat{\zeta} u)\|_{L_{2,s,\alpha}(Q_\rho,\Pi)} \leqslant C'_{\alpha,m-s}(1+\eta)\|\tilde{\xi}\Delta u\|_{L_{2,s,\alpha}(Q_\rho,\Pi)} + C\|u\|_{W^{(1)}_{2,s,\beta}(Q_\rho,\Pi)}. \tag{1.33}$$

In the estimates (1.31)–(1.33) *$\eta > 0$ is an arbitrary constant, $\beta \geqslant \alpha$, and β may be taken arbitrarily large if b in* (0.8) *is sufficiently close to* 1.

PROOF. We establish (1.31) first. It is obvious that

$$\tilde{\xi} D_{ij} u = D_{ij}(\tilde{\xi} u) - D_i \tilde{\xi} D_j u - D_j \tilde{\xi} D_i u - u D_{ij} \tilde{\xi}.$$

So,

$$\begin{aligned}|\tilde{\xi}D_{ij}u|^2 &= |D_{ij}(\tilde{\xi}u)|^2 - 2\tilde{\xi}D_{ij}u(D_i\tilde{\xi}D_ju + D_j\tilde{\xi}D_iu + uD_{ij}\tilde{\xi}) + |D_i\tilde{\xi}|^2|D_ju|^2 \\ &+ |D_j\tilde{\xi}|^2|D_iu|^2 + |D_{ij}\tilde{\xi}|^2|u|^2 + 2D_i\tilde{\xi}D_juD_j\tilde{\xi}D_iu + 2uD_i\tilde{\xi}D_{ij}\tilde{\xi}D_ju \\ &+ 2uD_j\tilde{\xi}D_{ij}\tilde{\xi}D_iu - 2(D_i\tilde{\xi}D_ju + D_j\tilde{\xi}D_iu + uD_{ij}\tilde{\xi})^2.\end{aligned}$$

Applying the "Cauchy inequality with η" (1.30) to the terms containing $\tilde{\xi}D_{ij}u$, we get

$$|\tilde{\xi}D_{ij}u|^2 \leqslant |D_{ij}(\tilde{\xi}u)|^2 + \eta|\tilde{\xi}D_{ij}u|^2 + C(S_1(\tilde{\xi},u) + S_2(\tilde{\xi},u) + S_3(\tilde{\xi},u)),$$

where

$$\begin{aligned}S_1(\tilde{\xi},u) &= |D_i\tilde{\xi}|^2|D_ju|^2 + |D_j\tilde{\xi}|^2|D_iu|^2 + |D_i\tilde{\xi}D_j\tilde{\xi}D_juD_iu|, \\ S_2(\tilde{\xi},u) &= |D_{ij}\tilde{\xi}|^2|u|^2, \\ S_3(\tilde{\xi},u) &= |uD_iuD_j\tilde{\xi}D_{ij}\tilde{\xi} + uD_juD_i\tilde{\xi}D_{ij}\tilde{\xi}|.\end{aligned}$$

Using (1.30), $S_3(\tilde{\xi},u)$ is estimated in terms of $S_1(\tilde{\xi},u)$ and $S_2(\tilde{\xi},u)$. Thus, we obtain

(1.34) $$|\tilde{\xi}D_{ij}u|^2 \leqslant (1+\eta)|D_{ij}(\tilde{\xi}u)|^2 + C(S_1(\tilde{\xi},u) + S_2(\tilde{\xi},u)).$$

Now on the basis of Lemma 1.10 we find that

(1.35) $$\begin{aligned}\|\tilde{\xi}D'^2u\|^2_{L_{2,s,\alpha}(Q_\rho,\Pi)} &\leqslant (1+\eta)C'^2_{\alpha,m-s}\|\Delta(\tilde{\xi}u)\|^2_{L_{2,s,\alpha}(Q_\rho,\Pi)} \\ &+ C\sum_{i,j=1}^m\{\|S_1(\tilde{\xi},u)\|_{L_{1,s,\alpha}(Q_\rho,\Pi)} + \|S_2(\tilde{\xi},u)\|_{L_{1,s,\alpha}(Q_\rho,\Pi)}\}.\end{aligned}$$

Similarly to (1.34) we establish also that

(1.36) $$|\Delta(\tilde{\xi}u)|^2 \leqslant (1+\eta)|\tilde{\xi}\Delta u|^2 + C(T_1(\tilde{\xi},u) + T_2(\tilde{\xi},u)),$$

where $T_1(\tilde{\xi},u) = \left(\sum_{i=1}^m D_i\tilde{\xi}D_iu\right)^2$, $T_2(\tilde{\xi},u) = |u\Delta\tilde{\xi}|^2$.

The estimates (1.35) and (1.36) imply

(1.37)
$$\begin{aligned}\|\tilde{\xi}D'^2u\|^2_{L_{2,s,\alpha}(Q_\rho,\Pi)} &\leqslant (1+\eta)C'^2_{\alpha,m-s}\|\tilde{\xi}\Delta u\|^2_{L_{2,s,\alpha}(Q_\rho,\Pi)} \\ &+ C\Big\{\sum_{i,j=1}^m[\|S_1(\tilde{\xi},u)\|_{L_{1,s,\alpha}(Q_\rho,\Pi)} + \|S_2(\tilde{\xi},u)\|_{L_{1,s,\alpha}(Q_\rho,\Pi)}] \\ &+ \|T_1(\tilde{\xi},u)\|_{L_{1,s,\alpha}(Q_\rho,\Pi)} + \|T_2(\tilde{\xi},u)\|_{L_{1,s,\alpha}(Q_\rho,\Pi)}\Big\}.\end{aligned}$$

Let us estimate the norms of $S_{1,2}$ and $T_{1,2}$ in (1.37). Note that these quantities do not contain second-order derivatives of $u(x)$. Consider, for example, $T_1(\tilde{\xi},u)$. Since $\hat{\nabla}\tilde{\xi} \equiv 0$,

$$\|T_1(\tilde{\xi},u)\|_{L_{1,s,\alpha}(Q_\rho,\Pi)} \leqslant C\sum_{i=m-s+1}^m\int_{Q_\rho}|D_i\tilde{\xi}|^2|D_iu|^2r_s^\alpha\,dx.$$

Taking (0.8) into account, we have

$$\|T_1(\tilde{\xi},u)\|_{L_{1,s,\alpha}(Q_\rho,\Pi)} \leqslant C\sum_{i=m-s+1}^m\int_{Q_\rho}\tilde{\xi}^{2b}|D_iu|^2r_s^\alpha\,dx.$$

Integrating by parts once and recalling that $D_i r_s^\alpha \equiv 0$ for $i = m - s + 1, \dots, m$, we obtain

$$\int_{Q_p} \xi^{2b} |D_i u|^2 r_s^\alpha \, dx \leqslant C \Big\{ \int_{Q_p} \xi^{2b} |u| \cdot |\Delta u| r_s^\alpha dx + \int_{Q_p} \xi^{2b-1} |D_i \xi| \cdot |u| \cdot |D_i u| r_s^\alpha \, dx \Big\}.$$

Again use $|D_j \xi| \leqslant C \xi^b$, and apply (1.30) to these integrals. We get

$$\int_{Q_p} \xi^{2b} |D_i u|^2 r_s^\alpha \, dx \leqslant \eta \int_{Q_p} \xi^2 |\Delta u|^2 r_s^\alpha \, dx + \eta \int_{Q_p} \xi^{2b} |D_i u|^2 r_s^\alpha \, dx + C \int_{Q_p} \xi^{4b-2} |u|^2 r_s^\alpha \, dx.$$

Consequently,

$$\int_{Q_p} \xi^{2b} |D_i u|^2 r_s^\alpha dx \leqslant \eta \|\xi \Delta u\|^2_{L_{2.s.\alpha}(Q_p, \Pi)} + C \|\xi^{2b-1} u\|^2_{L_{2.s.\alpha}(Q_p, \Pi)}.$$

The inequality $|D_{ij} \xi| \leqslant C \xi^{2b-1}$ gives us the following estimate for $T_2(\xi, u)$:

$$\|T_2(\xi, u)\|_{L_{1.s.\alpha}(Q_p, \Pi)} \leqslant C \|\xi^{2b-1} u\|^2_{L_{2.s.\alpha}(Q_p, \Pi)}.$$

In a similar way we estimate the norms of $S_{1,2}(\xi, u)$. Thus (1.37) and the inequalities already obtained give

$$\|\xi D'^2 u\|^2_{L_{2.s.\alpha}(Q_p, \Pi)} \leqslant C'^2_{\alpha, m-s} (1 + \eta) \|\xi \Delta u\|^2_{L_{2.s.\alpha}(Q_p, \Pi)} + C \|\xi^{2b-1} u\|^2_{L_{2.s.\alpha}(Q_p, \Pi)}. \tag{1.38}$$

In order to conclude the proof of (1.31) it remains to estimate the last term in (1.38). Apply the Hölder inequality with exponents $1/(2b-1)$ and $1/(2(1-b))$ and the Young inequality to get

$$\begin{aligned} \|\xi^{2b-1} u\|^2_{L_{2.s.\alpha}(Q_p, \Pi)} &= \int_{Q_p} \xi^{4b-2} |u|^2 r_s^\alpha \, dx \\ &= \int_{Q_p} \{ \xi^{4b-2} |u|^{4b-2} r_s^{(\alpha-2)(2b-1)} \} \{ |u|^{4-4b} r_s^{\alpha(2-2b)+2(2b-1)} \} dx \\ &\leqslant \Big[\int_{Q_p} \xi^2 |u|^2 r_s^{\alpha-2} dx \Big]^{2b-1} \Big[\int_{Q_p} |u|^2 r_s^{\alpha+(2b-1)/(1-b)} dx \Big]^{2(1-b)} \\ &\leqslant \eta \|\xi u\|^2_{L_{2.s.\alpha-2}(Q_p, \Pi)} + C \|u\|^2_{L_{2.s.\beta_1}(Q_p, \Pi)}, \end{aligned} \tag{1.39}$$

where $\eta > 0$ is arbitrary and $\beta = \alpha + (2b-1)/(1-b) \geqslant \alpha$ for $b \in [\frac{1}{2}, 1)$. The function u is compactly supported near the lateral area of the cylinder Q_p. Therefore (1.17) implies

$$\|\xi u\|^2_{L_{2.s.\alpha-2}(Q_p, \Pi)} \leqslant C \|\xi D'^2 u\|^2_{L_{2.s.\alpha}(Q_p, \Pi)}. \tag{1.40}$$

Finally, note that $\beta_1 \to +\infty$ as $\beta \to 1 - 0$; so, it follows from (1.38)–(1.40) that

$$\|\xi D'^2 u\|^2_{L_{2.s.\alpha}(Q_p, \Pi)} \leqslant C'^2_{\alpha, m-s} (1 + \eta) \|\xi \Delta u\|^2_{L_{2.s.\alpha}(Q_p, \Pi)} + C \|u\|^2_{L_{2.s.\beta}(Q_p, \Pi)},$$

where $\beta \geqslant \alpha$ may be chosen arbitrarily under suitable choice of the cut-off function ξ. This implies (1.31) immediately.

The inequality (1.33) follows from (1.31), which is valid for $\hat{\zeta}u$, since

$$\Delta(\hat{\zeta}u) = \hat{\zeta}\Delta u + 2\Delta\nabla\hat{\zeta}\nabla u + u\Delta\hat{\zeta}, \qquad 0 \leqslant \hat{\zeta} \leqslant 1,$$

and the function r_s^α is bounded in the region where $\nabla\hat{\zeta} \not\equiv 0$.

Finally, (1.32) follows from (1.33), since

$$\|\tilde{\xi}D'^2 u\| \leqslant \|\tilde{\xi}D'^2(\hat{\zeta}u)\| + \|\tilde{\xi}D'^2((1-\hat{\zeta})u)\|,$$

and r_s^α is bounded where $\hat{\zeta} \not\equiv 1$. □

§2. Weak regularity of generalized solutions of elliptic systems

Consider the elliptic system (0.1) with the boundary condition (0.5). Since the solution is always Hölder in the case $m = 2$ [**2**], we suppose that $m \geqslant 3$. For the generalized solution $u \in W_2^{(1)}(\Omega)$ of our problem it follows from the imbedding theorems that

$$u \in L_q(\Gamma_{m-1} \cap \Omega),$$

where $q < 2(m-1)(m-2)^{-1}$ and Γ_{m-1} is an arbitrary smooth $(m-1)$-dimensional manifold. The imbedding theorems guarantee no statements about the integrability of u over manifolds of lower dimensions.

The following result was proved in [**4**].

THEOREM 2.1. *Suppose $m \geqslant 3$, conditions 1 and 2 are fulfilled, $s \in [1, m-2]$ is an integer, $\omega \Subset \Omega$, Γ_s is an s-dimensional $C^{(1)}$ manifold, and*

$$K^2\left[1 + \frac{(m-s-2)^2}{m-s-1}\right] < 1, \tag{2.1}$$

where the constant K is defined in (0.4) *and $u \in W_2^{(1)}(\Omega)$ is a generalized solution of the problem* (0.1), (0.5). *Then*

$$u \in L_1(\Gamma_s \cap \omega)$$

and for each $\varphi \in \overset{\circ}{C}{}^\infty(\omega)$ the norm $\|u\varphi\|_{L_1(\Gamma_s\cap\omega)}$ is a Hölder function of the coordinates orthogonal to Γ_s. The condition (2.1) *is sharp.*

Theorem 2.1 can be strengthened. Namely, the following result holds.

THEOREM 2.2. *Under the conditions of Theorem* 2.1

$$u \in L_2(\Gamma_s \cap \omega) \tag{2.2}$$

and $\|u\varphi\|_{L_2(\Gamma_s\cap\omega)}$ is a Hölder function of the coordinates orthogonal to Γ_s.

The proof of this theorem is the same as that of Theorem 2.1; however, one has to use Lemma 1.6 instead of Lemma 2 in [**4**]. Note that the proof of Theorems 2.1 and 2.2 is based on the study of the convergence of the iterative procedure (0.6), (0.7) in the spaces $H^{(1)}_{2,s,\alpha}(\omega)$.

Let us discuss the sharpness of Theorem 2.2. The examples of systems given in [**4**] show that the inequality (2.1) alone cannot guarantee even the boundedness of solutions on manifolds of dimension less than s. The question of the best value of the exponent of integrability of the solution on s-dimensional manifolds is not clarified by these examples: if (2.1) is not fulfilled, then the solutions of these systems become unbounded everywhere on Γ_s.

We show here that in (2.2) one may not increase the exponent of integrability at least in the case $s = m - 2$ and $m \geqslant 3$. The condition (2.1) holds automatically in this case. On the other hand, in the famous example by De Giorgi the solution has the form $x|x|^{-\beta}$. This solution belongs to $W_2^{(1)}$ for $\beta < m/2$. At the same time the function $x_m|x|^{-\beta}$ is integrable over the $(m-2)$-dimensional manifold $x_1 = x_2 = 0$ with exponent no greater than $(m-2)(\beta-1)^{-1}$. Since $(m-2)(\beta-1)^{-1} \to 2$ as $\beta \to m/2$, the exponent 2 in (2.2) cannot be replaced by a greater one (when $s = m - 2$).

Now we turn to the study of the procedure (0.6), (0.7) in the space $H^{(2)}_{2,s,\alpha}$. We suppose that the coefficients of the system (0.1) are differentiable with respect to x. For $m = 2, 3, 4$ the solution of (0.1), (0.5) is Hölder without any conditions on the dispersion of the spectrum of the matrix A. For $m = 2, 3$ this follows from the inclusion $u \in W^{(2)}_{2,\mathrm{loc}}(\Omega)$, which holds under the above conditions. For $m = 4$ by use of the "inverse Hölder inequality" it was proved in **[5]** that $u \in W^{(2)}_{2+\delta,\mathrm{loc}}(\Omega)$ for some $\delta > 0$, and the regularity of the solution follows from imbedding theorems, as in the cases $m = 2, 3$. Therefore we shall assume that $m \geqslant 5$.

THEOREM 2.3. *Suppose that* $m \geqslant 5$, $s \in [1, m-4]$ *is an integer,* $\alpha_0 \in (s - m + 3, 0]$; *suppose that conditions* 1, 2 *are fulfilled as well as condition* 3 *for* $q > q_0 = 2(m - s)(m - s + \alpha_0)^{-1}$ *and the inequality*

$$KC'_{\alpha_0,m-s}C'_{-\alpha_0,m-s} < 1 \tag{2.3}$$

holds with the constant $C'_{\alpha,m}$ *defined in Lemma* 1.10. *Then the sequence* $\{u_n\}$, *defined by the iterative procedure* (0.6), (0.7), *is bounded in the norm of* $H^{(2)}_{2,s,\alpha_0}(\Omega')$, $\Omega' \Subset \Omega$, *if the initial approximation* $u_0 \in W^{(2)}_{q,\mathrm{loc}}(\Omega) \cap W_2^{(1)}(\Omega)$ *for* $q > q_0$.

PROOF. Let ω and ω' be domains such that $\omega \Subset \omega' \Subset \Omega$. Under conditions 1–3 for $u \in W^{(2)}_{q,\mathrm{loc}}(\Omega)$ the inclusion $L(u) \in L_{q,\mathrm{loc}}(\Omega)$ holds. Since $u_0 \in W^{(2)}_{q,\mathrm{loc}}(\Omega)$ and u_n satisfies (0.6), we have $u_n \in W_q^{(2)}(\omega')$ for all n. The Hölder inequality implies that $u_n \in W^{(2)}_{2,s,\alpha}(\omega', \Pi)$ for any s-dimensional linear manifold Π and $\alpha \geqslant \alpha_0$. Fix a manifold Π and assume that the coordinate system agrees with Π. Let $x^0 \in \Pi \cap \omega$, and let $\hat{\omega}$ be the projection of ω onto $\Pi^\perp$ as before. Consider the ball B_ρ and also the cylinder Q_ρ, the domain $K_\rho = Q_\rho \cap \omega'$, and the "cut-off in the direction of Π" function $\tilde{\xi}(\tilde{x})$ introduced before. Suppose that ρ is so small that $\hat{\nabla}\tilde{\xi} \equiv 0$.

Let $v(x) \in W_2^{(2)}(\omega')$ be compactly supported near the lateral area of the cylinder Q_ρ. Multiplying (0.6) by $\tilde{\xi}^2(x)\Delta v$ and integrating over K_ρ we obtain

$$\begin{aligned}\int_{K_\rho} \tilde{\xi}^2 \Delta u_{n+1} \Delta v\, dx &= \int_{K_\rho} \tilde{\xi}^2 (\Delta u_n - \varepsilon D_i a_i(x, Du_n)) \Delta v\, dx \\ &\quad + \int_{K_\rho} \tilde{\xi}^2 (u_{n+1} - u_n + \varepsilon a_0(x, Du_n)) \Delta v\, dx.\end{aligned} \tag{2.4}$$

In the first integral on the right-hand side of (2.4) integrate twice by parts. All the boundary terms vanish due to the properties of ξ and v; thus

$$\begin{aligned}\int_{K_\rho} \xi^2 \Delta u_{n+1} \Delta v\, dx &= \int_{K_\rho} \xi^2 \Big(D_{ij} u_n^{(k)} - \varepsilon \frac{\partial a_i^{(k)}}{\partial p_t^{(l)}} D_{jt} u^{(l)} \Big) D_{ij} v^{(k)} dx + S \\ &= \sum_{j=1}^m \int_{K_\rho} \xi^2 ((I - \varepsilon A) D(D_j u_n), D(D_j v))\, dx + S,\end{aligned} \tag{2.5}$$

where A is the $(L \times L)$-matrix (0.3), I is the unit matrix, $Dz \in \mathbb{R}^L$ is the vector with components $D_i z^{(k)}$, $i = 1, \ldots, m$, $k = 1, \ldots, N$; $(\cdot, \cdot)$ denotes the scalar product in $\mathbb{R}^L$;

$$\begin{aligned} S = \int_{K_\rho} \Big\{ \xi^2 \Big[(u_{n+1} - u_n - \varepsilon a_0(x, Du_n)) \Delta v - \varepsilon \Big(\frac{\partial a_i}{\partial x_j} + \frac{\partial a_i}{\partial p_0} D_j u_n \Big) D_{ij} v \Big] \\ + 2\xi (D_i u_n - \varepsilon a_i(x, Du_n))(D_j \xi D_{ij} v - D_i \xi \Delta v) \Big\}\, dx. \end{aligned}$$

Now multiply and divide by $r_s^{\alpha/2}$, $\alpha \in [\alpha_0, 0]$, in the integrand on the right-hand side of (2.5), use the Cauchy-Bunyakovskiĭ inequality (for sums and for integrals) and take Lemma 2.5 [**2**, p. 59] into account to get

$$\begin{aligned}&\Big| \int_{K_\rho} \xi^2 \Delta u_{n+1} \Delta v\, dx \Big| \\ &\quad\leqslant K \Big(\int_{K_\rho} \xi^2 |D'^2 u_n|^2 r_s^{\alpha} dx \Big)^{1/2} \Big(\int_{K_\rho} \xi^2 |D'^2 v|^2 r_s^{-\alpha} dx \Big)^{1/2} + |S|,\end{aligned} \tag{2.6}$$

where the constant K is determined by (0.4).

Next, we estimate the sum S. It evidently satisfies the inequality

$$|S| \leqslant C \left(\|u_n\|_{W^{(1)}_{2.s.\alpha}(K_\rho, \Pi)} + \|u_{n+1}\|_{L_{2.s.\alpha}(K_\rho, \Pi)} + \|\varphi\|_{L_{2.s.\alpha}(K_\rho, \Pi)} \right) \|\xi D'^2 v\|_{L_{2.s.\alpha}(K_\rho, \Pi)}, \tag{2.7}$$

where φ is the function from condition 3. It follows easily from the Hölder inequality that

$$\|\varphi\|_{L_{2.s.\alpha}(K_\rho, \Pi)} \leqslant C\rho^{\delta} \|\varphi\|_{L_q(\omega)},$$

where $\delta = \alpha - \alpha_0 + (m - s)(1/q_0 - 1/q) > 0$.

The estimates (2.6) and (2.7) lead us to the relation

$$\begin{aligned}\Big| \int_{K_\rho} \xi^2 \Delta u_{n+1} \Delta v\, dx \Big| \leqslant \Big[K s_n(\alpha, \omega') + C \Big(\|u_n\|_{W^{(1)}_{2.s.\alpha}(K_\rho, \Pi)} \\ + \|u_{n+1}\|_{L_{2.s.\alpha}(K_\rho, \Pi)} + \rho^{\delta} \Big) \Big] \|\xi D'^2 v\|_{L_{2.s.-\alpha}(K_\rho, \Pi)},\end{aligned} \tag{2.8}$$

where $s_n(\alpha, \omega') = \|\xi D'^2 u_n\|_{L_{2.s.\alpha}(K_\rho, \Pi)}$. Let $s_n^{(1)}(\alpha, \omega') = \|D'^2 u_n\|_{L_{2.s.\alpha}(K_\rho, \Pi)}$.

Applying (1.15), (1.20), and (1.21) we obtain

$$\|u_n\|_{W^{(1)}_{2.s.\alpha}(K_\rho, \Pi)} \leqslant C \|D'^2 u_n\|_{L_{2.s+\tau}(K_\rho, \Pi)} + \|u_n\|_{W_2^{(2)}(\omega')}$$

for all $\tau \in (0,2]$. Thus, (2.8) takes the form

(2.9)

$$\left|\int_{K_\rho} \tilde{\xi}^2 \Delta u_{n+1} \Delta v\, dx\right| \leqslant \big[K s_n(\alpha,\omega') + C\big(s_n^{(1)}(\alpha+\tau,\omega') + s_{n+1}^{(1)}(\alpha+\tau,\omega') + \|u_n\|_{W_2^{(2)}(\omega')} + \|u_{n+1}\|_{W_2^{(1)}(\omega')} + 1\big)\big] \|\tilde{\xi} D'^2 v\|_{L_{2,s,-\alpha}(K_\rho,\Pi)}.$$

Set $v = \hat{\zeta} v_1$. Apply the inequality (1.32) with $\beta = 0$ to the norm $s_n(\alpha,\omega')$ and apply the inequality (1.33) to the norm $\|\tilde{\xi} D'^2(\hat{\zeta} v_1)\|_{L_{2,s,\alpha}(K_\rho,\Pi)}$ with β temporarily not fixed. The estimates thus obtained together with (2.9) give

$$\left|\int_{K_\rho} \tilde{\xi}^2 \Delta u_{n+1} \Delta v\, dx\right| \leqslant K C'_{\alpha,m-s}(1+\eta) C'_{-\alpha,m-s}(1+\eta)$$

$$\times \big[\sigma_n(\alpha,\omega') + C\big(s_n^{(1)}(\alpha+\tau,\omega') + s_{n+1}^{(1)}(\alpha+\tau,\omega') + \|u_n\|_{W_2^{(2)}(\omega')} + \|u_{n+1}\|_{W_2^{(1)}(\omega')} + 1\big)\big] \cdot \big[\|\tilde{\xi}\Delta v_1\|_{L_{2,s,-\alpha}(K_\rho,\Pi)} + C\|v_1\|_{W^{(1)}_{2,s,\beta}(K_\rho,\Pi)}\big], \tag{2.10}$$

where $\sigma_n(\alpha,\omega') = \|\tilde{\xi}\Delta u_n\|_{L_{2,s,\alpha}(K_\rho,\Pi)}$.

Now we choose the test function $v_1(x)$ in this inequality in a proper way. Let v_1 be some solution in K_ρ of the equation

$$\Delta v_1 = \chi(x) r_s^\alpha \Delta u_{n+1}, \tag{2.11}$$

where $\chi(x)$ is a smooth cut-off function near Π, i.e., $\chi(x) = \chi(\hat{x})$, $0 \leqslant \chi \leqslant 1$ and $\chi(\hat{x}) \equiv 0$ for $|\hat{x}| \leqslant d$, $\chi(\hat{x}) \equiv 1$ for $|\hat{x}| \geqslant 2d$, $2d < \rho$.

Substitute this test function v_1 into (2.10) and note that the sequence $\{u_n\}$ is bounded in $W_2^{(2)}(\omega')$ (see **[2,** Ch. 4, Sect. 3]); this gives

(2.12)

$$\int_{K_\rho} \tilde{\xi}^2 \chi r_s^\alpha |\Delta u_{n+1}|\, dx \leqslant K C'_{\alpha,m-s} C'_{-\alpha,m-s}(1+\eta)$$
$$\times \big(\sigma_n(\alpha,\omega') + C(s_n^{(1)}(\alpha+\tau,\omega') + s_{n+1}^{(1)}(\alpha+\tau,\omega') + 1)\big)$$
$$\times \big(\sigma_{n+1}(\alpha,\omega') + C\|v_1\|_{W^{(1)}_{2,s,\beta}(K_\rho,\Pi)}\big).$$

Now take as v_1 the solution of (2.11) representable with the help of the parametrix of the Laplace operator:

$$v_1(x) = -\frac{1}{(m-2)S_m}\int_{K_\rho} \frac{f(y)}{|x-y|^{m-2}}\, dy, \qquad f(x) = \chi(x) r_s^\alpha \Delta u_{n+1},$$

where S_m is the area of the unit sphere in $\mathbb{R}^m$.

Here v_1 and ∇v_1 are integrals with weak polar singularity of order $m-\varepsilon$ for any $\varepsilon \in (0,1]$. By Lemma 1.9 they are bounded as operators acting from $L_{2,s,-\alpha+\tau}(K_\rho,\Pi)$ to $L_{2,s,-\alpha+\tau+2s-2\varepsilon}(K_\rho,\Pi)$ for all $\tau \in (0,2]$ and for α and s in the intervals specified above. Therefore,

$$\begin{aligned}\|v_1\|^2_{W^{(1)}_{2,s,-\alpha+\tau+2s-2\varepsilon}(K_\rho,\Pi)} &\leqslant C\|f\|^2_{L_{2,s,-\alpha+\tau}(K_\rho,\Pi)} \\ &= C\|\chi r_s^\alpha \Delta u_{n+1}\|^2_{L_{2,s,-\alpha+\tau}(K_\rho,\Pi)} \\ &\leqslant C(s_{n+1}^{(1)}(\alpha+\tau,\omega'))^2.\end{aligned}$$

In (2.12) one can set $\beta = -\alpha + \tau + 2s - 2\varepsilon$, which gives the estimate

(2.13)
$$\int_{K_\rho} \xi^2 \chi r_s^\alpha |\Delta u_{n+1}|^2\, dx \leqslant K C'_{\alpha.m-s} C'_{-\alpha.m-s}(1+\eta)(\sigma_n(\alpha,\omega') + C(s_n^{(1)}(\alpha+\tau,\omega') + s_{n+1}^{(1)}(\alpha+\tau,\omega') + 1))(\sigma_{n+1}(\alpha,\omega') + C s_{n+1}^{(1)}(\alpha+\tau,\omega')).$$

It is clear that the constants on the right-hand side of (2.13) do not depend on the function χ. Due to absolute continuity of the integral, for each $\eta > 0$ there exists a function χ such that

$$\int_{K_\rho} (1-\chi)\xi^2 |\Delta u_{n+1}|^2 r_s^\alpha\, dx \leqslant \eta \int_{K_\rho} \xi^2 |\Delta u_{n+1}|^2 r_s^\alpha\, dx. \tag{2.14}$$

Assuming (2.14) to hold, we obtain

$$\sigma_n^2(\alpha,\omega') \leqslant K C'_{\alpha.m-s} C'_{-\alpha.m-s} \frac{1}{1-\eta}\Big(\sigma_n(\alpha,\omega') + C(s_n^{(1)}(\alpha+\tau,\omega') + s_{n+1}^{(1)}(\alpha+\tau,\omega') + 1)\Big)(\sigma_{n+1}(\alpha,\omega') + C s_{n+1}^{(1)}(\alpha+\tau,\omega')).$$

Finally, applying the "Cauchy inequality with η" (1.30), we obtain

(2.15)
$$\sigma_n^2(\alpha,\omega') \leqslant K C'_{\alpha.m-s} C'_{-\alpha.m-s}(1+\eta)\sigma_n(\alpha,\omega')\sigma_{n+1}(\alpha,\omega') + \eta_1 \sigma_{n+1}^2(\alpha,\omega') + C\Big((s_n^{(1)}(\alpha+\tau,\omega'))^2 + (s_{n+1}^{(1)}(\alpha+\tau,\omega'))^2 + 1\Big).$$

Estimate (2.15) is established for all $\alpha \in [\alpha_0, 0]$. Note that the product $C'_{\alpha.m-s} C'_{-\alpha.m-s}$ decreases as a function of α for $\alpha \in (-m+s+3, 0]$. Therefore, under the conditions of the theorem, the inequality

$$K C'_{\alpha.m-s} C'_{-\alpha.m-s}(1+\eta) < 1 \tag{2.16}$$

holds for all $\alpha \in [\alpha_0, 0]$ if η is sufficiently small.

Given α_0, choose a finite sequence $\{\alpha_k\}$, $k = 1, \dots, k_0$, such that $\alpha_1 = 0$, $0 < \alpha_k - \alpha_{k+1} \leqslant 2$, $\alpha_{k_0} = \alpha_0$, and a sequence of domains $\Omega' = \Omega_k \Subset \cdots \Subset \Omega_1 \Subset \omega' \Subset \Omega$. Substituting $\alpha = \alpha_1$, $\tau = -\alpha_1$, $\omega = \Omega_1$ in (2.15) and taking the boundedness of the sequence $\{u_n\}$ in $W_2^{(2)}(\omega')$ into account we obtain, according to Lemma 1.7 **[2]** and inequalities (2.15), (2.16), the boundedness of the sequence $\sigma_n(\alpha_1, \omega')$. Lemma 1.11 gives the boundedness of the sequence $s_n^{(1)}(\alpha_1, \Omega_1)$. Next, by setting $\alpha = \alpha_2$, $\tau = \alpha_1 - \alpha_2$, $\omega' = \Omega_1$, $\omega = \Omega_2$, in a similar way we establish the boundedness of the sequences $\sigma_n(\alpha_2, \Omega_1)$ and $s_n^{(1)}(\alpha_2, \Omega_2)$. It is clear that after a finite number of such steps we obtain the boundedness of the sequence $s_n^{(1)}(\alpha_0, \Omega')$ and then the inequality (1.16) (with $p = 2$ and $\beta = 0$) implies the boundedness of the sequence $\|u_n\|_{W^{(2)}_{2.s.\alpha_0}(Q_\rho.\Pi)}$. Since the manifold Π is arbitrary, the theorem is proved. □

Theorem 2.3 produces the following result on the weak regularity of the solution of the problem (0.1), (0.5).

THEOREM 2.4. *Suppose* $u(x)$ *is the solution of* (0.1), (0.5), *the conditions* 1, 2, *and* 3 *hold for* $q > q_0 = (m-s)/2$, $m \geqslant 5$, s *is an integer in* $[1, \dots, m-4]$, $\omega \Subset \Omega$, Γ_s *is an* s*-dimensional* $C^{(1)}$ *manifold, and*

$$K^2\left(1 + \tfrac{1}{4}(m-s-4)(m-s-1)\right) < 1, \tag{2.17}$$

where K *is defined by the relations* (0.4). *Then* $u \in L_2(\Gamma_s \cap \omega)$ *and the norm* $\|u\varphi\|_{L_2(\Gamma_s \cap \omega)}$ *for each* $\varphi \in \overset{\circ}{C}{}^{\infty}(\omega)$ *is a Hölder function of the coordinates orthogonal to* Γ_s.

PROOF. Let Γ_s be a linear manifold. It is not hard to see that for $\alpha = s - m + 4 - 2\gamma$,

$$(C'_{\alpha,m-s} C'_{-\alpha,m-s})^2 = 1 + \tfrac{1}{4}(m-s-4)(m-s-1) + 2\eta,$$

where $\eta \to 0$ as $\gamma \to 0$. Therefore, under the condition (2.17), the sequence of iterations $\{u_n\}$ is bounded in the norm of $H^{(2)}_{2,s,\alpha}(\omega)$ with $\alpha = s - m + 4 - 2\gamma$. Thus, Lemma 1.7 implies that the sequence of norms $\|u_n\|_{L_2(\Gamma_s \cap \omega)}$ is bounded in $C^{(\gamma_1)}(\hat{\omega})$ for some $\gamma_1 > 0$. Due to the compactness of imbedding of Hölder spaces, some subsequence $\{u_{n_k}\}$ (we suppose that it is the sequence $\{u_n\}$ itself) converges in $C^{(\gamma)}(\hat{\omega})$ for all γ, $0 < \gamma < \gamma_1$. Denote by $u^*(x)$ the corresponding limit. Then

$$\|u^*\|_{L_2(\Gamma_s \cap \omega)} \in C^{(\gamma)}(\hat{\omega}).$$

Since the solution of (0.1), (0.5) is unique, $u = u^*$ and our theorem is proved for the case of a linear manifold Γ_s. The proof for arbitrary $\Gamma_s \in C^{(1)}$ could, as usual, be obtained by the well-known method of the local straightening of Γ_s. □

REMARK. 1) For $m \geqslant 5$ and $s = m - 4$ the condition (2.17) is satisfied automatically. Thus, the solution is square-integrable over an $(m-4)$-dimensional manifold only under the natural conditions 1–3.

2) The condition (2.17) is weaker than (2.1) only for $m - s = 4, 5, 6$. For such m and s, Theorem 2.4 strengthens Theorem 2.2 (for the case of differentiable coefficients). If $m - s \geqslant 7$, Theorem 2.4, in general, follows from Theorem 2.2. However, the method of proving Theorem 2.4 could be (in contrast to the proof of Theorem 2.2) extended to elliptic systems of arbitrary order.

References

1. H. O. Cordes, *Über die erste Randwertaufgabe bei quasilinearen Differentialgleichungen zweiter Ordnung in mehr als zwei Variablen*, Math. Ann. **131** (1956), 278–312.
2. A. I. Koshelev, *Regularity of solutions of elliptic equations*, "Nauka", Moscow, 1986. (Russian)
3. A. I. Koshelev and S. I. Chelkak, *Regularity of solutions of quasilinear elliptic systems*, Teubner-Texte Math., Vol. 77, Teubner-Verlag, Leipzig, 1985.
4. S. I. Chelkak, *Weak regularity of solutions of quasilinear elliptic systems of second order*, Izv. Vyssh. Uchebn. Zaved. Mat. **1989**, no. 11, 74–84; English transl. in Soviet Math. (Iz. VUZ) **33** (1989).
5. M. Giaquinta, *Multiple integrals in the calculus of variations and nonlinear elliptic systems*, Ann. of Math. Stud., Vol. 105, Princeton Univ. Press, Princeton, NJ, 1983.
6. L. Nirenberg, *On nonlinear elliptic partial differential equations and Hölder continuity*, Comm. Pure Appl. Math. **6** (1953), 103–156.
7. S. I. Chelkak, *Regularity of solutions of quasilinear elliptic systems of higher order*, Izv. Vyssh. Uchebn. Zaved. Mat. **1984**, no. 1, 33–41; English transl. in Soviet Math. (Iz. VUZ) **28** (1984).

8. V. G. Mazja and B. A. Plamenevskiĭ, *Weighted spaces with inhomogeneous norms and boundary value problems in domains with conical points*, Elliptische Differentialgleichungen, Vol. 10, Silhelm Pieck Univ., Rostock, 1977, pp. 10–16; 1978, pp. 161–190.
9. G. H. Hardy, G. E. Littlewood, and G. Polya, *Inequalities*, Cambridge Univ. Press, Cambridge, 1941.
10. V. G. Mazja, *Sobolev spaces*, Izdat. Leningrad. Univ., Leningrad, 1985; English transl., Springer-Verlag, Berlin and New York, 1985.
11. E. M. Stein, *Singular integrals and differentiability properties of functions*, Princeton Univ. Press, Princeton, NJ, 1970.
12. V. P. Glushko, *Some properties of the operators of potential-type and their applications*, Izv. Vyssh. Uchebn. Zaved. Mat. **1961**, no. 3, 3–13. (Russian)

St. Petersburg State Electrotechnical University, 5, ul. prof. Popova, St. Petersburg, 197022, Russia

Translated by G. ROZENBLUM

Amer. Math. Soc. Transl.
(2) Vol. **159**, 1994

Regularity of Statistical Solutions of the Transport Equations

N. B. Maslova

§1. Introduction

The transport equation

$$f_t + v \cdot f_x + f = g, \qquad t \in \mathbb{R}^1, \quad x \in \mathbb{R}^d, \quad v \in \mathbb{R}^d, \tag{1.1}$$

with a random velocity field v and random perturbation g plays a significant role in a number of problems of statistical physics. In the simplest cases v is a random variable independent of t, x with distribution given by some probability measure $\mathbf{P}$ on $\mathbb{R}^d \equiv \mathbb{R}^d_v$. The regularity of the velocity mean values $\mathbf{E}f = \int_{\mathbb{R}^d} f\,\mathbf{P}(dv)$ was investigated in **[1–3]**. The basic result in **[1]** is the following. Let f be a solution of the problem

$$v \cdot f_x + f = g, \qquad x \in \mathbb{R}^d, \quad v \in \mathbb{R}^d. \tag{1.2}$$

Suppose that there exists a positive constant C such that

$$\operatorname*{ess\,sup}_{e \in S^{d-1}} \mathbf{P}\left\{v \in \mathbb{R}^d \mid |v \cdot e| \leqslant \varepsilon\right\} \leqslant C\varepsilon, \qquad \varepsilon \in [0, 1].$$

Then the operator $g \to \mathbf{E}f$ from $L_p(L_p(\mathbb{R}^d), \mathbf{P})$ to $W_p^s(\mathbb{R}^d)$ is continuous if

$$\begin{aligned} s &< \inf(p^{-1}, 1 - p^{-1}), && 1 < p < \infty, \\ s &\leqslant 1/2, && p = 2. \end{aligned} \tag{1.3}$$

The condition (1.3) was derived from the case $p = 2$ with the help of interpolation. Counterexamples in **[1]** show the result to be sharp.

According to the basic results of the present paper, Mikhlin's theorem on Fourier multipliers allows one to prove that the exponent s in (1.3) may be far from optimal for measures with smooth density. In particular, if the measure $\mathbf{P}$ is compactly supported and has density in L_q, then

$$\mathbf{E}f \in W_p^{1-1/q}, \qquad 1 < p < \infty, \quad 1 \leqslant q < \infty. \tag{1.4}$$

An analogous result is valid for the nonstationary problem. For isotropic distribution $\mathbf{P}$ the relation (1.4) is fulfilled in the stationary problem without suppositions of density smoothness.

1991 *Mathematics Subject Classification*. Primary 35M99, 35Q72, 82B31; Secondary 35Q35, 82B40.

The corresponding assertions are more precisely formulated and proved in §§2 and 3. In §4 other manifestations of the smoothing properties of the mean velocities are discussed, such as the compactness of integral operators related to the kinetic equations

(1.5) $$f_t + v \cdot f_x + \nu f = Kf + g,$$

where K is an integral operator in the space of velocities $\nu = \nu(v)$.

§2. Stationary problem

We consider (1.2) under the condition $g = g(x) \in L_p(\mathbb{R}^d)$. This problem has a unique solution in $L_p(L_p(\mathbb{R}^d), \mathbf{P})$, which can be represented by the formula

$$f(x, v) = \int_0^\infty e^{-\tau} g(x - v\tau)\, d\tau \equiv (Ug)(x, v).$$

Let $\hat{f} = Ff$ be the Fourier transform of f with respect to x. Denoting by k the dual variable to x, we obtain

$$\hat{f} = \hat{U}\hat{g}, \qquad \hat{U} = (1 + ik \cdot v)^{-1}, \qquad \mathbf{E}\hat{f} = \mathbf{E}\hat{U}\hat{g}.$$

THEOREM 1. *Suppose that* $\mathbf{P}$ *has density in* W_q^L *and*

$$\int_{\mathbb{R}^d} |v|^{l+L} \mathbf{P}(dv) < \infty.$$

Then the operator $\mathbf{E}U$ *is continuous from* $L_p(\mathbb{R}^d)$ *to* $W_p^s(\mathbb{R}^d)$ *for*

$$[L] > d/2, \qquad s \leqslant 1 - \frac{1}{q} - r, \qquad 1 < p < \infty,$$
$$r = m(m + l)^{-1}(L + 1/q'), \qquad m = L + (d - 1)/q'.$$

PROOF. Set $\rho(k) = (1 + |k|^2)^{s/2} \mathbf{E}\hat{U}$. It suffices to check the inclusion $\rho \in M_p$, where M_p is the Fourier multiplier space in L_p [4]. By Mikhlin's theorem one has $\rho \in M_p$ if

(2.1) $$\sup_k |k|^{|\alpha|} |D^\alpha \rho(k)| < \infty, \qquad |\alpha| \leqslant L,$$

where $\alpha = (\alpha_1, \dots, \alpha_d)$ is a multi-index, $|\alpha| = \alpha_1 + \dots + \alpha_d$. The relation (2.1) is fulfilled if $J = \mathbf{E}\hat{U}$ satisfies

(2.2) $$\sup_k |k|^{|\alpha| + s} |D^\alpha J| < \infty, \qquad |\alpha| \leqslant L.$$

The last formula is verified below. Let us fix a positive R and consider a function ζ smoothly depending on $|v|$ such that

(2.3) $$\zeta \equiv 1, \quad |v| \in [0, R]; \qquad 0 \leqslant \zeta \leqslant 1, \qquad \operatorname{supp} \zeta = [0, 2R].$$

Set

(2.4) $$J = J_1 + J_2, \qquad J_1 = \mathbf{E}\hat{U}\zeta, \quad J_2 = \mathbf{E}\hat{U}(1 - \zeta).$$

We show that for a proper R each function J_1, J_2 satisfies (2.2). Introduce in the velocity space $\mathbb{R}_v^d$ a coordinate system with basis vectors

$$e_1 = |k|^{-1} k, e_2, \dots, e_d, \qquad (e_j, e_l) = \delta_{jl},$$

and set $w = Sv$, where

$$v_1 = |k|(v, e_1), \qquad w_j = (v, e_j), \qquad j \geqslant 2.$$

Denoting by $P(v)$ the density of $\mathbf{P}$, we obtain

$$J_1 = |k|^{-1} \int_{\mathbb{R}^d} [1 + iw_1]^{-1} P(S^{-1}w)\zeta(|S^{-1}w|)\, dw. \tag{2.5}$$

Applying the Hölder inequality to (2.5) one has

$$\begin{aligned} |J_1| &\leqslant |k|^{-1} \left(\int_{|v| \leqslant 2R} |1 + iw_1|^{-q'} dw \right)^{1/q'} \left(\int_{\mathbb{R}^d} |P(S^{-1}w)|^{-q} dw \right)^{1/q} \\ &\leqslant C R^{(d-1)/q'} |k|^{-1/q'} \qquad (1/q + 1/q' = 1). \end{aligned} \tag{2.6}$$

(From now on C denotes positive constants independent of the parameters involved.) Differentiation of (2.5) gives

$$\begin{aligned} \frac{\partial J_1}{\partial k_j} =& |k|^{-1} \int_{\mathbb{R}^d} (1 + iw_1)^{-1} \sum_l \frac{\partial v_l}{\partial k_j} \left[\frac{\partial}{\partial v_l} \left[P(S^{-1}w)\zeta(|S^{-1}w|) \right] \right] dw \\ &+ |k| \left(\frac{\partial}{\partial k_j} |k|^{-1} \right) J_1. \end{aligned} \tag{2.7}$$

Since $S^{-1}w$ linearly depends on the components of $|k|^{-1}k$, we have

$$\left| \frac{\partial v_l}{\partial k_j} \right| \leqslant C|k|^{-1} \sum_{j \geqslant 1} |(v, e_j)|.$$

The last inequality and (2.6), (2.7) lead to

$$\left| \frac{\partial J_1}{\partial k_j} \right| \leqslant C|k|^{-1-1/q'} R^{1+(d-1)/q'} \|P\|_{W_q^1}.$$

Estimating higher derivatives in an analogous way, we obtain

$$|D^\alpha J_1| \leqslant C|k|^{-|\alpha|-1/q'} \|P\|_{W_q^L} R^m \tag{2.8}$$

with $m = L + (d-1)/q'$. On the other hand, by virtue of (2.3), (2.4)

$$|D^\alpha J_2| \leqslant C \int_{|v| \geqslant R} |v|^{|\alpha|} P(v)\, dv \leqslant C_1 R^{-l}. \tag{2.9}$$

Since $D^\alpha J$ are bounded, it suffices to prove (2.2) for $|k| > 1$. Using (2.8), (2.9) one has

$$|k|^{|\alpha|+s} |D^\alpha J| \leqslant C \left[|k|^{s-1/q'} R^m + |k|^{L+s} R^{-l} \right]. \tag{2.10}$$

Choosing R so that

$$|k|^{s-1/q'} R^m = |k|^{L+s} R^{-l},$$

we obtain $R = |k|^{l_1}$, $l_1 = (L + 1/q')(m + l)^{-1}$. For such R the inequality (2.2) holds in view of (2.10). □

Now suppose that $\mathbf{P}$ has density depending only on $|v|$. Then taking into account (2.5) and passing to the variables $u_j = (v, e_j)$, we obtain

$$
\begin{aligned}
J_1 &= \int_{\mathbb{R}^d} [1 + i|k|u_1]^{-1} P(|u|)\zeta(|u|)\, du, \\
|D^\alpha J_1| &\leqslant C \int_{\mathbb{R}^d} |u_1|^{|\alpha|} \left|1 + i|k|u_1\right|^{-|\alpha|-1} P(|u|)\zeta(|u|)\, du \\
&\leqslant C|k|^{-|\alpha|} \int_{\mathbb{R}^d} \left|1 + i|k|u_1\right|^{-1} P(|u|)\zeta(|u|)\, du \qquad (2.11)\\
&\leqslant C|k|^{-|\alpha|} \left(\int_{|u|<2R} \left|1 + i|k|u_1\right|^{-q'} \right)^{1/q'} du \\
&\leqslant C|k|^{-|\alpha|-1/q'} R^{(d-1)/q'}.
\end{aligned}
$$

The relations (2.11) and (2.9) lead to the following.

THEOREM 2. *If the measure* $\mathbf{P}$ *has isotropic density in* L_q $(1 < q < \infty)$ *and* $(L + l)$*th-order moment* $([L] > d/2)$, *then the operator* $\mathbf{E}U$ *is continuous from* L_p *to* W_p^s *for all* p, s *subject to*

$$
s \leqslant 1 - \frac{1}{q} - r, \qquad 1 < p < \infty,
$$
$$
r = m(m + l)^{-1}(L + 1/q'), \qquad m = (d - 1)/q'.
$$

In the one-dimensional case the estimates analogous to (2.11) give

$$
|D^\alpha J| \leqslant C|k|^{-|\alpha|-1/q'} \|P\|_{L^q}.
$$

Hence, the following assertion holds.

THEOREM 3. *For* $d = 1$ *the operator* $\mathbf{E}U$ *is continuous from* L_p *to* $W_p^{1/q'}$ *if the measure* $\mathbf{P}$ *has density on* L_q, $1 < q < \infty$.

§3. The nonstationary problem

Now we turn to the nonstationary problem (1.1) and suppose that $g = g(t, x) \in L_p(\mathbb{R}^{d+1})$, $p \geqslant 1$. The problem has a unique solution in $L_1(L_p(\mathbb{R}^{d+1}), \mathbf{P})$ given by

$$
f(t, x, v) = \int_0^\infty e^{-\tau} g(t - \tau, x - v\tau) \equiv Vg(t, x, v).
$$

Denoting by $\hat{f} = Ff$ the Fourier transform of f with respect to t, x, we obtain

$$
\hat{f} = \hat{V}\hat{g}, \qquad \hat{V} = [1 + i(\tau + k \cdot v)]^{-1}, \qquad \mathbf{E}\hat{f} = E\hat{V}\hat{g},
$$

where $\tau \in \mathbb{R}^1$, $k \in \mathbb{R}^d$ are the variables dual to t, x.

THEOREM 4. *Under the conditions of Theorem* 1 *with* $[L] > (d+1)/2$ *the operator* $\mathbf{E}V$ *is continuous from* $L_p(\mathbb{R}^{d+1})$ *to* $W_p^s(\mathbb{R}^{d+1})$ *for*

$$s \leqslant 1 - \frac{1}{q} - r, \tag{3.1}$$
$$r = m(l+m)^{-1}(L + 1/q'), \qquad m = 2L + d/q'.$$

PROOF. As in the stationary problem the proof may be reduced to verifying the relation $\rho \in M_p$, where $\rho = [1 + (\tau^2 + |k|^2)^{s/2}]\mathbf{E}\hat{V}$, M_p being the Fourier multiplier space in $L_p(\mathbb{R}^{d+1})$. In view of Mikhlin's theorem it suffices to prove that

$$\sup_{\tau, k}(|\tau| + |k|)^{|\alpha|+s}|D^\alpha J| < \infty,$$

where $J = \mathbf{E}\hat{V} = \int_{\mathbb{R}^d}[1 + i(\tau + k \cdot v)]^{-1}\mathbf{P}(dv)$.

Set again

$$J_1 = \mathbf{E}\hat{V}\zeta, \qquad J_2 = \mathbf{E}\hat{V}(1 - \zeta),$$

where ζ is a cut-off function satisfying (2.4). Passing to the variables

$$w_1 = \tau + |k|(v, e_1), \quad w_j = (v, e_j), \quad j \geqslant 2, \tag{3.2}$$

in the integral that defines J_1 we obtain (2.8) for J_1, S being the transformation (3.2). Arguing as in §2 we derive

$$|D_k^\alpha J_1| \leqslant C|k|^{-|\alpha|-1/q'} R^{|\alpha|+(d-1)/q'}.$$

If $|\tau| \leqslant 4R|k|$, then from this estimate it follows that

$$|D_k^\alpha J_1| \leqslant C|\tau|^{-|\alpha|-1/q'} R^{2|\alpha|+d/q'}.$$

In the case $|\tau| > 4R|k|$ we use the equality

$$D_k^\alpha J_1 = \int_{\mathbb{R}^d}\{D_k^\alpha[1 + i(\tau + k \cdot v)]^{-1}\}P(v)\zeta(|v|)\, dv,$$

which implies

$$|D_k^\alpha J_1| \leqslant C|\tau|^{-|\alpha|-1}\int_{\mathbb{R}^d}|v|^{|\alpha|}\mathbf{P}(dv),$$

since $|\tau + k \cdot v| > |\tau| - 2R|k| > |\tau|/2$ for $|v| \leqslant 2R$, $|\tau| > 4R|k|$.

Let us estimate the derivatives with respect to τ. Taking into account (2.5) we obtain

$$D_\tau J_1 = |k|^{-1}\int_{\mathbb{R}^d}[1 + iw_1]^{-1}\sum_j \frac{\partial v_j}{\partial \tau}\frac{\partial}{\partial v_j}P\zeta\, dw. \tag{3.3}$$

By virtue of (3.2) one has $\left|\frac{\partial v_j}{\partial \tau}\right| \leqslant |k|^{-1}$. Therefore, from (3.3) it follows that

$$\begin{aligned} |D_\tau J_1| &\leqslant C|k|^{-2}|k|^{1/q}\left(\int_{|v|\leqslant 2R}|1 + iw_1|^{-q'}\right)^{1/q'} \\ &\quad \times \sum_j\left(\int_{\mathbb{R}^d}\left|\frac{\partial}{\partial v_j}P(v)\zeta(|v|)\right|^q\right)^{1/q} \\ &\leqslant C|k|^{-1-1/q'}\|P\|_{W_q^1}R^{(d-1)/q'}. \end{aligned}$$

Higher-order derivatives with respect to τ and mixed ones (with respect to k and τ) can be estimated in a similar way. In this fashion one proves that

$$|D^\alpha J_1| \leqslant C|k|^{-|\alpha|-1/q'} R^{|\alpha|+(d-1)/q'} \|P\|_{W_q^{|\alpha|}}. \tag{3.4}$$

If $|\tau| < 4R|k|$, then (3.4) implies

$$|D^\alpha J_1| \leqslant C|\tau|^{-|\alpha|-1/q'} R^{2|\alpha|+d/q'} \|P\|_{W_q^{|\alpha|}}.$$

In the case $|\tau| > 4R|k|$ we use the relations

$$D^\alpha J_1 = \int_{\mathbb{R}^d} \{D^\alpha[1 + i(\tau + k \cdot v)]^{-1}\} P(v)\zeta(|v|)\, dv,$$

$$|\tau + k \cdot v| > |\tau|/2,$$

and obtain

$$|D^\alpha J_1| \leqslant C|\tau|^{-|\alpha|-1}.$$

Thus the estimate

$$\left[|\tau|^{|\alpha|+s} + |k|^{|\alpha|+s}\right] |D^\alpha J_1| \leqslant C R^m (|\tau| + |k|)^{s-1/q'} \tag{3.5}$$

holds for $m = 2|\alpha| + d/q'$, $|\tau| > 1$, $|k| > 1$. Setting $R = (|\tau| + |k|)^{l_1}$, $l_1 = (L + 1/q')(m + l)^{-1}$ we derive (3.1) by means of (3.5), (2.9), and (2.4). □

§4. Compactness theorems

1. Stationary problem. Let U be the solving operator of the problem (1.2). Then

$$\mathbf{E}Ug = \int_0^\infty e^{-\tau} \left(\int_{\mathbb{R}^d} g(x - v\tau)\mathbf{P}(dv) \right) d\tau.$$

The change of variables $v, \tau \to |v|, y = x - v\tau$ for $d > 1$ leads to

$$\mathbf{E}Ug = K * g, \tag{4.1}$$

with $K(x) = |x|^{-d+1} \int_{\mathbb{R}^d} \exp\{-|x| \cdot |v|^{-1}\} P(|v| \cdot |x|^{-1}x)|v|^{d-2} d|v|$.

If $P \in L_\infty$ and is compactly supported, then $K(x) \leqslant C|x|^{-d+1}$; hence **[5,** Theorem 8.5] $\mathbf{E}U$ is a compact operator from $L_{p,\mathrm{loc}}$ to $L_{p_1,\mathrm{loc}}$ for $1 < p < d$ and $p_1 < dp(d-p)^{-1}$. Besides this, if P is continuous then $\mathbf{E}U$ is compact from $L_{1,\mathrm{loc}}$ to $L_{d(d-1)^{-1},\mathrm{loc}}$ and from L_d to L_{p_1}.

Moreover, from the results of §2 and (4.1) one obtains the following assertion.

THEOREM 5. *If the density of the measure* $\mathbf{P}$ *has a majorant* P_0 *subject to the conditions of Theorems* 1, 2, *or* 3, *then* $\mathbf{E}U$ *is compact from* $L_{p,\mathrm{loc}}$ *to* $L_{p_1,\mathrm{loc}}$ *for* $p > 1$, $p_1 < pd(d - ps)^{-1}$.

In particular, if $\mathbf{P}$ has a compact support and density in W_q^L, $[L] > d/2$, then the operator $\mathbf{E}U$ is compact from $L_{p,\mathrm{loc}}$ to $L_{p_1,\mathrm{loc}}$ for $p_1 < pd(d - p(1 - 1/q))^{-1}$.

One of the applications of the above theorems is connected with analyzing stationary boundary value problems for equation (1.5) in a bounded domain Ω with smooth boundary.

Let v be a continuous function subject to $\inf v > 0$, K a compact symmetric operator in $L_2(\mathbb{R}^d, \mathbf{P})$, $\mathbf{P}$ being the measure with Gaussian density

$$\mathbf{P}(dv) = (2\pi)^{-d/2} \exp\{-|v|^2/2\}\, dv.$$

Define the operator

$$(Ug)(x, v) = \int_0^\infty \exp\{-v\tau\}\chi(x - v\tau)g(x - v\tau, v)\, d\tau\, \chi(x),$$

where χ is the indicator of Ω.

In the proof of solvability of boundary value problems for the equation (1.5) the following theorem is crucial.

THEOREM 6. *The operator UK is compact in $L_2(L_2(\Omega), \mathbf{P})$.*

PROOF. Since K is compact in $L_2(\mathbb{R}^d, \mathbf{P})$, it suffices to prove the compactness of the operator UK, where K is the operator with real kernel

$$\tilde{K}(v, v_1) = \sum_{i=1}^N a_i(v)b_i(v_1),$$

a_i, b_i being smooth compactly supported functions. Theorem 5 provides the compactness of $\tilde{K}U^*$ and, hence, that of $U\tilde{K}$. □

Besides, from Theorem 5 it follows that KUK is a compact operator from $L_p(\Omega)$ to $L_{p_1}(\Omega)$ for $p < d$ and $p_1 < pd(d-p)^{-1}$. This operator property is important for the analysis of nonlinear problems.

2. Nonstationary problems. Let V be the solving operator of the problem (1.1) for $g \in L_p(\mathbb{R}^{d+1})$. Then

$$\mathbf{E}Vg = \int_0^\infty e^{-\tau}g(t - \tau, x - v\tau)\, d\tau.$$

The change of variables $s = t - \tau$, $y = x - v\tau$ gives

$$\mathbf{E}V = K * g, \tag{4.2}$$

with

$$K(\tau, y) = \tau^{-d}P(y\tau^{-1})e^{-\tau}\chi, \tag{4.3}$$

χ being the indicator of $(0, \infty)$.

Applying the theorem on operators with compact majorants, we obtain the following assertion.

THEOREM 7. *Let $\mathbf{P} \leqslant P_0$, where P_0 is a density satisfying the conditions of Theorem 4. Then $\mathbf{E}V$ is a compact operator from $L_p(\mathbb{R}^{d+1})$ to $L_{p_1,\mathrm{loc}}(\mathbb{R}^{d+1})$ for $1 < p < (d+1)s^{-1}$, and $\mathbf{E}V$ is a bounded operator from L_p to $L_{\infty,\mathrm{loc}}$ in the case of $p \geqslant (d+1)s^{-1}$.*

The Kantorovich criterion [**5**, Theorem 7.1] allows one to get the compactness conditions under weaker restrictions on the measure $\mathbf{P}$.

THEOREM 8. *Let Ω be a bounded domain in $\mathbb{R}^d$, χ the indicator of $\Omega \times [0, T]$. If $P \in L_q$, $1 < q < (d+1)d^{-1}$, then the operator $\mathbf{E}\chi V\chi$ is compact from $L_p(\mathbb{R}^{d+1})$ to $L_{p_1}(\mathbb{R}^{d+1})$ for $1 < p < q'$, $p_1 < pq'(q'-p)^{-1}$.*

Indeed, the function K defined by (4.3) satisfies

$$\|K\|_{L_q}^q \leqslant \|P\|_{L_q}^q \int_0^\infty e^{-q\tau}\tau^{-qd+d}\,d\tau.$$

Hence,

$$K \in L_q \quad \text{for} \quad q < (d+1)d^{-1}. \tag{4.4}$$

Theorem 8 is an evident consequence of this condition and the Kantorovich criterion.

On the other hand, by virtue of the Hölder inequality and (4.2), (4.4)

$$\|\mathbf{E}Vg\|_{L_\infty} \leqslant \|K\|_{L_q}\|g\|_{L_{q'}}.$$

By virtue of the Minkowski and Yang inequalities

$$\|\mathbf{E}Vg\|_{L_{p_1}} \leqslant \|K\|_{L_q}\|g\|_{L_p}, \qquad 1 \leqslant p < q',\ 1/p_1 = 1/p - 1/q'.$$

Thus, we obtain the following result.

THEOREM 9.

(i) *For* $1 \leqslant p < q'$, $1/p_1 = 1/p - 1/q'$ *the operator* $\mathbf{E}V$ *is continuous and weakly compact from* $L_p(\mathbb{R}^{d+1})$ *to* $L_{p_1,\mathrm{loc}}(\mathbb{R}^{d+1})$ *if*

$$P \in L_q(\mathbb{R}^d), \qquad q > 1.$$

(ii) *The operator* $\mathbf{E}V$ *maps every weakly compact set in* $L_1(\mathbb{R}^{d+1})$ *to a compact set in* $L_{1/q,\mathrm{loc}}$ $(1 \leqslant q < \infty)$.

One derives (ii) in an obvious way from (i) and the regularity of $\mathbf{E}V$ **[5,** Theorem 5.5].

The above results are related to perturbations g independent of a random parameter. Passing to the general case we consider

$$g \in L_j(L_p(\mathbb{R}^{d+1}), \mathbf{P}), \qquad 1 \leqslant p, j \leqslant \infty.$$

For $p = j = 2$ we have

$$|\mathbf{E}\hat{V}\hat{g}| \leqslant \left(\mathbf{E}|\hat{V}|^2\right)^{1/2}\left(\mathbf{E}|\hat{g}|^2\right)^{1/2}.$$

Applying to $\mathbf{E}|\hat{V}|^2$ the estimates from the proof of Theorem 4, we obtain that $\mathbf{E}V$ is continuous from $L_2(L_2(\mathbb{R}^{d+1}), \mathbf{P})$ to $W_2^{s/2}$ in the case

$$2s < 1/q' - r, \qquad r = m(l+m)^{-1}/q', \qquad m = d/q'.$$

Interpolation theorems imply that $\mathbf{E}V$ is continuous from $L_p(L_p(\mathbb{R}^{d+1}), \mathbf{P})$ to W_q^m for m satisfying $m < \inf(p^{-1}, 1 - p^{-1})s/2$.

One has also another estimate for the limit exponent m:

$$m < \inf(p^{-1}, 1 - p^{-1}, l/(l+1))$$

if $\sup_k |k|^{-1}\int_{\mathbb{R}^d}(1 + |w_1|)^l P(S^{-1}w)\,dw < \infty$, where S is the transformation (3.2).

From Theorem 9 it follows that $\mathbf{E}V$ is continuous from $L_\infty(L_1(\mathbb{R}^{d+1}), \mathbf{P})$ to L_q. The weak compactness of $\mathbf{E}V$ from $L_j(L_1(\mathbb{R}^{d+1}), \mathbf{P})$ to $L_{p_1,\mathrm{loc}}(\mathbb{R}^{d+1})$, $1 \leqslant j \leqslant \infty$, $p_1 < qj(q-1+j)^{-1}$, $1 < q < (d+1)d^{-1}$, is provided by the interpolation theorems. In the case $q = 1$ the operator $\mathbf{E}V$ sends every weakly compact sequence in $L_1(L_1(\mathbb{R}^{d+1}), \mathbf{P})$ to a compact sequence in $L_{1,\mathrm{loc}}(\mathbb{R}^{d+1})$ **[1]**.

We now state some simple consequences of Theorems 8, 9.

COROLLARY 1. *Under the conditions of Theorem* 8 *the operator* $\mathbf{E}\chi V\chi\mathbf{E}$ *is compact from* $L_p(L_r(\mathbb{R}^{d+1}), \mathbf{P})$ *to* $L_{r_1}(L_{p_1,\mathrm{loc}}(\mathbb{R}^{d+1}), \mathbf{P})$, $1 < r,\ r_1 < \infty$.

COROLLARY 2. *Under the conditions of Theorem* 9 *the operator* $\mathbf{E}V\mathbf{E}$ *is continuous from* $L_p(L_r(\mathbb{R}^d, \mathbf{P}), \mathbb{R}^{d+1})$ *to* $L_{r_1}(L_{p_1,\mathrm{loc}}(\mathbb{R}^{d+1}), \mathbf{P})$.

By virtue of (1.1) the relation

$$\|V\mathbf{E}g\|_2^2 = (\mathbf{E}g, V\mathbf{E}g)_2 = (g, \mathbf{E}V\mathbf{E}g)_2 \tag{4.5}$$

holds, where $\|\cdot\|_2$ and $(\cdot,\cdot)_2$ are the norm and inner product in $L_2(L_2(\mathbb{R}^{d+1}), \mathbf{P})$.

Let g_n be an arbitrary sequence weakly convergent to zero. In view of (4.5) and Corollary 1 one has $\|\chi V\chi\mathbf{E}g^n\|_2 \to 0$. Thus, the operator $\chi V\chi\mathbf{E}$ is compact in L_2. Now let K be a compact symmetric operator in $L_2(\mathbb{R}^d{}_v, \mathbf{P})$. Then the above relations imply the compactness of $\chi V\chi K$ and the continuity of $K\chi V\chi K$ from $L_2(L_p(\mathbb{R}^{d+1}), \mathbf{P})$ to $L_2(L_{p_1,\mathrm{loc}}(\mathbb{R}^{d+1}), \mathbf{P})$ for p, p_1 satisfying the conditions of Theorems 8, 9. This allows one to describe the solutions of the nonlinear equations (1.1) with $g = \Gamma(f)$ for a broad class of operators Γ [7].

References

1. F. Golse, P.-L. Lions, B. Perthame, and R. Sentis, *Regularity of the moments of the solution of a transport equation*, J. Funct. Anal. **76** (1988), 110–125.
2. F. Golse, *Quelques résultats de moyennisation pour les équations aux dérivées partielles*, Rend. Sem. Mat. Univ. Politec. Torino **1988**, Special Issue, 101–123.
3. R. J. Di Perna and P.-L. Lions, *Global weak solutions of Vlasov-Maxwell systems*, Comm. Pure Appl. Math. **42** (1989), 729–757.
4. J. Bergh and J. Löfström, *Interpolation spaces. An introduction*, Grundlehren Math. Wiss, Band 223, Springer-Verlag, Berlin and New York, 1976.
5. M. A. Krasnoselskiĭ, P. P. Zabreiko, E. I. Pustilnik, and P. E. Sobolevskiĭ, *Integral operators in spaces of summable functions*, "Nauka", Moscow, 1966; English transl., Noordhoff, Leiden, 1976.
6. N. B. Maslova, *Solvability theorems for the nonlinear Boltzmann equations*, Supplement to C. Cercignani, "Theory and Applications of the Boltzmann Equation", "Mir", Moscow, 1978, pp. 461–480. (Russian)
7. ———, *Global solutions of nonstationary kinetic equations*, Zap. Nauchn. Sem. Leningrad. Otdel. Mat. Inst. Steklov. (LOMI) **119** (1982), 169–177; English transl. in J. Soviet Math. **27** (1984).

DEPARTMENT OF PHYSICS, ST. PETERSBURG STATE UNIVERSITY, STARY PETERHOF, 1ST ULYANOVA STR., ST. PETERSBURG, 198904, RUSSIA

Translated by B. A. PLAMENEVSKIĬ

Amer. Math. Soc. Transl.
(2) Vol. **159**, 1994

On the Symmetric Envelope of Averaging of Symmetric Ideals

A. A. Mekler

In this paper a full presentation of results announced in [**1**] and related facts is given. We consider the symmetric (i.e., rearrangement invariant and order ideal) envelope $N_{E(X|\mathscr{F})}$ of the image $E(X|\mathscr{F})$ of a symmetric ideal X of measurable functions by the averaging operator $E(\cdot|\mathscr{F})$ over a countable partition $\mathscr{F}$ of the segment $[0,1]$. Besides the symmetric envelope the majorant envelope $M_{E(X|\mathscr{F})}$ of $E(X|\mathscr{F})$ (in other words, the interpolational envelope between $L^1(0,1)$ and $L^\infty(0,1)$) is analyzed. It is shown that if X is a principal majorant ideal M_f (i.e., generated by some summable function f), then $M_{E(X|\mathscr{F})}$ is a principal majorant ideal too, i.e., $M_{E(M_f|\mathscr{F})} = M_g$ for a proper average g of the function f. An analogous assertion for the principal symmetric ideal N_f is only true under some conditions for f; the basic result (Theorems 2.9 and 2.10) contains a list of properties of f equivalent to this condition.

Note that Theorem 2.9 as well as Theorem 2.2 may be related to problems connected with the dominated convergence of martingales and ergodic means.

Preliminaries are given in §1, basic results in §2. The simple assertions in §1 are given without explanations.

§1. Definitions, notation, and preliminaries

As usual, by $L^1 = L^1(0,1)$ one denotes the space of all summable (classes of) functions on the Lebesgue segment $[0,1] = ([0,1], \Lambda, \lambda)$. The indicator function of a set $A \in \Lambda$ is denoted by $\mathbf{1}_A$. The contraction τ_s, $s > 0$, is defined by $(\tau_s f)(t) = f(st)$ for a measurable function f, $t \in [0,1]$ (one assumes f to be extended by zero over $(1,\infty)$). Two functions f and g are called equivalent, $f \simeq g$, if $f \leqslant C_1\tau_{s_1}g \leqslant C_2\tau_{s_2}f$ for proper positive constants C_i, s_i, $i = 1, 2$. By definition, f and g are equimeasurable, $f \sim g$, if $\lambda\{f > r\} = \lambda\{g > r\}$ for any real r. A nonincreasing function f equimeasurable with $|f|$ is denoted by f^*; for $f \in L^1$ one writes $f^{**}(t) = t^{-1}\int_0^1 f^* d\lambda$, $t \in (0,1]$.

LEMMA 1.1. *Let $f \in L^1$, $A \in \Lambda$, $\lambda A > 0$. For every measurable $B \subset A$, $\lambda B > 0$, there exists $\tilde{f}$ equimeasurable with f, $\tilde{f} = f$ outside A, such that*

$$(\lambda B)^{-1}\int_B \tilde{f}\,d\lambda = (\lambda A)^{-1}\int_A f\,d\lambda.$$

1991 *Mathematics Subject Classification*. Primary 16S60; Secondary 16W99.

LEMMA 1.2 ([2]). *For any* $f \in L^1$,

$$0 \leqslant f^*(t) \leqslant f^{**}(t) = (f^{**})^*(t) \geqslant t^{-1} \int_0^t f\, d\lambda, \qquad t \in (0,1]; \tag{1}$$

$$f^* \leqslant \tau_s f^*, \quad \tau_s f^{**} \leqslant s^{-1} f^{**}, \qquad s \in (0,1]. \tag{2}$$

A vector variety X in L^1 is called an order ideal if $|y| \leqslant |x|$, $x \in X$, implies $y \in X$; a symmetric ideal (s.i.) if $y^* \leqslant x^*$, $x \in X$, implies $y \in X$; a majorant ideal (m.i.) if $y^{**} \leqslant x^{**}$, $x \in X$, implies $y \in X$.

REMARK. 1) For an s.i. X, set $X^* = \{x^* : x \in X\}$. It is evident that $X^* \subset Y^*$ implies $X \subset Y$ for s.i. X, Y.

2) Every majorant ideal is symmetric, and every symmetric ideal is an order ideal (from now on we omit the word "order").

3) Majorant ideals can be thought of as vector varieties in L^1 that are interpolational between L^1 and L^∞, [2].

LEMMA 1.3 ([3]). $\tau_s X \subset X$ *for any s.i.* X *and* $s > 0$.

Let $\{X_\gamma,\ \gamma \in \Gamma\}$ be a family of ideals in L^1. We set

$$\sum_{\gamma\in\Gamma} X_\gamma = \left\{ \sum_{i=1}^n x_{\gamma_i} : x_{\gamma_i} \in X_{\gamma_i}, i = 1, \dots, n; (\gamma_1, \dots, \gamma_n) \in \Gamma^n, n = 1, 2, \dots \right\}.$$

LEMMA 1.4. *The sum* $\sum_{\gamma\in\Gamma} X_\gamma$ *is an ideal in* L^1. *If* X_γ *are symmetric or majorant, the same is true for the sum.*

Since the space L^1 is a majorant ideal, for any $Z \subset L^1$ there exists a minimal s.i. (minimal m.i.) containing Z, which we denote by N_Z (respectively, M_Z). For a singleton set $Z = \{f\}$, $f \in L^1$, we write N_f (M_f) instead of $N_{\{f\}}$ $(M_{\{f\}})$; N_f (M_f) is called the principal symmetric (majorant) ideal generated by f.

Lemmas 1.2 and 1.3 allow us to describe N_f and M_f in a unified fashion.

LEMMA 1.5. *For any* $f \in L^1$ *the following equalities hold*:

$$\begin{aligned} N_f &= \left\{ z \in L^1 : \text{ there exists } q = q(z) \text{ such that } z^* \leqslant q\tau_{q^{-1}} f^* \right\}; \\ M_f &= \left\{ z \in L^1 : \text{ there exists } q = q(z) \text{ such that } z^{**} \leqslant q\tau_{q^{-1}} f^{**} \right\} \\ &= \left\{ z \in L^1 : \text{ there exists } q = q(z) \text{ such that } z^{**} \leqslant q f^{**} \right\}. \end{aligned} \tag{3}$$

(In the sequel we shall also use the notation N_f in the sense of the first equality in (3) for measurable nonsummable functions.)

The last equality shows that the principal m.i. M_f contains the same elements as the Marcinkiewicz space M_ψ [2], ψ being a concave continuous nondecreasing function such that $\psi(0) = 0$ and $\psi(t) = t f^{**}(t)$, $t \in (0,1]$.

LEMMA 1.6. *The equalities*

$$N_{\sum_{\gamma\in\Gamma} X_\gamma} = \sum_{\gamma\in\Gamma} N_{X_\gamma}, \quad M_{\sum_{\gamma\in\Gamma} X_\gamma} = \sum_{\gamma\in\Gamma} M_{X_\gamma} \tag{4}$$

hold for any family $\{X_\gamma, \gamma \in \Gamma\}$ *of ideals in* L_1.

We say that a σ-subalgebra $\mathscr{F}$ in Λ is a countable partition of $[0, 1]$ if it is generated by a countable collection of pairwise disjoint sets $F_n : \lambda F_n > 0$ $(n \geqslant 1)$, $\sum_{n=1}^{\infty} \lambda F_n = 1$. We denote the algebra $\mathscr{F}$ by $\sigma(F_n)$, the vector with components λF_n, $n \geqslant 1$, by $\overline{\mathscr{F}}$ and introduce the class $[\overline{\mathscr{F}}]$ of all vectors obtained by permuting the components of $\overline{\mathscr{F}}$. Two partitions $\mathscr{F}_1$ and $\mathscr{F}_2$ are said to be equimeasurable, $\mathscr{F}_1 \sim \mathscr{F}_2$, if $[\overline{\mathscr{F}}_1] = [\overline{\mathscr{F}}_2]$. It is well known that for any equimeasurable partitions $\mathscr{F}_1$ and $\mathscr{F}_2$ there exists a mod 0-automorphism π of $[0, 1]$ sending one partition to the other; this will be denoted by $\mathscr{F}_2 = \mathscr{F}_1 \circ \pi$. We say that $\mathscr{F}_1$ is coarser than $\mathscr{F}_2$ if the inclusion $\mathscr{F}_1 \subset \mathscr{F}_2$ holds for the σ-algebras $\mathscr{F}_1$, $\mathscr{F}_2$.

Let $\{t_n\}_{n=0}^{\infty}$ be a sequence strictly decreasing to zero, $t_0 = 1$. The partition $\mathscr{T} = \sigma((t_n, t_{n-1}])$ is called an interval partition (i.p.), denoted $\mathscr{T} = (t_n)$. An i.p. $\mathscr{F}$ is said to be monotone if $t_n - t_{n+1} \leqslant t_{n-1} - t_n$, $n \geqslant 1$. It is evident that for any partition $\mathscr{F}$ there exists an equimeasurable monotone i.p. $\mathscr{F}^*$.

The i.p. (2^{-m_n}) with integers m_n, $0 = m_0 < m_1 < m_2 < \cdots$, is called binary. The binary i.p. (2^{-n}) is denoted by $\mathscr{D}$, and the interval $(2^{-n}, 2^{-n+1}]$ by D_n, $n \geqslant 1$. To every i.p. $\mathscr{T} = (t_n)$ we associate its (right) binary projection that is the binary i.p. $\mathscr{T}_{(2)}$ consisting of all pairwise distinct points of the form $2^{-[\log_2 t_n]+1}$, $[\alpha]$ being the integer part of the real α. Note that $\mathscr{T}_{(2)}$ is always monotone. For a partition $\mathscr{F}$ we write $\mathscr{F}^*_{(2)}$ instead of $(\mathscr{F}^*)_{(2)}$.

Given an i.p. $\mathscr{T} = (t_n)$, we define the following two functions:

$$\omega_{\mathscr{T}}(k) = \sup_{n \geqslant 0} t_{n+k}/t_n, \quad q_{\mathscr{T}}(k) = \overline{\overline{\{t_n\}_{n=0}^{\infty} \cap D_k}}, \qquad k \geqslant 1,$$

for positive integers k.

LEMMA 1.7. *The following conditions are equivalent*:
(1) *there is an integer $k \geqslant 1$ such that $\omega_{\mathscr{T}}(k) < 1$;*
(2) *the sequence $\{q_{\mathscr{T}}(k)\}$ is bounded.*

Let a nondecreasing function ψ on $[0, 1]$ be continuous at 0, $\psi(0) = 0$, and let $\mathscr{T} = (t_n)$ be an i.p.. By $\psi(\mathscr{T})$ we denote the i.p. formed by all pairwise distinct points of the sequence $\{\min[1, \psi(t_n)]\}_{n=0}^{\infty}$ (adding 1, if necessary). The notations $\omega_{\psi(\mathscr{T})}$ and $q_{\psi(\mathscr{T})}$ are obvious.

An i.p. $\mathscr{S} = (s_n)$ is coarser than an i.p. $\mathscr{T} = (t_n)$ if and only if $\{s_n\}_{n=0}^{\infty} \subset \{t_n\}_{n=0}^{\infty}$. $\mathscr{S}$ and $\mathscr{T}$ are said to be equivalent, $\mathscr{S} \simeq \mathscr{T}$, if $C^{-1}s_n \leqslant t_n \leqslant Cs_n$, $n \geqslant 0$, with a proper constant $C > 0$.

LEMMA 1.8 (**[4]**).
(1) *Every constant $\alpha \geqslant (\sqrt{5}+1)/2$ possesses the following property. Let $\mathscr{T} = (t_n)$ be an arbitrary i.p., and let (t_n^*) denote the i.p. $\mathscr{T}^*$. For any integer $n \geqslant 0$ there is an integer $m \geqslant 0$ such that*

$$\alpha^{-1} t_m^* \leqslant t_n \leqslant \alpha t_m^*.$$

(2) *$(\sqrt{5}+1)/2$ is the infimum for the numbers α with the above property.*

COROLLARY 1.9. *Let $\mathscr{T}$ be any i.p. and $\mathscr{S} \subset \mathscr{T}$. For every i.p. $\mathscr{U}$ equimeasurable with $\mathscr{S}$ there exists an i.p. $\mathscr{V} \subset \mathscr{T}^*$ equivalent to $\mathscr{U}$.*

For any partition $\mathscr{F} = \sigma(F_n)$ the averaging operator $E(\cdot|\mathscr{F}) : L^1 \to L^1$ is defined by

$$E(f|\mathscr{F}) = \sum_{n=1}^{\infty} \left[(\lambda F_n)^{-1} \int_{F_n} f \, d\lambda \right] \mathbf{1}_{F_n}. \tag{5}$$

The next assertion follows from Lemma 1.1.

LEMMA 1.10. *Let a partition $\mathscr{F}_1$ be coarser than $\mathscr{F}_2$. For every $f \in L^1$ there is $\tilde{f} \sim f$ such that*

$$E(\tilde{f}|\mathscr{F}_2) = E(f|\mathscr{F}_1).$$

LEMMA 1.11. *Let $\mathscr{F}$ be a partition, $f \in L^1$. There exist $\tilde{f} \sim f$ and i.p. $\mathscr{T} \sim \mathscr{F}$ such that $E(f|\mathscr{F})^* = E(\tilde{f}|\mathscr{T})$.*

LEMMA 1.12. *Let π be a* mod 0 *automorphism of* $[0,1]$, $f \in L^1$, $\mathscr{F}$ *a partition. Then*

$$E(f \circ \pi | \mathscr{F} \circ \pi) = E(f|\mathscr{F}) \circ \pi. \tag{6}$$

LEMMA 1.13. *Let $\mathscr{F}$ be a partition, $\{X_\gamma, \gamma \in \Gamma\}$ a family of ideals in L^1. Then*

$$E(\textstyle\sum_{\gamma\in\Gamma} X_\gamma|\mathscr{F}) = \displaystyle\sum_{\gamma\in\Gamma} E(X_\gamma|\mathscr{F}). \tag{7}$$

LEMMA 1.14. *Let $\mathscr{F}_1$ and $\mathscr{F}_2$ be partitions, X an s.i., $\mathscr{T}$, $\mathscr{T}_1$, and $\mathscr{T}_2$ i.p.. Then*
(1) $N_{E(X|\mathscr{F}_1)} \subset N_{E(X|\mathscr{F}_2)}$, $M_{E(X|\mathscr{F}_1)} \subset M_{E(X|\mathscr{F}_2)}$ *in the case* $\mathscr{F}_1 \subset \mathscr{F}_2$;
(2) $N_{E(X|\mathscr{F}_1)} = N_{E(X|\mathscr{F}_2)}$, $M_{E(X|\mathscr{F}_1)} = M_{E(X|\mathscr{F}_2)}$ *in the case* $\mathscr{F}_1 \sim \mathscr{F}_2$;
(3) $N_{E(X|\mathscr{T}_1)} = N_{E(X|\mathscr{T}_2)}$, $M_{E(X|\mathscr{T}_1)} = M_{E(X|\mathscr{T}_2)}$ *if* $\mathscr{T}_1 \simeq \mathscr{T}_2$;
(4) $N_{E(X|\mathscr{T})} = N_{E(X|\mathscr{T}_{(2)})}$.

LEMMA 1.15. *Let $\mathscr{F}$ be a partition, $f \in L^1$. Then*

$$N_{E(N_f|\mathscr{F})} = \sum_{\tilde{f}\sim f, \tilde{\mathscr{F}}\sim\mathscr{F}} N_{E(\tilde{f}|\tilde{\mathscr{F}})}, \qquad M_{E(N_f|\mathscr{F})} = \sum_{\tilde{f}\sim f, \tilde{\mathscr{F}}\sim\mathscr{F}} M_{E(\tilde{f}|\tilde{\mathscr{F}})}. \tag{8}$$

For $\mathscr{T} = (t_n)$ and a monotone $f = f^*$ we denote $f_{\mathscr{T}} = \sum_{n=1}^{\infty} f(t_{n-1}) 1_{(t_n, t_{n-1}]}$. It is clear that $f_{\mathscr{T}} = (f_{\mathscr{T}})^*$.

LEMMA 1.16. *For any $f \in L^1$ and any i.p. $\mathscr{T}$*

$$(f^*)_{\mathscr{T}} \leqslant E(f^*|\mathscr{T}) = E(f^*|\mathscr{T})^* = E(f^*|\mathscr{T})_{\mathscr{T}} \leqslant (E(f^*|\mathscr{T})^{**})_{\mathscr{T}})(f^{**})_{\mathscr{T}}. \tag{9}$$

From now on the right-hand side and the left-hand side in (9) will be denoted by $f^{**}_{\mathscr{T}}$ and $f^*_{\mathscr{T}}$, respectively.

LEMMA 1.17. *Let $\mathscr{T}$ be an i.p., $f \in L^1$, and $\alpha \geqslant (\sqrt{5}+1)/2$. Then*

$$f^{**}_{\mathscr{T}} \leqslant \alpha f^{**}_{\mathscr{T}^*}. \tag{10}$$

REMARK. In fact, the minimal α in (10) is equal to $4/3 < (\sqrt{5}+1)/2$ [**4**].

We say that a partition $\mathscr{F}$ averages an s.i. X if $E(X|\mathscr{F}) \subset X$. Lemma 1.14 implies

LEMMA 1.18.
(1) *If one of two equimeasurable partitions (or equivalent i.p.) averages an s.i. X, so does the other.*
(2) *If $\mathcal{F}_2$ averages an s.i. X and $\mathcal{F}_1$ is coarser than $\mathcal{F}_2$, then $\mathcal{F}_1$ averages X.*
(3) *averages an s.i. X if and only if the binary projection $\mathcal{T}_{(2)}$ averages X.*

LEMMA 1.19 (**[2]**). *Every m.i. is averaged by any partition.*

LEMMA 1.20 (**[5]**). *Let $\mathcal{T}$ be an i.p., $f \in L^1$, and $g = g^*_{\mathcal{T}} \in M_f$. There exists $\tilde{f} \sim f^*$ and $C > 0$ such that $g \leqslant CE(\tilde{f}|\mathcal{T})$.*

REMARK. From this result it immediately follows that for s.i., Lemma 1.19 can be reversed. It was shown in **[6]** that such reversion is possible not only for s.i., but for any ideals.

A partition $\mathcal{F}$ is called a checking partition if $\mathcal{F}$ averages only majorant ideals among the s.i.

LEMMA 1.21 (**[7]**). *$\mathcal{F}$ is a checking partition if and only if the condition*

(11)
$$\sup_{n\geqslant 0} t^*_n/t^*_{n+1} < \infty$$

holds for the i.p. $\mathcal{F}^* = (t^*_n)$.

REMARK. It is easy to see that an i.p. $\mathcal{T}$ is a checking partition if and only if the same is true for $\mathcal{T}_{(2)}$.

LEMMA 1.22. *If $\mathcal{F}$ is a checking partition, then $E(f^*|\mathcal{F}^*) \simeq f^*$ for any $f \in L^1$.*

COROLLARY 1.23. *If $\mathcal{F}$ is a checking partition, then the relations*

(12)
$$X^* = \left(N_{E(X^*|\mathcal{F}^*)}\right)^* \subset \left(N_{E(X|\mathcal{F})}\right)^*$$

are valid for every s.i. X.

A function f is said to be $\mathcal{F}$-regular if the partition $\mathcal{F}$ averages the principal s.i. N_f. (True, Part 1 of Lemma 1.18 shows that f should rather be called $[\bar{\mathcal{F}}]$-regular.)

LEMMA 1.24 (**[8]**). *$f \in L^1$ is $\mathcal{F}$-regular if and only if $f^{**}_{\mathcal{F}^*} \simeq f^*_{\mathcal{F}^{**}}$.*

REMARK. For any i.p. $\mathcal{T}$ a function f is $\mathcal{T}$-regular if and only if f is $\mathcal{T}_{(2)}$-regular.

§2. Results

PROPOSITION 2.1. *Let $\mathcal{T}$ be an i.p., $f \in L^1$. Then*

(13)
$$f^{**}_{\mathcal{T}} = \sup\{E(f^*|\mathcal{S}) \colon \mathcal{S} \subset \mathcal{T}\}.$$

PROOF. Let us take any number $u : 0 < u \leqslant 1$ and suppose that $\mathcal{S} = (s_n) \subset (t_n) = \mathcal{T}$. Choose a number n such that $s_{n+1} < u \leqslant s_n$. By virtue of (9)

$$E(f^*|\mathcal{S})(u) = E(f^*|\mathcal{S})(s_n) \leqslant f^{**}(s_n) = f^{**}_{\mathcal{S}}(s_n) = f^{**}_{\mathcal{T}}(s_n) = f^{**}_{\mathcal{T}}(u);$$

so the right-hand side in (13) does not exceed the left-hand side. To prove the reverse inequality we fix $\varepsilon > 0$ and choose a positive integer k such that $t_{k+1} < u \leqslant t_k$. Next, take $m > k$ so that

$$(t_k - t_m)^{-1} \int_{t_m}^{t_k} f^* d\lambda \geqslant f^{**}(t_k) - \varepsilon = f^{**}_{\mathscr{T}}(u) - \varepsilon.$$

This inequality means that

$$E(f^*|\mathscr{S}_\varepsilon)(u) \geqslant f^{**}_{\mathscr{T}}(u) - \varepsilon \tag{14}$$

provided $\mathscr{S}_\varepsilon = (s_n)$ is a coarser partition than $\mathscr{T}$ such that $s_i = t_k$, $s_{i+1} = t_m$ for some integer $i \geqslant 0$. □

THEOREM 2.2.
(1) *For any* $f \in L^1$ *and any partition* $\mathscr{F}$

$$E(f|\mathscr{F})^* \leqslant \tfrac{4}{3} f^{**}_{\mathscr{F}^*}. \tag{15}$$

(2) *The constant* $\frac{4}{3}$ *in* (15) *is minimal.*

PROOF. By Lemma 1.11, $E(f|\mathscr{F})^* = E(\tilde{f}|\mathscr{T})$ with $\mathscr{T} \sim \mathscr{F}$, $\tilde{f} \sim f$. Thus, according to (1), (9), and the remark to Lemma 1.17,

$$E(f|\mathscr{F})^* = E(\tilde{f}|\mathscr{T}) \leqslant (E(\tilde{f}|\mathscr{T})^{**})_{\mathscr{T}} \leqslant E(f^*|\mathscr{T})^{**}_{\mathscr{T}} = f^{**}_{\mathscr{T}} \leqslant \tfrac{4}{3} f^{**}_{\mathscr{F}^*}. \tag{16}$$

That gives the first part of the theorem. We turn to the second part. Assume that for $f \in L^1$ and every partition $\mathscr{F}$,

$$E(f|\mathscr{F})^* \leqslant b f^{**}_{\mathscr{F}^*},$$

where $0 < b < \frac{4}{3}$. By virtue of Lemmas 1.10 and 1.11, $E(f|\mathscr{F}_2)^* \leqslant b f^{**}_{\mathscr{F}^*}$ holds for any partition $\mathscr{F}_2$ that is coarser than $\mathscr{F}_1$, equimeasurable with $\mathscr{F}$. This, along with Proposition 2.1, implies $f^{**}_{\mathscr{T}} \leqslant b f^{**}_{\mathscr{F}^*}$. But this contradicts the remark to Lemma 1.17. □

Theorem 2.2 and Lemma 1.15 lead to the following assertion.

COROLLARY 2.3. *For any partition* $\mathscr{F}$ *and* $f \in L^1$

$$N_{E(N_f|\mathscr{F})} \subset N_{f^{**}_{\mathscr{F}^*}}.$$

LEMMA 2.4. *The equality*

$$N_{E(N_f|\mathscr{F})} = N_{E(M_f|\mathscr{F})}$$

holds for any partition $\mathscr{F}$ *and* $f \in L^1$.

PROOF. The inclusion $N_{E(N_f|\mathscr{F})} \subset N_{E(M_f|\mathscr{F})}$ is trivial owing to $N_f \subset M_f$. The reverse inclusion follows from

$$E(M_f|\mathscr{F}) \subset N_{E(N_f|\mathscr{F})}, \tag{17}$$

which is a direct consequence of Lemmas 1.11, 1.12, 1.20, and 1.18.1. □

THEOREM 2.5. *Any partition* $\mathscr{F}$ *averages the s.i.* $N_{E(X|\mathscr{F})}$ *with every s.i.* X.

PROOF. We start with $X = N_f$. It suffices to prove $E(N_{E(N_f|\mathscr{F})}|\mathscr{F}) \subset N_{E(N_f|\mathscr{F})}$. Since $N_f \subset M_f$, Lemma 1.19 implies $E(N_f|\mathscr{F}) \subset M_f$. Hence $N_{E(N_f|\mathscr{F})} \subset M_f$. It remains to take (17) into account.

We pass to the general case. Using Lemma 1.13, Lemma 1.6, and the above case one has

$$E(N_{E(X|\mathcal{F})}|\mathcal{F}) = E(\textstyle\sum_{x\in X} N_{E(N_x|\mathcal{F})}|\mathcal{F}) = \displaystyle\sum_{x\in X} E(N_{E(N_x|\mathcal{F})}|\mathcal{F})$$

$$\subset \sum_{x\in X} N_{E(N_x|\mathcal{F})} = N_{E(\sum_{x\in X} N_x|\mathcal{F})} = N_{E(X|\mathcal{F})}. \qquad \square$$

COROLLARY 2.6. *Let $\mathcal{F}$ be a checking partition, X an s.i.. Then $M_X = N_{E(X|\mathcal{F})}$. In particular, $M_f = N_{E(N_f|\mathcal{F})}$.*

PROOF. The inclusion $X \subset N_{E(X|\mathcal{F})}$ follows from Corollary 1.23 and Remark 1 to the definition of s.i.. Everything else is evident. $\square$

THEOREM 2.7. *For any partition $\mathcal{F}$ and $f \in L^1$,*

$$M_{E(M_f|\mathcal{F})} = M_{E(N_f|\mathcal{F})} = M_{E(f^*|\mathcal{F}^*)}.$$

PROOF. The first equality immediately follows from Lemma 2.4. Taking into account Lemma 1.15, Lemma 1.11, the last inequality in (1), and Lemma 1.17 one has

$$M_{E(M_f|\mathcal{F})} = \sum_{\tilde{f}\sim f, \tilde{\mathcal{F}}\sim\mathcal{F}} M_{E(\tilde{f}|\tilde{\mathcal{F}})} = \sum_{\mathcal{T}\sim\mathcal{F}, \bar{f}\sim f} M_{E(\bar{f}|\mathcal{T})}$$

$$= \sum_{\mathcal{T}\sim\mathcal{F}} M_{E(f^*|\mathcal{T})} \subset M_E(f^*|\mathcal{F}^*).$$

The reverse inclusion is evident. $\square$

The s.i. spanned by $E(N_f|\mathcal{F})$, unlike the m.i. spanned by the same set, is not always generated by the function $E(f^*|\mathcal{F}^*)$ even when principal. The following theorem describes a favorable case.

THEOREM 2.8. *Let $\mathcal{F}$ be a partition and $f \in L^1$. The equality $N_{E(N_f|\mathcal{F})} = N_{E(f^*|\mathcal{F}^*)}$ is valid if and only if $E(f^*|\mathcal{F}^*)$ is $\mathcal{F}$-regular.*

PROOF. Suppose that the above equality is valid. Then by Theorem 2.5 the function $E(f^*|\mathcal{F}^*)$ is $\mathcal{F}$-regular.

Conversely, let $E(f^*|\mathcal{F}^*)$ be $\mathcal{F}$-regular. If we prove that $E(\tilde{f}|\tilde{\mathcal{F}}) \in N_{E(f^*|\mathcal{F}^*)}$ for each $\tilde{f} \sim f$, $\tilde{\mathcal{F}} \sim \mathcal{F}$, we shall obtain $N_{E(N_f|\mathcal{F})} \subset N_{E(f^*|\mathcal{F}^*)}$ by Lemma 1.15, while the reverse inclusion is trivial.

Applying Lemmas 1.11, 1.16, and 1.17, and the relation (1), we obtain for appropriate $\bar{f} \sim f$ and i.p. $\tilde{\mathcal{F}} \sim \mathcal{F}$

$$E(\tilde{f}|\tilde{\mathcal{F}})^* = E(\bar{f}|\mathcal{T}) \leqslant E(\bar{f}|\mathcal{T})^{**}_{\mathcal{T}} \leqslant E(f^*|\mathcal{T})^{**}_{\mathcal{T}} = f^{**}_{\mathcal{T}}$$

$$\leqslant \tfrac{4}{3} f^{**}_{\mathcal{F}^*} = \tfrac{4}{3} E(f^*|\mathcal{F}^*)^{**}_{\mathcal{F}^*} \leqslant \tfrac{4}{3} C\tau_s E(f^*|\mathcal{F}^*),$$

where C and s are the equivalence constants in Lemma 1.24. Therefore $E(\tilde{f}|\tilde{\mathcal{F}}) \in N_{E(f^*|\mathcal{F}^*)}$. $\square$

Let us state our basic result concerning the case in which $N_{E(N_f|\mathcal{F})}$ (or $N_{E(M_f|\mathcal{F})}$) is a principal s.i..

THEOREM 2.9. *For any partition $\mathscr{F}$ and $f \in L^1$ the following conditions are equivalent*:

(1) *there exists* $g \in L_1$ *such that* $N_{E(N_f|\mathscr{F})} = N_g$;

(2) $N_{E(N_f|\mathscr{F})} = N_{f^{**}_{\mathscr{F}^*}}$;

(3) $f^{**}_{\mathscr{F}^*}$ *is an* $\mathscr{F}$*-regular function*;

(4) $M_{f^{**}_{\mathscr{F}^*}} = M_{E(f^*|\mathscr{F}^*)}$;

(5) $f^{**}_{\mathscr{F}^*} \in M_f$.

PROOF. 1) $\Rightarrow$ 2). If $N_{E(N_f|\mathscr{F})} = N_g$, then by Theorem 2.5, g is $\mathscr{F}$-regular; hence by Lemma 1.24, $g^{**}_{\mathscr{F}^*} \simeq g^*_{\mathscr{F}^*}$. Besides, by virtue of Theorem 2.7, $M_g = M_{E(f^*|\mathscr{F}^*)}$; so $g^{**} \simeq E(f^*|\mathscr{F}^*)^{**}$, whence $g^{**}_{\mathscr{F}^*} \simeq f^{**}_{\mathscr{F}^*}$ by (9). Thus $g^* \geqslant g^*_{\mathscr{F}^*} \simeq f^{**}_{\mathscr{F}^*}$, which gives $N_{f^{**}_{\mathscr{F}^*}} \subset N_g$. Lemma 1.15 and Theorem 2.2 provide the reverse inclusion.

2) $\Rightarrow$ 3). This follows directly from Theorem 2.5.

3) $\Rightarrow$ 4). In view of (9), $f^{**}_{\mathscr{F}^*}$ and $E(f^*|\mathscr{F}^*)^{**}_{\mathscr{F}^*}$ coincide. Therefore, by Lemma 1.24, $(f^{**}_{\mathscr{F}^*})^{**}_{\mathscr{F}^*} \simeq E(f^*|\mathscr{F}^*)^{**}_{\mathscr{F}^*}$. Since $E(f^*|\mathscr{F}^*)$ and $f^{**}_{\mathscr{F}^*}$ are $\mathscr{F}^*$-step functions, this implies $(f^{**}_{\mathscr{F}^*})^{**} \simeq E(f^*|\mathscr{F}^*)^{**}$, i.e., $M_{f^{**}_{\mathscr{F}^*}} = M_{E(f^*|\mathscr{F}^*)}$.

4) $\Rightarrow$ 5). This is trivial owing to $M_{E(f^*|\mathscr{F}^*)} \subset M_f$.

5) $\Rightarrow$ 1). The inclusion $f^{**}_{\mathscr{F}^*} \in M_f$ can be written as $f^{**}_{\mathscr{F}^*} \in E(M_f|\mathscr{F}^*)$. Now from Lemma 2.4 and Lemma 1.14, 2) it follows that

$$N_{f^{**}_{\mathscr{F}^*}} \subset N_{E(M_f|\mathscr{F}^*)} = N_{E(N_f|\mathscr{F}^*)} = N_{E(N_f|\mathscr{F})}.$$

Lemma 1.15 and Theorem 2.2 give the reverse inclusion. □

REMARK. 1) The $\mathscr{F}$-regularity of f implies the $\mathscr{F}$-regularity of $E(f^*|\mathscr{F}^*)$, which, in turn, gives the $\mathscr{F}$-regularity of $f^{**}_{\mathscr{F}^*}$. Indeed, if f is $\mathscr{F}$-regular, then by (9) and (1) one has

$$E(f^*|\mathscr{F}^*)^{**}_{\mathscr{F}^*} = f^{**}_{\mathscr{F}^*} \simeq f^*_{\mathscr{F}^*} \leqslant E(f^*|\mathscr{F}^*) = E(f^*|\mathscr{F}^*)_{\mathscr{F}^*} \leqslant E(f^*|\mathscr{F}^*)^{**}_{\mathscr{F}^*},$$

and so $E(f^*|\mathscr{F}^*)$ is $\mathscr{F}$-regular as well. Further, if $E(f^*|\mathscr{F}^*)$ is $\mathscr{F}$-regular, then by Theorem 2.8, $N_{E(N_f|\mathscr{F})} = N_{E(f^*|\mathscr{F}^*)}$. Now, $f^{**}_{\mathscr{F}^*}$ is $\mathscr{F}$-regular in accordance with Theorem 2.9.

2) The reverse implications to the ones indicated in 1) are, generally speaking, wrong. A corresponding example was given in **[9]**. It was shown there that the $\mathscr{F}$-regularity of $g = f^{**}_{\mathscr{F}^*}$ is equivalent to the $\mathscr{F}$-regularity of $g^{**}_{\mathscr{F}^*}$.

A function $f \in L^1$ is called regular if it is $\mathscr{F}$-regular for any partition $\mathscr{F}$. It is known **[10]** that the regularity of f is equivalent to the regularity of f^{**}. Hence f^{**} and $(f^{**})^{**}$ can be regular only simultaneously. This fact can be viewed as an analogy between the regularity of f^{**} and the $\mathscr{F}$-regularity of $f^{**}_{\mathscr{F}^*}$. Let us analyze this analogy in more detail.

Set $\psi(t) = tf^{**}(t)$, $t \in (0, 1]$. Applying the aforementioned result in **[10]** and Lemma 1.7, one can obtain that f^{**} is regular if and only if the sequence $\{q_{\psi(\mathscr{D})}(k)\}_{k=1}^{\infty}$ is bounded (see **[11**, Theorem 2**]**).

THEOREM 2.10. $f^{**}_{\mathscr{F}^*}$ *is* $\mathscr{F}$*-regular if and only if the sequence* $\{q_{\psi(\mathscr{F}^*_{(2)})}(k)\}_{k=1}^{\infty}$ *is bounded.*

PROOF. Taking into account Theorem 2.9, Lemmas 1.17 and 1.14, the remark to Lemma 1.24, and the fact that $f^{**}_{\mathscr{F}^*_{(2)}}$ and $E(f^*|\mathscr{F}^*_{(2)})$ are $\mathscr{F}^*_{(2)}$-step functions, it suffices to prove the equivalence of the following two conditions:

1) There exists $C > 0$ such that

$$\int_0^{s_n} f_{\mathcal{S}}^{**}\,d\lambda \leqslant C\psi(s_n), \qquad n \geqslant 0; \tag{18}$$

2) There exist positive integers p and α, $0 < \alpha < 1$, such that

$$\psi(s_{n+p}) \leqslant \alpha\psi(s_n), \qquad n \geqslant 0, \tag{19}$$

where $\mathcal{S} = (s_n)$ denotes the i.p. $\mathcal{F}^*_{(2)}$.

Let us prove that 2) implies 1). The inequalities (19) can be written in the form

$$\psi(s_{n+kp+i}) \leqslant \alpha^k \psi(s_{n+i}), \qquad n \geqslant 0,\ k \geqslant 0,\ i = 0, 1, \dots, p-1.$$

Now from 2) for any $n \geqslant 0$ it follows that

$$\begin{aligned}\int_0^{s_n} f_{\mathcal{S}}^{**}\,d\lambda &= \int_0^{s_n} \sum_{k=n}^{\infty} f^{**}(s_k)\mathbf{1}_{(s_{k+1},s_k]}\,d\lambda \\ &= \sum_{k=0}^{\infty}\sum_{i=0}^{p-1} f^{**}(s_{n+kp+i})(s_{n+kp+i} - s_{n+kp+i+1}) \\ &\leqslant \sum_{k=0}^{\infty} \alpha^k \sum_{i=0}^{p-1} \psi(s_{n+i}) \leqslant \frac{p}{1-\alpha}\psi(s_n).\end{aligned}$$

It remains to set $C = \frac{p}{1-\alpha}$.

Check that 1) implies 2). Regarding $C \geqslant 1$ in 1), one obtains

$$\frac{\int_0^{s_{n+1}} f_{\mathcal{S}}^{**}\,d\lambda}{\int_0^{s_n} f_{\mathcal{S}}^{**}\,d\lambda} = 1 - \frac{\psi(s_n)}{\int_0^{s_n} f_{\mathcal{S}}^{**}\,d\lambda}\left(1 - \frac{s_{n+1}}{s_n}\right) \leqslant 1 - \frac{1}{2C} := \alpha \in (0,1)$$

for any $n \geqslant 0$.

Thus, for every positive integer p,

$$\psi(s_{n+p}) \leqslant \int_0^{s_{n+p}} f_{\mathcal{S}}^{**}\,d\lambda \leqslant \alpha \int_0^{s_{n+p-1}} f_{\mathcal{S}}^{**}\,d\lambda \leqslant \cdots \leqslant \alpha^p \int_0^{s_n} f_{\mathcal{S}}^{**}\,d\lambda \leqslant C\alpha^p \psi(s_n).$$

It remains to choose p such that $C\alpha^p < 1$. □

Theorem 2.10 can be useful for verifying the conditions of Theorem 2.9. Besides, it is of independent interest in connection with the possibility (indicated in **[11,** Theorem 1]) of constructing $f = f^* \in L^1$ such that $q(k) \doteq q_{\psi(\mathcal{D})}(k)$, $k \geqslant 1$, for an arbitrary sequence $\{q(k)\}_{k=1}^{\infty}$ consisting of positive integers; here $\psi(t) = tf^{**}(t)$, $t \in (0,1]$.

References

1. A. A. Mekler, *Addition to "The averaging operator over a countable partition on a minimal symmetric ideal of $L^1(0,1)$"*, Zap. Nauchn. Sem. Leningrad Otdel. Mat. Inst. Steklov. (LOMI) **149** (1986), 137–141; English transl. in Soviet J. Math. **42** (1988).
2. S. G. Krein, Yu. I. Petunin, and E. M. Semenov, *Interpolation of linear operators*, "Nauka", Moscow, 1978; English transl., Transl. Math. Monographs, vol. 54, Amer. Math. Soc., Providence, RI, 1982.
3. M. S. Braverman and A. A. Mekler, *The Hardy-Littlewood property for symmetric spaces*, Sibirsk. Mat. Zh. **18** (1977), no. 3, 522–540; English transl. in Siberian Math. J. **18** (1977).

4. E. V. Abakumov and A. A. Mekler, *Shifting segments and the problem of identification of symmetric spaces*, Izv. Vyssh. Uchebn. Zaved. Mat. **1988**, no. 2, 3–9; English transl. in Soviet Math. (Iz. VUZ) **32** (1988).
5. A. A. Mekler, *Intermediate spaces and bistochastic projections*, Mat. Issled. **10** (1975), no. 1, 270–275. (Russian)
6. ———, *Averageable ideal spaces and interpolation between L_∞ and L_1*, Studies in the Theory of Several Variables, Yaroslavl. Gos. Univ., Yaroslavl′, 1980, pp. 126–139. (Russian)
7. ———, *Averageable ideal spaces and interpolation between L_∞ and L_1*. II, Qualitative and Approximate Methods for Investigating Operator Equations, Yaroslavl. Gos. Univ., Yaroslavl′, 1984, pp. 56–70. (Russian)
8. ———, *Averaging operators over countable partition on a minimal symmetric ideal of $L^1(0,1)$*, Zap. Nauchn. Sem. Leningrad Otdel. Mat. Inst. Steklov. (LOMI) **107** (1982), 136–149; English transl. in J. Soviet Math. **36** (1987).
9. ———, *A weakening of Muckenhoupt's A_1-condition for positive non-increasing functions*, Operators and their Applications, Leningrad. Gos. Ped. Inst., Leningrad, 1988, pp. 65–71. (Russian)
10. A. A. Mekler and N. F. Sokolovskaya, *On some properties of functions generating the Marcinkiewicz spaces*, Uporyadochennye prostranstva i operatornye uravneniya. Syktyvkar and Perm, 1982, pp. 136–143. (Russian)
11. A. A. Mekler, *On natural sequences related to concave functions*, Approximation of Functions by Special Classes of Operators, Vologd. Gos. Ped. Inst., Vologda, 1987, pp. 121–128. (Russian)

St. Petersburg State Academy of Aerospace Instrumentation, 67, Bolshaya Morskaya, St. Petersburg, 190000, Russia

Translated by B. A. PLAMENEVSKIĬ

Amer. Math. Soc. Transl.
(2) Vol. **159**, 1994

On the History of Mathematics at Leningrad State University at the End of the 1920s

S. C. Mikhlin

1. At the end of January 1927, I received a letter from Leningrad University, which informed me that I had been admitted to the University as a second-year student of Mathematics. Vera Nikolaevna Zamyatina (subsequently Faddeeva, by marriage) received a similar letter. Before that both of us were second-year mathematics students at Hertzen Pedagogical Institute. Due to the great difference in curricula, it turned out that we were far behind our new fellow students; we had to work hard, but by the beginning of the third year, in the fall of 1927, we caught up with them.

2. At that time there was no Department of Mathematics and Mechanics at the University; there was a Department of Physics and Mathematics, large in scope but with a comparatively small number of students. It consisted of five divisions dealing with Mathematics and Mechanics, Physics and Geophysics, Chemistry, Biology, and Geology. It was a four-year course. In fact, the Department itself was an administrative body, each division being independent, each with its own staff of professors and teachers, its own scientific Council and its own disciplines. There was a dean of the Department, but as far as I remember he did not interfere with the activities of the divisions. When I entered the University, Professor Kamenshikov, an astronomer, was the dean of the Department of Physics and Mathematics. I do not remember how long he remained in this post, nor do I remember who was the dean of the Department when I graduated from the University in 1929.

At the division of Mathematics and Mechanics four disciplines were taught, namely, Mathematics, Mechanics, Mathematical Statistics, and Astronomy. Most students studied Mechanics. This could be explained by the fact that in the 1920s technical institutes were most popular with young people, but not everyone who wished to study there could pass the exams; therefore, studying Mechanics was considered as an acceptable alternative to studying in a technical institute.

3. I do not remember the names of the students who studied at the Astronomy division for I studied Mathematics, and we had no courses in common with these students. A few students studied Statistics and there were about ten students in Mathematics.

Gradually we became acquainted with some junior and senior students who later made a significant contribution to mathematical science: Isidor Pavlovich Natanson

1991 *Mathematics Subject Classification*. Primary 01A73; Secondary 01A60, 01A70.

(1906–1964, admitted to the University in 1923), Dmitri Konstantinovich Faddeev (1907–1989, admitted to the University in 1923), Gennadi Mikhailovich Goluzin (1905–1950, admitted to the University in 1924), Leonid Vitalievich Kantorovich (1912–1988, admitted to the University in 1926), Viktor Amazaspovich Ambartsumyan (born in 1908, admitted to the University in 1924), an astronomer.

Of my fellow students the most worthy of note is one of the great mathematicians of the twentieth century, Sergei Lvovich Sobolev (1908–1989), who introduced into science the fundamental notion of generalized functions and derivatives; he also developed a theory for a very important class of function spaces, which are now called Sobolev spaces. Sobolev's works on dynamic elasticity (jointly with V. I. Smirnov), on the theory of hyperbolic equations, and on optimal cubic formulae are world famous. It is worth mentioning that S. L. Sobolev wrote his first scientific paper when he was a third-year student. He became a correspondent member of the Academy of Science of the USSR when he was not yet 25; at the age of thirty he was elected academician.

The next to be mentioned is Sergei Alekseevich Khristianovich (born in 1908), an outstanding scientist in the field of Mechanics and the author of many profound works on hydrodynamics and plasticity. At the age of thirty he became a correspondent member of the Academy of Sciences of the USSR; at the age of thirty-five he was elected academician.

Boris Borisovich Davidson (1908–?) was of English origin. Not long before graduating from the University he reinstated his English citizenship, but for some time continued living in Leningrad and studying at the University. He studied some mathematical problems of hydrodynamics and wrote a book, "Some New Problems of Mechanics of Continua", together with Khristianovich and me. In the 1930s (I do not remember the exact date) Davidson left for England; then, according to unconfirmed information, he went to Canada where he died relatively soon thereafter.

Vera Nikolaevna Faddeeva (Zamyatina) (1906–1983) became an outstanding specialist in numerical methods of linear algebra. Her book on this subject was the first of its kind in this field. Later, together with Dmitri Konstantinovich Faddeev, she greatly expanded this work, and a new monograph was translated into many languages. Apart from this unique book, Faddeeva wrote a number of prominent works on numerical methods of linear algebra.

In particular, I should like to mention our fellow student Nina Arcadievna Rozenson (1910–1942?) whose tragic death prevented her great talent from being fully realized. Her field of scientific interests was Geometry. After graduating from the University she worked in Leningrad Polytechnical Institute where she defended her candidate thesis (to my regret, I do not remember its subject). It is interesting to note the fact that professor Onufri Konstantinovich Zhitomirski, a well-known Leningrad geometrist, who examined N. A. Rozenson at her candidate examination said later that it would have been more fair if she had examined him for she knew Geometry much better than he did. At the beginning of the Great Patriotic War, the staff of Leningrad Polytechnical Institute was evacuated to the Northern Caucasus, to the town of Kislovodsk, as far as I remember. The town was later occupied by German Nazis. As I was told, the Nazis killed Nina Arcadievna in a horrible fashion – she was crushed under a steam roller; in the same manner they killed another mathematician, Tuvi Naumovich Blinchikov, assistant professor at Leningrad Polytechnical Institute.

My fellow students, the late Goarik Amazaspovna Ambarzumyan (born in 1907, her married name is Petrosyan) and Gersh Isaakovich Egudin (born in 1908) became candidates of science in Physical Mathematics. Both of them specialized in probability and after graduation worked in the Institute of Financial Economics. Goarik Amazaspovna spent the last years of her life in Erevan.

I should like to say a few words about my own work in the field of Mathematics. My first paper was my diploma on a generalization to double power series of the well-known Cauchy-Hadamard theorem on the radius of convergence of a power series. This paper was written in 1929 but was published only in 1932 because it was lost twice in the editorial office of "Matematicheski Sbornik". Anyway, while it was lost I supplemented it with a number of new results. In the years that followed I was dealing with elasticity and plasticity, integral equations, both Fredholm and singular, and their applications, operator theory, mathematical physics, numerical analysis, and the theory of errors.

4. In my graduation certificate there is a list of exams I have passed; from it one can judge with some certainty the curriculum for mathematics at the end of the 1920s. I must emphasize that I cannot always differentiate the major disciplines from special courses. Our mathematical education was based on a three-year course in calculus (differential and integral calculus of one and several variables, theory of series, ordinary differential equations, first-order partial differential equations, calculus of variations). In addition to this we had three more basic courses: 1) the theory of functions of complex variables (3rd year), 2) the theory of functions of real variables (4th year), 3) equations of mathematical physics (4th year). The first of these supplementary courses included, apart from the usual things such as Cauchy-Riemann equations, Cauchy integral, conformal mapping, and power series, a large section on elliptical functions. The second of the aforementioned courses was devoted to the Cantor theory of transfinite numbers, elements of set theory, and the Lebesgue integral. The mathematical physics course was quite elementary. It mostly dealt with the string oscillation equation written in different forms under different initial and boundary conditions. The wave equation on two and three coordinates, heat equation, elementary properties of harmonic functions, and properties of Newton potential were briefly covered.

There was also a course on approximate calculations, but I have only a vague idea of what it was about.

The Geometry course included classes on analytical geometry, projective geometry, and the theory of surfaces. There was also drawing, which could be thought of as a curiosity since it was taught in the first year on a very primitive level.

Our mathematical education was reinforced by courses in higher algebra, number theory, and probability. To my regret, I do not remember what these courses contained.

Our general education was reinforced by courses in physics, foreign languages, social politics, and military training.

According to our curriculum, we had to write a diploma paper, but this was cancelled at the end of 1929 or at the beginning of 1930. Unless my memory fails me, of my fellow students in Mathematics, I was the only one who managed to present my diploma paper, although S. L. Sobolev, S. A. Khristianovich, and N. A. Rozenson had their papers ready.

5. Among our teachers there were a number of brilliant scientists and educators. Nikolai Maksimovich Gunter, corresponding member of the Academy of Sciences of the USSR, gave lectures throughout the four-year course. In the first year he lectured in calculus; in the fourth year he taught a special course on analytical theory of ordinary differential equations. He was a man of principle, although very kind and

gentle. His lectures were brilliant in form and deep in substance. In four years he only once stumbled while giving a lecture, and I should like to describe this episode. It was in 1929 when we were third-year students. Nikolai Maksimovich lectured on first-order partial differential equations. On that day he was talking about Professor Sultikov's work. Suddenly he got stuck at some point, which had never happened to him before. He thought for a minute, then excused himself and said that he would have to think it over and would report the results at the next lecture. In a few days, Nikolai Maksimovich told us that there seemed to be an error in Sultikov's work and suggested that those who wished should analyze this work. Naturally, S. L. Sobolev took up this task. He studied Sultikov's work in detail, clarifying errors as well as conditions under which Sultikov's assertions were valid. The results of Sobolev's analysis were published in the Proceedings of the Academy of Sciences in 1929; it was the first scientific publication of this great, future mathematician.

Of the many significant works by N. M. Gunter I should like to focus on two series. One of them is devoted to the theorems on the existence of a solution to the main problems of hydrodynamics. The other deals with an attempt to solve a particular contradiction which arises when stating and solving classical problems of mathematical physics. On the one hand, these solutions should satisfy differential equations of mathematical physics and therefore possess certain smoothness, but, on the other hand, solutions derived in practice do not possess such smoothness. In N. M. Gunter's work this contradiction is solved by introducing special functionals – domain functions – in place of the usual functions, and differential equations of mathematical physics as well as the notions of their solutions are formulated in terms of these functionals.

It should be noted that another approach to this problem was later developed by S. L. Sobolev. This approach is now generally accepted; it is based on the notion of generalized derivatives and generalized solutions of differential equations.

Vladimir Ivanovich Smirnov (1887–1974), professor, correspondent member of the Academy of Sciences since 1932, academician since 1943, mainly taught at the division of Physics (later it became an independent department). At the division of Mathematics he lectured on the theory of functions of a complex variable. He is famous for his studies of boundary properties of complex variables, as well as for his five-volume work, "A Course on Higher Mathematics", which was translated into many languages. It is worth mentioning that V. I. Smirnov, together with S. L. Sobolev, wrote a great number of works on dynamical elasticity from 1930 to 1934.

Vladimir Ivanovich had many students; one of the most talented was Gennadi Mikhailovich Goluzin (1905–1950). I consider it a great honor that I also can count myself among his students.

I think that the course on the theory of real variables was very important for us as students in Mathematics at the end of the 1920s. It was taught by Professor Grigori Mikhailovich Fikhtengoltz (1888–1959), later the Honored Scientist of the USSR. In general, our education could be considered as a classical one, but the lectures given by Grigori Mikhailovich, which included, in particular, the theory of measure and the theory of the Lebesgue integral, gave us an idea of relatively up-to-date trends in the field of calculus. These lectures were of great help to me (and I think to many students in Mathematics at that time) when I had to study functional analysis several years after graduating from the University.

Functional analysis was the field of personal interest of G. M. Fikhtengoltz. Particularly, a number of his works are devoted to the study of a general form of linear functionals (sometimes operators) in various abstract spaces. He was also interested in the history of science. In 1927, he gave a series of brilliant public lectures on the evolution of analysis. He was also the author of several textbooks, among them "A Course on Differential and Integral Calculus", a lengthy and, in my opinion, very interesting work. It is worth noting that among his students was one of the great scientists of our time, mathematician and economist, academician Leonid Vitalievich Kantorovich, a Nobel Laureate.

I should like to pay tribute to two of our teachers, namely, to Professor Nadezhda Nikolaevna Gernet (to my regret, I do not know the dates of her birth and death—I only know that she died in Leningrad during the siege) and Assistant Professor Olga Andreevna Polosukhina (1883–1958). Nadezhda Nikolaevna gave a series of lectures on the calculus of variations. She loved her subject very much and it was very interesting to listen to her. Olga Andreevna lectured on integral equations; it was a special course. There were not many students but quite a lot of special courses, and it so happened that I was her only student. This did not ruffle Olga Andreevna—we used to meet strictly according to the timetable, sat down side by side and talked. It is due to these talks that integral equations became one of the major subjects of my scientific work.

Unfortunately, I know little about the Geometry course at that time, because it was already finished when I was admitted to the University. I passed my exams in Analytical Geometry when I studied at the Pedagogical Institute. At the University I passed my exams in Projective Geometry and in the theory of surfaces, as well as my final test in drawing. In these subjects I was examined by the same teacher. I do not want to provoke the dead and give his name as he was no scientist.

Unless my memory fails me, Professor Andrei Mitrofanovich Zhuravski lectured in Higher Algebra and Probability. To my regret, I have no memories of the higher algebra course; I was interested in probability and was even thinking that I might choose it as the subject of my narrow specialization. In the end, however, I settled upon the theory of complex variables.

Ivan Matveevich Vinogradov, one of the most prominent mathematicians of the twentieth century, member of many foreign academies, member of the Academy of Sciences of the USSR since 1929, lectured on number theory. His lectures were interesting, but looking back I cannot help expressing my regret that he did not tell us anything about his own work, which contributed greatly to number theory, but concentrated on the classical topics.

7. It goes without saying that the curricula of the 1920s differed greatly from modern times. Moreover, even at this time curricula were not up-to-date. Here are a few examples.

Picard's theorem in the theory of ordinary differential equations was not included in the program; we learned about its existence from a paper presented at an optional seminar.

By the end of the 1920s great developments had taken place in such important branches of mathematics as functional analysis. By this time the notions of abstract spaces, among them the spaces that we now call Banach, had been studied, the theory as well as the application of Fredholm integral equations had been greatly developed, the fundamentals of the theory of one-dimensional singular integral equations had been lain down, and the theory of compact operators and equations with such operators had been developed. Surely, there was enough material for a course such as, for example, "An Introductory Course in Functional Analysis", but there was no such course.

At one of the lectures given by G. M. Fikhtengoltz we learned that there existed a new mathematical science called topology, but we learned (at least many of us) about the subject itself many years after graduating from the University.

One could give more examples of that kind. But at the same time one cannot help recognizing the fact that there were people among the students of the 1920s who later became scientists, some of them great scientists, such as the late G. M. Goluzin, L. V. Kantorovich, S. L. Sobolev, and D. K. Faddeev. It seems to me that of great importance was what and how our teachers taught us, and they taught us two simple things: to work and to think, which helped us overcome the limitations of the university education of that time. For this we are eternally grateful to our teachers and will remember them forever.

S. C. Mikhlin passed away at the age of 83, on August 29, 1991. We repeat the last sentence of his article: " ... we are eternally grateful to our teachers and will remember them forever".

Translated by K. BERESOVSKAYA

Amer. Math. Soc. Transl.
(2) Vol. **159**, 1994

On the History of the St. Petersburg and Petrograd Mathematical Societies

N. S. Ermolaeva

In 1890 three new mathematical societies started their activities in Kazan, Kiev, and St. Petersburg in addition to the already functioning Moscow Society (1867) and Kharkov Society (1879).

The activities of the St. Petersburg Mathematical Society during the first decade of its existence, 1890–1899, are covered in the collection [**1**] of concise records of its meetings. This collection was extensively used in the lengthy and detailed paper by I. Y. Depman [**2**] and in the note by A. A. Kiselev [**3**]. Some information on the Society is also contained in [**4**]. In the present paper all these sources have been used and also supplemented. The dates of the events that took place in Russia before 1918 are given according to the Old Style (the Julian calendar).

The initiative for founding the Society came from Academician V. G. Imshenetzky (1832–1892), who already had experience in organizing scientific societies. After graduating from Kazan University he founded a scientific circle of mathematics teachers and young scientists from the University [**4,** p. 252], and during his stay in Kharkov (from 1872) he played a significant part in setting up the Mathematical Society at the local University, and from 1880 to 1881—the year he left for St. Petersburg owing to his election to the Academy—he was its Chairman.

The constituent meeting of the St. Petersburg Mathematical Society took place on October 20, 1890, at the apartment of V. I. Shiff, a teacher of mathematics at St. Petersburg Higher Women's Courses ("Bestuzhevsky Courses"), and her husband P. A. Shiff, a professor at Mikhailovskaya Artillery Academy whose scientific interests centered around mathematics, elasticity, hydromechanics, and artillery. Besides the hosts present were Academicians V. G. Imshenetskiĭ, O. A. Backlund, A. A. Markov, Professors Yu. V. Sokhotskiĭ, N. A. Zabudskiĭ, V. V. Vitkovskiĭ, D. A. Gravé, I. I. Ivanov, I. A. Kleiber, N. P. Kolomeytsev, I. V. Meshcherskiĭ, P. M. Novikov, I. L. Ptashitskiĭ, D. F. Selivanov, and V. I. Stanevich. Professors K. A. Possé, A. N. Korkin, D. K. Bobylev, and A. M. Zhdanov notified in writing their consent to be on the list of the founders of the Society.

The majority of the participants of the meeting supported the founding of the Mathematical Society, at the meetings of which lectures on Pure Mathematics, Theoretical Mechanics, Theoretical Astronomy, and Mathematical Physics would be given. At the same time the Bureau of the Society was elected, including V. G. Imshenetskiĭ

1991 *Mathematics Subject Classification*. Primary 01A74; Secondary 01A70, 01A55, 01A60.

(Chairman), Y. V. Sokhotzky (Vice-chairman), and P. A. Shiff (Secretary). It was planned to meet once a month for reading scientific papers. New members were to be admitted by vote on the recommendation of the members of the Society.

The Society started its activities after receiving an oral permission from the Minister of Education who said that there was no hurry about formalities. During the first two years the meetings took place in the Study of Physics of the Academy of Sciences and after that at the University.

From 1890 to 1899 the meetings were held regularly. Only once, on February 17, 1899, the meeting was cancelled "on the occasion of riots in the University". The Society had a library at its disposal replenished mostly by donated books and reprints.

On January 17, 1891, the Society started work on its Charter. For this purpose a committee under the chairmanship of Academician A. V. Gadolin (1828–1892) was elected including A. N. Krylov (1863–1945). The petition, dated February 5, 1892, for confirming the Charter was signed by 43 founding members, including A. V. Gadolin, A. A. Markov, D. K. Bobylev, Yu. V. Sokhotskiĭ, D. F. Selivanov, O. A. Backlund, N. A. Zabudskiĭ, N. V. Maĭevskiĭ, and H. S. Golovin. However, the Ministry of Education confirmed the Charter only on December 27, 1892.

The membership of the Society was rapidly increasing. While there were only 18 persons at the founding meeting this number doubled by the time of the second meeting and by 1897 it had increased to 98.

The Society was rather an eclectic one. First of all, there were mathematicians and experts in mechanics from St. Petersburg University and university graduates working at secondary and higher schools of the city. The other part of the Society consisted of the teachers from higher military schools: N. V. Maĭevskiĭ (1823–1892), P. A. Shiff (1849–1910), N. A. Zabudskiĭ (1853–1917) from Mikhailovskiĭ Artillery Academy, V. V. Vitkovskiĭ (1856–1924), the geodesist from the General Staff Academy, and E. S. Fedorov (1853–1919), the geometrist and crystallographer from the Mining Institute. There was a small group of astronomers: O. A. Backlund (1846–1916), S. P. Glazenap (1848–1937), A. M. Zhdanov (1858–1914), A. A. Ivanov (1867–1939), and I. A. Kleiber (1863–1892). There were also lovers of mathematics working in other fields of science and engineering.

Among Society members there were women, most of them graduates from the Higher Women's Courses. Some of them were astronomers-calculators at the Pulkovo Observatory. Besides V. I. Shiff, one of the founding members of the Society was E. F. Litvinova (1845–1919(?)) who had received her mathematical education at the Zurich Polytechnicum and had defended her thesis, "The solution of one mapping problem", in Bern (1878). The thesis was written under the supervision of H. A. Schwarz and L. Schläfli. Later she became well known as an outstanding teacher and the author of a series of scientific biographies of mathematicians.

The graduate of Higher Women's Courses, L. N. Zapolskaya (1871–1943), defended her thesis "On the theory of relatively Abelian number fields" (1902) in Göttingen, where she worked under the supervision of D. Hilbert. Having returned to Russia she defended her master's thesis, "The theory of algebraic domains of rationalities appearing in solving third-degree equations". After 1923 Zapolskaya was a professor at Saratov University and at the Yaroslavl and Kuban teacher-training institutes.

The first change in the management of the Society took place in 1892. V. G. Imshenetzky suddenly died and after that Yu. V. Sokhotskiĭ become the permanent

Chairman of the Society. Up to 1903 the post of the Secretary was held by P. A. Shiff. It was due to his efforts that the records of 72 Society meetings during the first ten years were published [**1**]. In 1903 the post of the Secretary went to D. F. Selivanov (1856–1932).

P. L. Chebyshev was elected honorary member on February 15, 1893, but even before his election he was present at the meeting on December 16, 1892, and by Y. V. Sokhotskiĭ's proposal was presiding there. This meeting was called "special" in the Records since Chebyshev's presence was probably caused by the arrival to St. Petersburg of the Swedish mathematician G. Mittag-Leffler (1846–1927), the founder and editor of "Acta Mathematica", who was elected member of the Society on March 21, 1894, correspondent member of St. Petersburg Academy of Sciences in 1896, and Honorary Member of the Academy of Sciences of the USSR in 1926. At this meeting two papers were read (in French): "On nonalgebraic singularities of a differential equation not depending on a constant" by G. Mittag-Leffler and "On approximate calculation of a certain definite integral" by P. L. Chebyshev. Mittag-Leffler read another paper at the Society's meeting on March 18, 1893: "On univalent regular functions".

The paper by Chebyshev, indirectly connected with the theory of orthogonal polynomials, is of special interest since Chebyshev never published it and its summary was not included in his Complete Works; it was published by I. V. Depman much later [**2**, pp. 20–22]. In his talk Chebyshev deduced an approximate formula for calculating the definite integral

$$\frac{1}{\pi}\int_{-\pi}^{\pi}\frac{\cos m\phi}{\sqrt{1+r^2-2r\cos\phi}}\,d\phi$$

and gave the upper bound for the error. The derivation was based on the formula for approximation of $x^{-1/2}$ by an expression of the form $A+\sum_{i=1}^{\infty}B_i/(C_i+x)$, where the coefficients are determined with the help of Jacobi elliptic functions (published by Chebyshev in 1889, [**5**, p. 247]). In the second part of the paper, Chebyshev gave a formula for approximation of the fraction $1/(H-x)$ on $(-h,h)$ by a polynomial of degree $n-1$ and showed how this formula may serve for approximation of integrals, in particular, of

$$\int_{-h}^{h}\frac{f(x)\,dx}{H-x},$$

where $f(x)>0$ on this interval. This result was also new to the audience, since Chebyshev had reported it among other results at the meeting of the Section of Physics and Mathematics of St. Petersburg Academy of Sciences on February 2, 1892, only shortly before the meeting of the Society, and had not published it yet (the corresponding paper appeared in 1893, see [**6**, p. 372]).

Another meeting connected with the name of Chebyshev was held on January 14, 1895, in memory of the great mathematician who had died on November 26, 1894. Yu. V. Sokhotskiĭ gave a speech dedicated to the life and work of P. L. Chebyshev which he concluded by the words " ... we express the conviction that his life's work has found a reliable and faithful sanctuary here, in his motherland: nothing has been missed and nothing will be forgotten. We study it thoroughly; having furnished it with necessary supplements and proper interpretation we shall pass it in good time to the next generation as a pledge of further independent development of mathematical sciences in Russia and as a pledge of unfailing gratitude to our compatriot scientist whose name will live as long as science itself lives." [**1**, p. 94].

After that an outline of Chebyshev's works was given in the talks by D. A. Gravé, I. I. Ivanov, V. I. Stanevich, D. F. Selivanov, I. L. Ptashitskiĭ, V. I. Shiff, N. Y. Sonin, V. A. Markov, D. K. Bobylev, and N. B. Deloné (the summary of the latter two talks is included in [**1**] and reproduced in [**2,** pp. 24–26]).

At the period under consideration the Society had three more meetings dedicated to R. Descartes, N. I. Lobachevskiĭ, and K. Weierstrass.

At the meeting in honor of the 300th anniversary of Descartes's birthday on March 3, 1896, talks were given by S. E. Savich, D. F. Selivanov, and Yu. V. Sokhotskiĭ. The meeting of the Society on the occasion of the Centenary of N. I. Lobachevskiĭ's birthday was held on October 20, 1893, and a talk on his life and works in Geometry was given by S. E. Savich. Two weeks after the death of K. Weierstrass, on February 19, 1897, the Society held a meeting in his memory where D. F. Selivanov gave a talk, "On the variational calculus of Weierstrass".

We can add that besides G. Mittag-Leffler among the members of the Society was the known Polish mathematician and historian of mathematics S. Dikstein (1851–1939), the founder and first Vice-President of the Warsaw Scientific Society, later Vice-President of the International Society for the History of Sciences. Among the members from other cities we note N. V. Bugaev (1837–1903), President of the Moscow Mathematical Society since 1893, and A. V. Vasilyev, a mathematician from Kazan, who will be spoken about later.

On October 19, 1895, and February 9, 1896, D. A. Gravé reported to the Society on the solution of the problem of determining integer roots of the equation $\pi/4 = m\,\mathrm{arctg}\,(1/p) + n\,\mathrm{arctg}\,(1/q)$ posed by him and solved by C. Stormer (1874–1957), a student of the University of Christiania (now Oslo) and a future Norwegian Academician.

According to I. Y. Depman, the records of the Society meetings during the years 1890–1899 contain the summaries of 172 talks given by 50 members, among them D. A. Gravé (20 talks), N. Y. Sonin (16), Yu. V. Sokhotskiĭ (14), P. A. Shiff (11), B. M. Koyalovich (11), A. A. Markov (5), V. A. Markov (6), N. B. Deloné (8), I. V. Meshcherskiĭ (4), D. K. Bobylev (3), and O. A. Backlund (4).

The talks fall into the following thematical categories: Calculus—50 talks, Differential Equations—35, Algebra and Number Theory—35, Mechanics—24, Geometry (mostly differential geometry and the theory of curves and surfaces)—12. As Depman notes, "it is somewhat strange that there are no papers on Probability, one of the main topics of P. L. Chebyshev's interest" [**1,** p. 17]—only three talks devoted to it were given, namely, "On the so-called Petersburg Paradox" by B. M. Koyalovich and the talks by N. N. Pirogov on applications of probability to physics and by D. F. Selivanov on P. L. Chebyshev's works in probability.

Now we focus on some papers presented to the Society, noting here that "the Records" do not contain summaries of all the papers. Some of the papers were published by the authors, the others were given in order to inform the members of the Society about new ideas in Mathematics, but there were also talks about which nothing is known except the title.

Yu. V. Sokhotskiĭ, as has already been said, gave fourteen talks during the first decade. His first lecture was given on March 16, 1891, and was devoted to geodesic lines although this topic was not in the mainstream of his interests. On December 21, 1892, the subject of Sokhotskiĭ's talk was "The principle of the greatest common divisor as applied to divisibility of algebraic numbers." This talk appeared later as a separate publication [**7**], which attracted the attention of Russian mathematicians. In this work Sokhotskiĭ gave a simple and clear presentation of his vision of the theory, developed by E. I. Zolotarev (1847–1878).

In 1897 Sokhotskiĭ gave two review lectures on group theory (their content is unknown), and three lectures in 1897 and 1898 were devoted to equations of degree four and five.

The necessity of presenting Weierstrass's theory of elliptic functions in a form understandable to engineers induced Sokhotskiĭ to give a series of talks on this subject in 1896–1898 (the summary of these talks is missing in **[1]**). The first lecture of this series was given by Sokhotskiĭ on March 17, 1896, at the meeting dedicated to Descartes. Having noted briefly the role of Descartes for subsequent development of calculus and geometry, Sokhotskiĭ said that the study of the curves leads to the study of new functions. In the same way as one can obtain all the curves of degree two by starting with one of them (say, the circle) and using appropriate transformations, all the curves of third degree can also be obtained from one of them, e.g.,

$$y^2 = 4x^3 - g_2 x - g_3$$

(recall that the folium of Descartes is a curve of the third degree too). Accordingly, from the study of analytical properties of the circle one can obtain the theory of circular functions while the analysis of the curves of third degree leads naturally to the elliptic functions. Then Sokhotskiĭ demonstrated a simple method of obtaining the elliptic functions by using descriptive-geometrical arguments. In the following lectures he proposed another way of presenting this theory. These talks by Sokhotskiĭ were made use of by another Society member, S. G. Petrovich (1868–1926), a professor at Mikhailovskaya Artillery Academy, who published in 1898 "An approach to the elementary theory of Weierstrass's functions" **[8]**.

This topic was also discussed by K. A. Possé in his talk given on September 23, 1918, in honor of the 50th anniversary of Sokhotskiĭ's professorship: "The greater part of these lectures was devoted to your elementary presentation of the theory of elliptic functions. This presentation was not published by you in Russian but it formed the basis of the thesis by S. G. Petrovich, one of the most diligent participants of the Society" [**9**, p. 1].

Sometimes results obtained by some members of the Society stimulated further research on the subject by the other members. Thus, on February 29, 1892, V. A. Markov (the brother of A. A. Markov) proved the theorem: "If two algebraic functions of degree n have only real and alternating roots, then their derivatives of equal order also have alternating roots." This paper, read by a young scientist (V. A. Markov (1871–1897) graduated from the University in 1892), was to be continued. On September 15, 1892, P. A. Shiff gave a lecture "Some corollaries of Rolle's theorem and an application of Rolle's theorem to the proof of V. A. Markov's theorem on the distribution of roots of two functions and their derivatives"; also he extended Markov's theorem to the complex domain (published in **[10]**); I. I. Ivanov gave another proof of V. A. Markov's theorem. Finally, at the next meeting on October 15, 1892, V. A. Markov himself reported on a further generalisation of this theorem.

These results are not contained in the published works of V. A. Markov, and they can be found only in **[1]** and **[2]**. Of the other talks given by V. A. Markov we mention his paper, "On the axioms lying in the foundation of geometrical systems", read at two meetings, which is highly noteworthy taking into account the lack of interest in Geometry on the part of St. Petersburg mathematicians.

Somewhat earlier the Society had discussed the results by P. A. Shiff published by him later in [**10**] where many relations for definite integrals were derived from the Green formula for functions of several variables by means of a suitable change of variables. Besides this, the author obtained the equation for which the equation applied by Beltrami in studying functions of a complex variable on a surface is a particular case. Finally, Shiff obtained the multi-dimensional analogue of the Cauchy integral formula and some generalizations of the integrals of Poisson, Parceval, and Kummer.

Only two talks were given by I. A. Kleiber, a gifted mathematician and astronomer, who died in his prime. One of them was "On a particular case of the problem of three bodies" (February 15, 1891) and the other "On some integrals of complete elliptic integrals" (March 16, 1891).

Results, which later became parts of their theses, were repeatedly reported by D. A. Gravé, B. M. Koyalovich, and others. The talks given by A. A. Markov (February 15, 1891) and K. A. Possé (October 20, 1893) were devoted to the proof of the transcendence of π.

Among the other papers were, for example, the following:

– O. A. Backlund: "Integrating a differential equation for determining the longitude in the motion of one group of minor planets" (April 23, 1897);

– D. K. Bobylev: "On the motion over the rough horizontal surface of a spherical shell containing a rotating gyroscope, with the axis fixed with respect to the sphere" (February 15, 1891);

– G. F. Voronoy: "On determining the sum of quadratic residues of the prime p of the form $4m+3$ using Bernoulli numbers" (January 15, 1891);

– N. M. Gunter: "On the integration of the second-order equations in hypergeometrical functions" (March 17, 1899);

– N. B. Deloné: "On the new ellipsograph" (March 17, 1893);

– V. D. von Dervis: "On a theorem in the theory of sets" (December 17, 1894);

– V. F. Kagan: "On some irreducible polynomials" (a generalization of Schönemann's theorem) (January 22, 1896);

– G. B. Kolosov: "On Greffe's method for numerical solution of equations" (February 19, 1897);

– A. A. Markov: "The cases when the integral $\int \frac{x\,dx}{(x^3+B)\sqrt{x^3-1}}$ is expressed through logarithms" (November 13, 1892);

– I. V. Meshcherskiĭ: "On the problem on n bodies" (January 15, 1893);

– D. F. Selivanov: "On periodic continued fractions" (December 15, 1890 and March 18, 1898);

– N. Y. Sonin: "Notes on two sections of Jordan's Course d'Analyse" (November 20, 1896); "On a series of Johann Bernoulli" (October 15, 1892);

– M. M. Filippov: "On the arithmetical differentiation" (December 17, 1894).

Almost nothing is known about the activities of the Society after 1900. In the first years of the twentieth century, both the number of members and the number of meetings were gradually decreasing (in the years 1904–1905 there were 3 meetings where 5 talks were given). Attempts to find the documentary data on the activities of the Society failed. I. V. Depman supposed that since the Society had no premises of its own, the documentation had been kept by its Secretary.

In 1902–1903, according to the official information, this post was held by D. F. Selivanov. It remains unknown if anybody held this post later. As for Selivanov, he emigrated to Prague in the 1920s. It is most probable that the Society ceased to exist in 1905 when all gatherings were banned.

Only two pieces of evidence on the activities of the Society came to light. Thus, G. V. Kolosov informs us in his master's thesis that in October, 1901, he read a paper on coincidence of Brun's and Weber's problems [**11,** p. 51] at the meeting of the Society but he does not mention the exact title of the paper. The other evidence is an unsigned typewritten notice dated March 20, 1910, that was found in the archive of V. A. Steklov. It read, "The members of St. Petersburg Mathematical Society are invited to the meeting on Wednesday, March 24, this year, to lecture-hall 3 at the University at 8 o'clock p.m. for the election of the members of the Council and discussion of the affairs of the Society" [**12,** p. 152]. This note may be considered as an effort to resume the activities of the Society, apparently unsuccessful. It is not completely clear what part Steklov played in this Society. Later he was the Honorary Member of the Leningrad Mathematical Society and also the founder (1926) and the first editor of the journal of the Society. The Editors of the journal wrote about it in the obituary article in memory of Steklov [**13,** p. 1]. It was also said there that Steklov "had been dreaming since 1910" about publishing such a mathematical journal. He might have expressed this wish in March, 1910, at the same meeting mentioned in the notice. Also, it might have been his idea to hold the aforementioned meeting.

The new Mathematical Society was founded in Petrograd on May 14, 1921, on the initiative of Professor A. V. Vasilyev (1853–1929) who had been elected a member of the St. Petersburg Society on November 13, 1895.

A graduate of St. Petersburg University, A. V. Vasilyev worked at Kazan University from 1874 to 1906 (as a professor since 1887). He was one of the founders of the Kazan Physical and Mathematical Society and its Chairman up to his departure for St. Petersburg (in 1906 he was elected member of the State Council). At St. Petersburg University, Vasilyev gave a course on Mathematics for Chemistry students, and at the same time he taught the Higher Women's Courses, then at the Teacher-Training Institute. Being a mathematician with a wide range of interests and great erudition, a mathematician of "philosophical and historical method of thinking" (as was said by Professor S. I. Shorokh-Trotzkiĭ (1853–1923) [**14,** p. 423], a known St. Petersburg teacher), Vasilyev did a lot for promoting new mathematical and physical theories. In particular, he was the editor of a collection of papers, "New Ideas in Mathematics", where the known philosopher P. S. Yushkevich actively participated.

In 1920 Vasilyev organized a mathematical circle at the Teacher-Training Institute affiliated with the University, later transformed into the Petrograd Physical and Mathematical Society. The Society mainly aimed at promoting scientific contacts among all interested in mathematics and mathematical education. There was a pedagogical section associated with the Society which united secondary school teachers and university instructors [**15,** pp. 18–19].

A. V. Vasilyev wrote to Moscow Academician P. P. Lazarev on May 15, 1921 about the first meetings of the Society: "Yesterday Petrograd Physical and Mathematical Society was founded, i.e., the presiding body was elected (my Vice-Presidents are Y. V. Uspenskiĭ and Y. A. Krutkov, the Learned Secretary is V. K. Frederics), and it begins its activities on May 25 by the meeting where, apart from your talk, there will be mine on 'The Geometry of the Universe'. As this day is Wednesday, and a session of the branch of the Academy of Sciences might be scheduled for this day, I intend to open the meeting at 8 o'clock. On May 26 (the anniversary of Chebyshev's birthday), the second meeting will be held where two lectures will be given: by myself on the history of Mathematics in Russia and by A. S. Bezikovich on some topic in Probability closely related to Chebyshev's works. You may have already known the convocation in his memory at the Academy is scheduled for June 10" [**16**, p. 6]. In the last sentence Vasilyev writes about the conference held by the Academy of Sciences in June 1921 on the occasion of the centenary of Chebyshev's birthday where V. A. Steklov gave a talk on "the theory and practice in Chebyshev's studies" [**17**].

By the way, the arrival at the conference in Petrograd of the group of Moscow mathematicians headed by N. N. Luzin contributed considerably to establishing scientific contacts between the two mathematical schools. Vasilyev's activities were conducive to it too. In an undated note addressed to Academician Lazarev he wrote "... I am very happy about the wonderful progress of our Moscow-Petrograd mathematical community ... What an enthusiastic and whole-hearted interest you took in all that!"

In the first two years of the Society's existence Y. D. Tamarkin, E. L. Nikolai, N. M. Gunter, A. A. Satkevich, and other scientists read their papers at its meetings.

After A. V. Vasilyev had left for Moscow in 1923, N. M. Gunter was elected Chairman (the exact date is unknown), A. A. Friedman was elected Secretary of the Society, and after his death this post went to A. F. Gavrilov.

In his talk at the meeting dedicated to the memory of A. A. Friedman on September 26, 1925, N. M. Gunter said, "Our Society is greatly indebted to A. A. Friedman; one may say that there would have been no Society without him; the Society was revived by A. A. The regular activities of the Society started the day when A. A. became its Secretary. He composed and got approved the new Charter of the Society. Busy as he was, he still found time to attend the meetings of the Board, made arrangements for our journal during his visits to Moscow and planned to participate in editing it" [**18**, pp. 8–9]. By "regular activities of the Society", N. M. Gunter probably meant its more scientific than educational nature, to which N. M. Gunter himself contributed later.

At present the activities of the St. Petersburg and Petrograd Mathematical Societies are carried on by the Leningrad (now again St. Petersburg) Mathematical Society founded in 1959 (see the paper by A. M. Vershik [**19**]).

References

1. *The Records of St. Petersburg Mathematical Society*, Kirschbaum's Printing House, St. Petersburg, 1899. (Russian)
2. I. Ya. Depman, *La société mathématique de St. Pétersbourg*, Istor. Mat. Issled. **13** (1960), 11–106. (Russian)
3. A. A. Kiselev, *St. Petersburg Mathematical Society at the end of the* 20*th Century*, The Topics on the History of Physical-Mathematical Sciences, Vysshaya Shkola, Moscow, 1963, pp. 41–42. (Russian)
4. I. Z. Stokalo, A. N. Bogolyubov, Yu. A. Mitropolskiĭ, I. B. Pogrebysskiĭ, E. Ya. Remez, K. A. Rybnikov, Yu. D. Sokolov, and V. S. Sologub (eds.), *History of national mathematics in four volumes. Vol.* II. 1801–1917, "Naukova Dumka", Kiev, 1967. (Russian)
5. P. L. Chebyshev, *Complete collected works*, vol. 2, pp. 240–255. (Russian)
6. ______, *On the polynomials of the best approximation for elementary rational functions on a given interval*, Complete Collected Works, vol. 2, pp. 363–372. (Russian)
7. Yu. V. Sokhotskiĭ, *Introduction to the greatest common factor applied to divisibility theory of algebraic numbers*, Academy Printing House, St. Petersburg, 1893. (Russian)
8. S. G. Petrovich, *On the elementary theory of Weierstrass's functions* sn u, cn u, *and* dn u, Kushner and Co., St. Petersburg and Moscow, 1898. (Russian)
9. *Leningrad State Historical Archives*, F. 14, Op. 1, D. 6646.
10. P. A. Shiff, *On some corollaries of Rolle's theorem*, Proc. Sect. Phys. Sci. Soc. Enthusiasts Nat. Sci., Antopol. and Etnogr. **5** (1893), no. 2, 45–59. (Russian)

11. G. V. Koslov, *On some modifications of the Hamilton principle in solving some problems in mechanics of solids*, Ehrlich, St. Petersburg, 1903. (Russian)
12. *Leningrad Branch of the Archives of the Acad. Sci. USSR*, F. 162, Op. 3, D. 121.
13. J. Leningrad Phys. Math. Soc. **1** (1926), 1. (Russian)
14. *The anniversary of Prof. A. V. Vasiliev*, Scient. Review **1900**, no. 2, 420–423. (Russian)
15. A. N. Bogolyubov, *The starting period in the development of the Soviet mathematics*, The History of National Mathematics in Four Volumes. Vol. III. 1917–1967, "Naukova Dumka", Kiev, 1968, pp. 9–56. (Russian)
16. *The Archives of the Acad. Sci. USSR (Moscow)*, F. 459, Op. 4, D. 17. (Russian)
17. V. A. Steklov, *Theory and practice in Chebyshev's studies*, The talk given at the celebration in honour of the 100th birthday anniversary of P. L. Chebyshev at the Russian Academy of Sciences, Akad. Sci. Publ., Petrograd, 1921. (Russian)
18. N. M. Gunther, *In memory of A. A. Friedman*, J. Leningrad Phys.-Math. Soc. **1** (1926), no. 1, 5–17. (Russian)
19. A. M. Vershik, *The Leningrad Mathematical Society*, Trudy Leningrad. Mat. Obshch. **1** (1990), 4–8; English transl. in Amer. Math. Soc. Transl. Ser. 2 **155** (1993), ix–xiii.

St. Petersburg State University for Architecture and Civil Engineering, 4, 2nd Krasnoarmeĭskaya ul., St. Petersburg, 198005, Russia

Translated by G. ROZENBLUM

Amer. Math. Soc. Transl.
(2) Vol. **159**, 1994

St. Petersburg Mathematical Society

A. M. Vershik

Since the publication of the first volume of "Trudy" in 1990, great changes have taken place in the life and structural organization of our society as, indeed, in the life of the whole country. In brief, these changes are the following.

In the winter of 1992, a decision was made to rename the Leningrad Mathematical Society to St. Petersburg Mathematical Society (PMS), and thus the Society got back its initial name—the name under which the first mathematical society was founded in St. Petersburg one hundred years ago (see the paper on this subject in the first volume of "Trudy"). At the same meeting of the Society (February 11, 1992) it was decided that PMS should be registered as an independent public scientific association, and a motion was carried to enter the European Mathematical Society. In the spring of the same year the new Charter of the Society was worked out, and, on August 4, the St. Petersburg Mathematical Society was registered with the Department of Justice in St. Petersburg as a voluntary public association.

In July 1992, at the meeting of the council of the European Mathematical Society, PMS was admitted to EMS as a collective member.

On December 15, 1992, an administrative meeting of PMS was held at which the new Board and other bodies of the society were elected.

President: O. A. Ladyzhenskaya.

Vice Presidents: A. M. Vershik, Yu. A. Davydov.

Members of the Board: A. D. Aleksandrov, M. I. Bashmakov, V. S. Buslaev, O. Ya. Viro, I. A. Ibragimov, S. V. Kerov, G. A. Leonov, A. A. Lodkin, Yu. A. Matiyasevich, A. S. Merkur′ev, G. I. Natanson (scientific secretary), N. Yu. Netsvetaev, M. A. Semenov-Tyan-Shanskiĭ.

Members of the Audition Committee: O. A. Ivanov, B. V. Lur′e, A. A. Semenov, V. N. Sudakov, V. N. Fomin, V. P. Khavin.

Members of the Scientific Council: V. M. Babich, M. S. Birman, V. S. Buslaev, A. M. Vershik, S. A. Vinogradov, S. V. Vostokov, I. A. Ibragimov, O. A. Ladyzhenskaya, G. A. Leonov, Yu. A. Matiyasevich, N. Yu. Netsvetaev (scientific secretary), A. P. Oskolkov, N. N. Ural′tseva, L. D. Faddeev, V. P. Khavin, V. A. Yakubovich.

Members of the Editorial Board: O. A. Ladyzhenskaya (editor-in-chief), A. M. Vershik (deputy editor-in-chief), V. P. Orevkov (secretary-in-chief), S. A. Vinogradov, G. A. Leonov, B. S. Pavlov, B. A. Plamenevskiĭ, M. A. Semenov-Tyan-Shanskiĭ.

In addition, several bodies continue their activities without any changes in membership: School Liaison Council (chairman M. I. Bashmakov), College Liaison

Council (chairman S. V. Vostokov), and the Bureau of the Mathematical Section in the House of Scientists (chairman A. M. Vershik, secretary S. V. Kerov).

Among others, the duties of the newly founded Scientific Council are: to provide coordination (under the agreement with the Foundation "Matematika" at the Russian Academy of Sciences, the American Mathematical Society Fund, and others) of financial support for St. Petersburg mathematicians, to provide scientific guidance for a special students' group at the Department of Mathematics and Mechanics of St. Petersburg University, and to promote other scientific initiatives.

Deceased members of the Society:

Mikhlin Solomon Grigor′evich (1908–1990)

Bondareva Ol′ga Nikolaevna (1937–1991)

Gavurin Mark Konstantinovich (1911–1992)

Milin Isaak Moiseevich (1919–1992)

Thirty-one persons were admitted to the society: S. I. Karpushev, Ya. I. Belopol′skaya, P. V. Sporyshev, V. K. Zakharov, N. S. Ermolaeva, A. A. Mekler, B. B. Pokhodzeĭ, D. Yu. Burago, A. L. Gromov, A. I. Nazarov, M. N. Gusarov, I. B. Fesenko, A. A. Semenov, È. A. Musaev, V. V. Makeev, I. N. Kostin, I. B. Zhukov, B. V. Budaev, V. L. Oleĭnik, A. G. Izergin, A. P. Karp, B. A. Lifshits, A. L. Chistov, V. M. Nezhinskiĭ, S. A. Evdokimov, A. D. Lisitskiĭ, I. N. Ponomarenko, N. V. Proskurin, G. A. Seregin, N. V. Smorodina, I. V. Denisova.

The Society Prize winners are:

1991. G. Ya. Perel′man. Alexandrov spaces with curvature bounded from below.

1992. D. Yu. Burago. Geodesic flows on manifolds with limit metric.

I. B. Fesenko. The class field theory.

Reports on the meetings of the society were published in the journal "Uspekhi matematicheskikh nauk": 1991, vol. 46, 2(278), pp. 233–234; 1992, vol. 47, 3(285), pp. 191–192; 1993, vol. 48, 2(290).

The following papers have been presented:

N. A. Vavilov. Maximal subgroups of finite simple groups.

A. P. Veselov (Moscow). Integrable congruences.

A. I. Barvinok. Methods of representation theory in optimization.

O. Ya. Viro. New invariants in the theory of three-dimensional manifolds.

A. M. Olevskiĭ (Moscow). Free interpolation in classical and harmonic analysis.

G. B. Shabat (Moscow). On the realization of Grothendieck's program.

G. L. Litvinov (Moscow). Rational approximations and error autocorrection.

A. G. Khovanskiĭ (Moscow). The number of lattice points in integer polytopes. The Riemann–Roch theorem and a multidimensional generalization of the Euler–MacLaurin theorem.

G. Ya. Perel′man. Alexandrov spaces with curvature bounded from below.

V. A. Yakubovich. Linear-quadratic problems of optimal control with quadratic restriction.

N. N. Ural′tseva. On the evolution of surfaces under the influence of the mean curvature.

A. M. Vershik. Hydrodynamic limits of classical Lie algebras.

I. A. Panin. Vector bundles on homogeneous spaces.

G. Ya. Perel′man. Convergence and collapse of Riemannian manifolds.

S. V. Kerov. Asymptotics of reciprocal separation.

M. I. Graev, V. S. Retakh (Moscow). Latest developments in the theory of hypergeometric functions.

I. B. Fesenko. On higher class field theory.

Marcel Berger (Paris). Packing by disks.

Yu. V. Matiyasevich. Interactive proofs (I shall prove it to you...).

A. Bolibruh (Moscow). The XXI Hilbert problem for Fuchsian linear systems.

A meeting in memory of Dmitri Konstantinovich Faddeev, president of the Society since 1985.

A meeting in memory of Mark Grigor′evich Kreĭn (1907–1989) and his works.

A meeting in memory of Yuriĭ Vladimirovich Linnik (1915–1972) on the occasion of his 75th birthday anniversary.

At one meeting, A. M. Vershik, O. Ya. Viro, N. K. Nikol′skiĭ, and others shared their views on how mathematical education and research are organized in American and European universities.

A meeting in memory of Solomon Grigor′evich Mikhlin (1908–1990).

A meeting in memory of Yuri Aleksandrovich Volkov, professor at the Leningrad State University (1930–1981).

A meeting in honor of Nikolaĭ Sergeevich Koshlyakov, Laureat of the State Prize of the USSR, University professor, correspondent member of the Academy of Sciences of the USSR, on the occasion of his 100th birthday anniversary (1891–1958).

The following lectures have been given within the framework of the Students Lecturing Bureau:

N. A. Vavilov. Classification of finite simple groups.

N. V. Ivanov. Discrete groups as geometrical objects.

A. P. Veselov (Moscow). Chebyshev polynomials, Lie algebras and integrable mappings.

O. I. Reĭnov. Bases and approximation problems in Banach spaces.

V. G. Osmolovskiĭ. Evolution of the notion of solution in classical problems of mathematical physics.

G. L. Litvinov (Moscow). Computer evaluation of functions, and rational approximations.

N. N. Petrov. New invariants in search problems on graphs.

A. N. Borodin. A mathematical solution of the problem of heat transfer.

D. Yu. Burago. Geometric problems in the theory of dynamic systems.

A. L. Smirnov. Diophantine geometry: problems and achievements.

N. Yu. Netsvetaev. The Poincaré conjecture: its 90th anniversary.

A. M. Vershik. What is a limit picture in geometric problems?

Yu. V. Matiyasevich. Interactive proofs (I shall prove it to you...).

S. Yu. Pilyugin. What is chaos?

E. P. Golubeva. The well-known unsolved Gauss class-one problem and continued fractions.

Recent Titles in This Series

(*Continued from the front of this publication*)

(See the AMS catalog for earlier titles)